AI System Support for Con‹

Springer
*London
Berlin
Heidelberg
New York
Barcelona
Budapest
Hong Kong
Milan
Paris
Santa Clara
Singapore
Tokyo*

John Sharpe (Ed.)

AI System Support for Conceptual Design

Proceedings of the 1995 Lancaster International Workshop on Engineering Design, 27-29 March 1995

With 169 Figures

 Springer

John E.E. Sharpe

Engineering Design Centre
Lancaster University
Lancaster LA1 4YR, UK

ISBN 3-540-76000-8 Springer-Verlag Berlin Heidelberg New York

British Library Cataloguing in Publication Data
AI System Support for Conceptual Design:
Proceedings of the 1995 Lancaster International Workshop on Engineering Design,
27-29 March 1995
 I. Sharpe, John
 620.0042028563
 ISBN 3-540-76000-8

Library of Congress Cataloging-in-Publication Data
A catalog record for this book is available from the Library of Congress

Apart from any fair dealing for the purposes of research or private study, or criticism or review, as permitted under the Copyright, Designs and Patents Act 1988, this publication may only be reproduced, stored or transmitted, in any form or by any means, with the prior permission in writing of the publishers, or in the case of reprographic reproduction in accordance with the terms of licences issued by the Copyright Licensing Agency. Enquiries concerning reproduction outside those terms should be sent to the publishers.

© Springer-Verlag London Limited 1996
Printed in Great Britain

The publisher makes no representation, express or implied, with regard to the accuracy of the information contained in this book and cannot accept any legal responsibility or liability for any errors or omissions that may be made.

Typesetting: Camera-ready by authors
Printed and bound at the Athenæum Press Ltd., Gateshead, Tyne and Wear
69/3830-543210 Printed on acid-free paper

INTRODUCTION

Conceptual design is often considered to be the most important step in the whole product development process, the reason being that more than half of the total life cycle cost is committed and the product quality determined at this stage. A poorly conceived design concept can never be compensated for at the manufacturing stage or via marketing policies. The design process at the conceptual stage is inherently a complex activity, due to the need to satisfy many requirements and comply with diverse types of constraints. Individual designers very often restrict themselves in a situation where they are not able to tackle large and complex systems partly because of a lack of sufficient suitable information and partly because of the limited availability of appropriate computer support tools.

There is a growing awareness that designers can benefit substantially from computer support during conceptual design and this is being driven by three factors:

- An understanding of the importance of early decisions in the overall success and profitability of a product or scheme,
- The growth in the complexity of products and their interdisciplinary nature especially in the design of mechatronic products,
- The need to provide AI support to the designer as the numbers of those trained and educated to such functions reduces.

This book consists of a collection of papers delivered to the Lancaster International Workshop on Engineering Design, all of which represent significant contributions to the advancement of computer support for conceptual design. The contributions are not "wish lists" of proposed research, which are becoming rather fashionable, but reports of solid achievements in the field. Also included are two major contributions from industry, which outline the need for conceptual design tools and their potential impacts. Many of the papers cover advanced application of artificial intelligence, 3D solid modelling, simulation and many aspects of on-line support.

The concept of this book and the associated workshop was born in the realisation of the need to produce a clear presentation of the formative research.

We would very much like to thank all those who have contributed to this book both from academia and industry. Our particular thanks go to the staff of the Lancaster EDC, especially Jan Anderson, who has worked so hard to ensure its success.

John E.E. Sharpe
Lancaster March 1994

CONTENTS

Introduction
John E.E. Sharpe

Adaptive Search and Optimisation

1. An Adaptive Machine Learning System for Computer Supported Conceptual Engineering Design
   M J Hague, A Taleb-Bendiab & M J Brandish 1
2. The Application of Genetic Algorithms to Conceptual Design
   M G Hudson & I C Parmee 17
3. Solution Clustering with Genetic Algorithms and DFA: An Experimental Approach
   S D Santillan-Gutierrez & I C Wright 37

Handling of Geometric Data and Knowledge

4. Handling of Positional Information in a System for Supporting Early Geometric Design
   X Guan & K J MacCallum 54
5. An Architecture for the Intelligent Support of Knitwear Design
   C Eckert & M Stacey 71

Knowledge Based Reasoning

6. Representing Conceptual Design Knowledge with Multi-Layered Logic
   K Clibbon, E Edmonds & L Candy 93
7. A Synthetic Reasoning Method Based on a Physical Phenomenon Knowledge Base
   M Ishii & T Tomiyama 109
8. Design Model: Towards an Integrated Representation for Design Semantics & Syntax
   L B Keat, C L Tan & K Mathur 124

Integrated Intelligent Support for Conceptual Design

9. Égide: A Design Support System for Conceptual Chemical Process Design
   R Bañares-Alcántara, J M P King & G H Ballinger 138

10. Development of an Integrated AI System for Conceptual Design Support
    M X Tang 153

11. Integrated Platform for AI Support of Complex Design (Part I): Rapid Development of Schemes from First Principles
    R H Bracewell, R V Chaplin, P M Langdon, M Li, V K Oh, J E E Sharpe & X T Yan 170

12. Integrated Platform for AI Support of Complex Design (Part II): Supporting the Embodiment Process
    R H Bracewell, R V Chaplin, P M Langdon, M Li, V K Oh, J E E Sharpe & X T Yan 189

13. A Computerized Tool to Create Concept Variants from Function Structures
    D Brady & N P Juster 208

Interactive Knowledge Support

14. Integrated Innovative Computer Systems for Decision Support in Bridge Design
    C J Moore & J C Miles 227

15. 'MODESSA', A Computer Based Conceptual Design Support System
    T Kersten 241

16. Interactive Knowledge Support to Conceptual Design
    L Candy, E A Edmonds & D J Patrick 260

17. A Support System for Building Design - Experiences and Convictions from the Fabel Project
    B Bartsch-Spörl & S Bakhtari 279

Intelligent Management of Design Procedures

18. Conflict Management in an Interdisciplinary Design Environment
    V Oh & J E E Sharpe 298

19. A Fuzzy Thesaurus for Semantic Integration of Design Schemes
    I Mirbel 319

20. Managing Design and Manufacturing Constraints in a Distributed Industrial Environment: The Creation of a Managed Environment for Engineering Design
   A Medland 336
21. Computer Support for Design Team Decisions
   D G Ullman & D Herling 349

AI Support of Detail Design

22. Use of Visualisation and Qualitative Reasoning in Configuring Mechanical Fasteners
   G Zhong & M Dooner 362
23. Conceptual Design of Polymer Composite Assemblies
   J K McDowell, T J Lenz, J Sticklen & M C Hawley 377
24. Conceptual Design for Mechatronics
   H P Hildre & K Aasland 390
25. Reasoning and Truth Maintenance of Casual Structures in Interdisciplinary Product Modelling and Simulation
   X-T Yan & J E E Sharpe 405

LIST OF CONTRIBUTORS

M J Hague, A Taleb Bendiab &
M J Brandish
Department of Mechanical Engineering,
Design and Manufacture
Concurrent Engineering Research Group
Manchester Metropolitan University
Manchester, UK

M G Hudson & I C Parmee
Engineering Design Centre
University of Plymouth
Plymouth, UK

S D Santillan-Gutierrez & I C Wright
Engineering Design Institute
Loughborough University of Technology
Loughborough, UK

X Guan & K J MacCallum
CAD Centre
Department of Design, Manufacture and
Engineering Management
University of Strathclyde
75 Montrose Street
Glasgow Gl 1XJ
Scotland UK

C Eckert
Design Discipline
The Open University
Milton Keynes
MK7 6AA, UK

M Stacey
Computing Department
The Open University
Milton Keynes
MK7 6AA, UK

K Clibbon, E Edmonds, L Candy &
D J Patrick
LUTCHI Research Centre
Department of Computer Studies
Loughborough LE11 3TU, UK

M Ishii & T Tomiyama
Department of Precision Machinery
Engineering
Faculty of Engineering
The University of Tokyo
Hongo 7-3-1
Bunkyo-ku,
Tokyo 113, Japan

L B Keat
School of Information Technology &
Applied Science
Temasek Polytechnic
51 Grange Road
Singapore 1024

T C Lim
Department of Information Systems &
Computer Science
National University of Singapore
10 Kent Ridge Crescent
Singapore 0511

K Mathur
School of Building & Estate Management
National University of Singapore
10 Kent Ridge Crescent
Singapore 0511

R Bañares-Alcántara, J M P King &
G H Ballinger
Department of Chemical Engineering,
University of Edinburgh
Edinburgh, Scotland

M X Tang
Engineering Design Centre
Department of Engineering
University of Cambridge
Trumpington Street
Cambridge, CB2 1PZ, UK

R H Bracewell, R V Chaplin,
P M Langdon, M Li & J E E Sharpe
Engineering Design Centre
Lancaster University
Lancaster LA1 4YR, UK

D Brady
Keyworth Institute
Department of Mechanical Engineering
University of Leeds
Leeds, LS2 9JT, UK

N P Juster
Department of Mechanical Engineering
University of Leeds
Leeds, LS2 9JT, UK

C J Moore & J C Miles
Cardiff School of Engineering
University of Wales
College of Cardiff
Cardiff CF2 1YF
Wales, UK

T Kersten
Technology Application Unit
Unilever Research Laboratory Vlaardingen
P.O. Box 114
3130 AC Vlaardingen
The Netherlands

B Bartsch-Spörl & S Bakhtari
BSR Consulting GmbH
Wirtstrasse 38
D-81539 Muenchen
Germany

V K Oh   *(currently at)*
Silicon Graphics pte Ltd
83 Science Park Drive # 0403/04
The Curie
Singapore 0511

I Mirbel
University of Nice - Sophia Antipolis
I3S, CNRS-URA 1376
250 Avenue Albert Einstein
Sophia-Antipolis 06560 Valbonne
France

A J Medland
School of Mechanical Engineering
University of Bath
UK

D G Ullman & D Herling
Department of Mechanical Engineering
Oregon State University
Corvallis
Oregon 97331
USA

G Zhong & M Dooner
Department of Engineering Design &
Manufacture
University of Hull
UK

J K McDowell, T J Lenz, J Sticklen &
M C Hawley
Composite Materials & Structures Center
and Intelligent Systems Lab
Michigan State University
East Lansing
MI48824-1326
USA

H P Hildre
NTH Department of Machine Design &
Materials Technology
N-7034 Trondheim
Norway

X-T Yan (*currently at*)
D.M.E.M. Dept
James Weir Building
University of Strathclyde
Glasgow G1 1XJ, Scotland

K Aasland
SINTEF Materials Technology
N-7034
Trondeim
Norway

AN ADAPTIVE MACHINE LEARNING SYSTEM FOR COMPUTER SUPPORTED CONCEPTUAL ENGINEERING DESIGN

M J Hague, A Taleb-Bendiab & M J Brandish

1. CONCURRENT CONCEPTUAL DESIGN

Recent research has shown that the highest return of investments made during various stages of product development is recorded during the design stage (Fig. 1) [1]. With a view to survival, a number of manufacturing companies are, as a result, now reviewing their design practices, and in particular, since it is recognised as the most critical and determinant factor in product competitiveness, the conceptual engineering design phase [2,3,4].

Furthermore, the generation of a "good" conceptual design solution does not necessarily lead to a product's successful penetration (or sustenance of its position) in the market place, this has to be achieved by the selection of appropriate production, marketing and distribution solutions. To this end, product developers have, at an early design stage, to take into account all the life-cycle concerns such as: manufacturing, reliability, marketing, distribution. It is also found that designers at early design stages are less willing to make any required changes, as opposed to in the later part of the product development process. Although no explanation for this is proposed, the authors believe that it could be related to the lack of design life-cycle information (facts) necessary to convince a designer to incorporate suggested changes.

To this end, there is a pressing need for the development of a methodology to support the evaluation of conceptual design alternatives from multiple life-cycle facets (design, manufacturing, assembly) throughout the generation of alternative solutions. This requires access to any necessary down-stream information throughout the early design stages. Inherent in this approach is the need for appropriate management of the large amount of information involved, and the development of an efficient method for presenting the required local design information that is required for the global optimisation, as well as a facility to enable the "preview" of design decisions before any commitment is made.

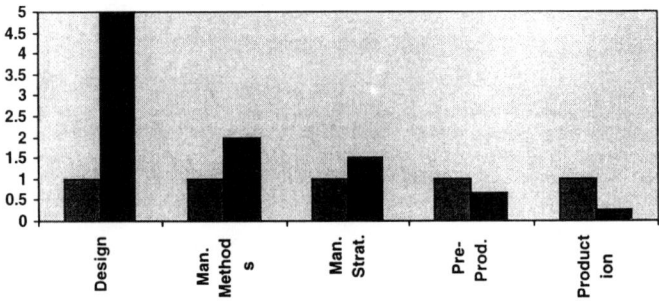

Figure 1: Investment yields of throughout different product stages [1].

The development of a computer supported conceptual design system, the Co-Designer, is motivated by the above mentioned product development requirements, stated more succinctly, these are, to assist a designer in the generation of conceptual design solutions and their evaluation from multi-perspective viewpoints, as well as to "preview" design decisions.

This paper will start by presenting a review of research works which focus on the development of computer-based conceptual aids, before describing the structure of the Computer-Aided Conceptual Design Tool "Co-Designer". A design example is used to illustrate the motivations for and use of this tool. This knowledge-based system is complemented by a decision consequence prediction module, which learns interactively from the user while solving a design problem. Such knowledge will enable the user to assess the suitability of a solution at an early design stage. This facility is vital for enabling designers to focus on a small set of potential solutions.

2. BACKGROUND

Research into computer supported conceptual design is continually gaining importance. A number of works have focused on the development of computer aides to support such activities as; conceptual solution generation [5,6,7], concept evaluation [8], decision-making [9,10], management of conceptual design processes by means of constraint networks [11], and learning [9].

Conceptual solution generation is a combinatorial search problem, in that, all combinations are evaluated against the requirements of a desired solution. Research into decision-making is an important area to formalise the evaluation of conceptual alternatives and the access of necessary information for critical decision-making points. Two distinct methods have been adopted to support evaluation of generated conceptual solutions, these are analytical and heuristic methods.

Under the analytical method umbrella, techniques such as genetic algorithms [12], simulated annealing, branch and bound and multi-criterion decision making [13] methods have been used to support early design decision making activities. Whilst, these techniques have shown potential in many design areas they require a "static" search space throughout the solution search process, an objective function will remain unchanged, even if a new piece of information is obtained.

Heuristic methods are characterised by the use of domain knowledge to guide the search for conceptual solutions. CADET [14], for example, uses heuristic knowledge stored as successful design cases to automatically produce potential solutions that match a desired product's specifications [9]. This mechanical engineering design assistant retrieves and re-uses previous successful designs whilst avoiding previous failures.

Machine learning techniques have been used to acquire background knowledge (heuristics) to guide search mechanisms [15]. The ARMS System [16], when failing to generate an assembly robot plan, uses an explanation-based learning technique to acquire domain knowledge through an example assembly run provided by a human assembly planner. The Co-Designer system uses an adaptive machine learning technique to interactively store a user's design episode and build a Decision Consequences Prediction network. The latter can then be accessed by the user to "predict" the consequences of any given design decision [9].

3. CO-DESIGNER

French [17] defines conceptual design as the phase that takes the statement of the problem and generates broad solutions to it in the form of schemes. It is in the conceptual design phase where greatest demand is made on the designer and where there is the most scope for striking improvements. Additionally, it is in this phase where engineering science, practical knowledge, production methods and commercial aspects need to be brought together, and where the most important decisions are made.

A scheme can be defined as an outline of a solution to a design problem, carried to a point where the means of performing each major function has been fixed, as have the spatial and structural relationships of the principal components. A scheme should be sufficiently worked out in detail for it to be possible to supply approximate costs, weights and overall dimensions, and the feasibility should have been assured as far as circumstances allow. A scheme should be relatively explicit about special features or components but need not go into much detail [17]. The points raised in the two preceding statements are fundamental considerations when commencing Computer Supported Conceptual Design (CSCD) development.

A key step towards developing any form of conceptual design tool is understanding and isolating the various stages comprising the conceptual design process. As shown in Figure 2, Pahl & Beitz [2] define these stages as involving *the establishment of function structures*, the search for suitable *solution principles* and their combination into *concept variants*. Concept variants that do not satisfy the *demands* (design specification) have to be eliminated and the rest must be judged by the systematic application of criteria based on the *wishes* (specifications). On the basis of this evaluation the best *solution concept* can now be selected.

3.1. Hybrid Conceptual Design Model

This example is a more specific form of the definition of the "Top-down" process by which humans solve problems, an essential part of which involves step by step *analysis* and *synthesis* [2] during which a transition from the qualitative to the quantitative occurs. However, there is a strong belief [18] that conceptual design practitioners do not adhere rigidly to the top-down approach. Instead, they draw on their own knowledge and experience, effectively a large collection of previously solved cases, from which they can evoke an appropriate solution principle to solve totally (or partly) a given conceptual design problem. Such gradual refinement can be termed "bottom-up" problem solving (Fig. 3), [18].

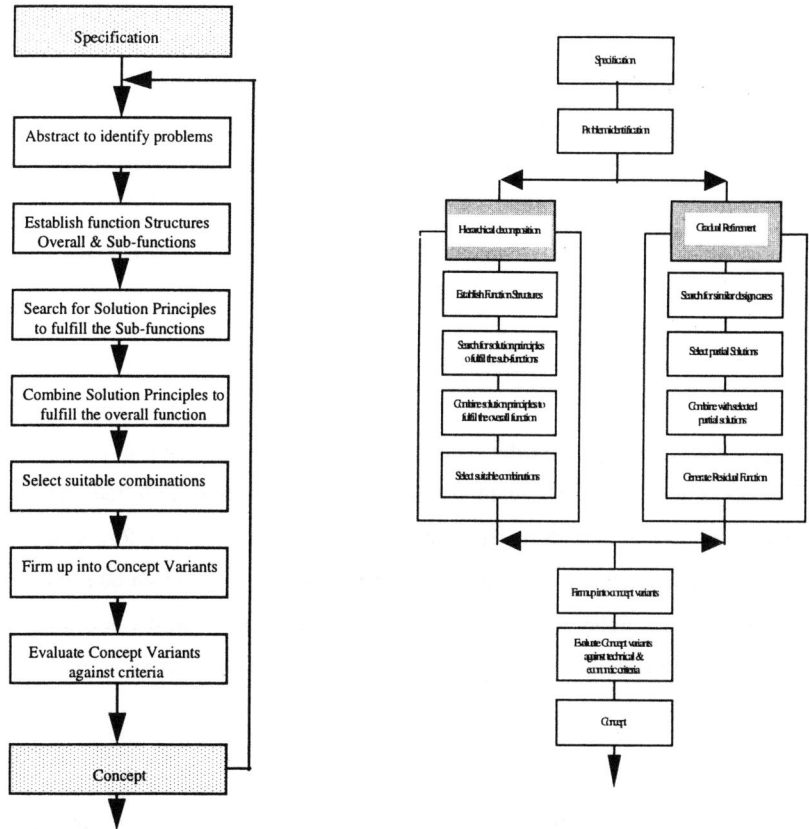

Figure 2: **Pahl & Beitz's Model of Conceptual Design.**   Figure 3: **Hybrid Conceptual Design Model.**

The Co-Designer system has been developed based on a hybrid conceptual design model, (i.e. top-down/bottom-up) (Fig. 3). Where the user is able to search freely for possible solutions to given design problems. If the evoked solution principle subsets can then undergo some form of optimisation, referenced against input criteria, then a useful facility can be created. Depending upon the power of the optimisation process implemented, it is envisaged that the output from this process may not only be the optimal relationship between the individual solution principles, but, may even generate new hybrid solution principles.

3.2. Co-Designer System Architecture

Figure 4 illustrates the Co-Designer's main components, which are briefly described below:

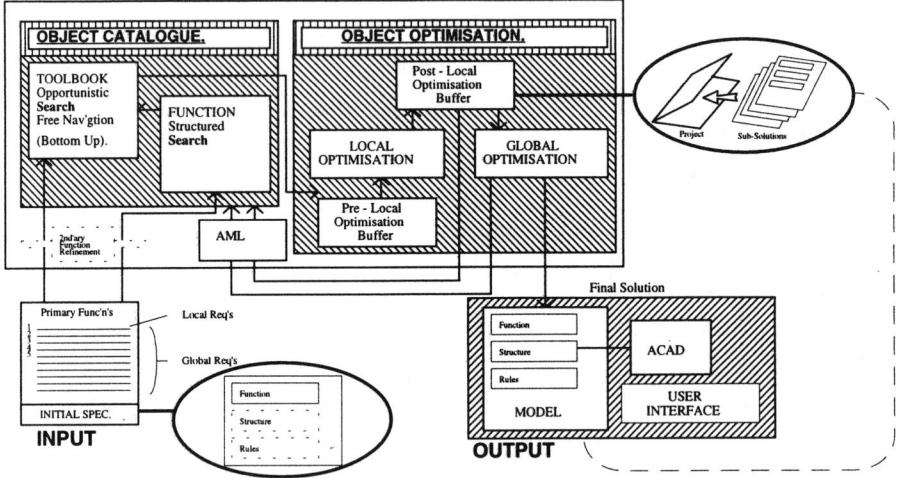

Figure 4: **Co-Designer System Architecture.**

1. **Function Specification Module:** This facilitates the inputting of a new product's functional requirements. The information input is structured into required design functions and any associated mathematical rules (constraints). This information is then used for the generation of conceptual solution variants. The solution search is undertaken as an opportunistic and/or structured search mechanisms. Opportunistic Searching is the means by which the designer can browse a multimedia level within the concept base with the aim of generating design solutions. The structured search mechanism uses the Concept Base's sub-level database of function structure and constraints to retrieve a required solutions, during this process, depending on the nature of the problem and the size of the

progressively growing Concept Base, numerous sub-solutions will be generated as different permutations of solution principles produce different final solutions. The full description of the conceptual solution generation is out of the scope of this paper and can be found elsewhere [9,18].

2. **Design Object Catalogue:** This contains design information structured into: (i) hypermedia form contained in the interactive Design Tool. The latter contains generic engineering description of a specific object/concept as well as rules, guidelines and principles related to a particular concept. (ii) Database form contained in the Concept Base. As mentioned in the previous paragraph, these two forms of knowledge representation are necessary to support both the opportunistic and the structured solution search mechanisms.

3. **Optimisation Module:** This enables the search of local optimum solutions for a given sub-problem, such solutions will be added to the global search space, from which a global conceptual solution will be selected using the concurrent global optimisation module. The global optimisation module is based on a developed concurrent global optimising algorithm (Sec. 6).

4. **Machine Learning Module:** This stores solved conceptual design cases as a weighted directed graph of users interactions. The latter can be used to predict the consequences of a given design decision before any commitment is made. The *design decision consequence* is based on the use of an adaptive machine learning method (Sec. 5). Co-Designer contains two basic learning modules, which are the *Concept Re-use* [18] and the *decision consequence prediction* module [9]. These modules are based on the rote learning approach, a basic machine learning technique that stores computed facts and results for future reuse in order to save recomputation time. Design decision consequences may be defined as a sequence of events that follow the selection of the considered decision (action).

5. **Design Object Catalogue Update Module:** Whilst the Adaptive Machine Learning Element will facilitate the updating of the catalogue with any sub-solution generated, the updating of the 'Multimedia' level within the catalogue will be undertaken by the user. This process is often carried out at the end of a design project.

6. **Product Model:** This contains all the information concerning a given design. A project file storage mechanism is used to store information related to a particular partial solution.

The structure of a final solution will contain information such as function, physical structure and rules, and any other required information for example cost, dimensions and weight.

7. **Design Product Management:** Depending on the size of the catalogue and the nature of the design problem the potential for the generation of a large amount of information is great (Sec. 4.2). This will be further compounded if the product design has been decomposed into various elements which are then designed concurrently. As a result the amount of information that could exist relating to each design element which together forms one design project could be vast and the likelihood of conflict arising between the various design groups is high.

4. ADAPTIVE MACHINE LEARNING

This multi-strategy machine learning technique uses case-based reasoning and incremental machine learning technique. The latter is based on two principles, these are:

1. **Parameter adjustment:** Under the same conditions, the more frequent a plan (sequence of actions) is utilised the more likely it will be used again in future problem-solving. This concept technique has been adopted in a number of techniques such as: neural-networks and plausible inference techniques.

2. **Conceptual Clustering:** This concept learning approach is in the present implementation based on a user-driven concept clustering technique to generalise and abstract generated plans' clusters [19]. For instance, consider the "toolRobotJoin" task that is used to select a Robotic clamping device from a given solution space, the possible solutions within this space could range from mechanical clamps to pneumatic clamps. A plans' cluster is a set of plans that are "close" to each other. In other words, as it is in our application, they are plans used in the same scenario (design context), (Fig. 5a).

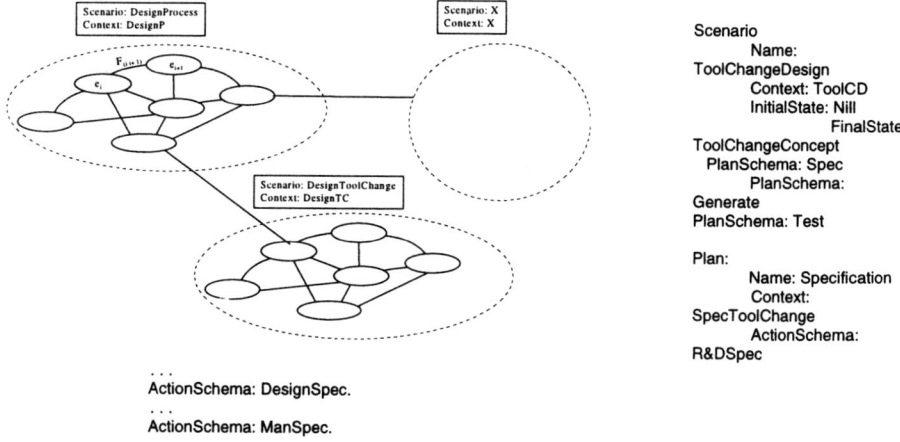

Figure 5a: **Plan Clusters.**   Figure 5b: **A script-based knowledge structure.**

As shown in Figure 5a, any event (action) e is an element of the set E of stored scenarios. A scenario S_i is defined by a triple (SA,C,P), SA is an ordered collection (sequence) of actions that will occur under a defined context C with a probability of occurrence P.

$$\forall e \in E$$
$$where\, E = \{S \mid S = (SA, C, P)\}$$

A sequence of actions SA_i will be described as a weighted directed graph, in which events e_i represent the nodes and arcs are represented by causal links. The strength of an arc between two consecutive nodes (events) e_i and e_{i+1} is calculated as:

$$F_{|i\,i+1|} = \frac{N_s}{N_t}$$

$F_{(i\ i+1)}$ is a statistical function that measures the total number N_t of utilisation's of a considered S_i over the number N_s of successful traversal from node e_i to e_{i+1}.

Bayes theorem is applied to estimate the probability for the prediction of a the occurrence of event B:

$$P(A_i|B) = \frac{P(A_i)P(B|A_i)}{\sum_{j=1}^{n} P(B|A_j)P(A_j)}$$

Similar to explanation-based learning, the implemented incremental machine learning technique runs as an underlying process throughout a user's concurrent engineering project planning, the system updates existing plans by updating the frequency factor (weight) of arcs.

Figure 6 shows a simplified design episode of a "robot tool change". Script-based knowledge representation is used to structure background (domain) knowledge [20] in terms of context, and a set of operators (actions) an agent (user) can apply to perform the required changes to a design world to change from the initial state to the goal (final) state.

The pseudocode below illustrates the major steps of the incremental machine learning algorithm. The user is first required to define the initial design context to select the most appropriate stored design scenario, and to initialise the inferential network. The next event (action) is selected based on the event prediction mechanism which selects the next node with the highest frequency factor[1]. The algorithm reads the next user's action and compares it to the predicted node. This test (in the present implementation) leads to the either the addition of a new action or increment of the selected node frequency factor. The algorithm ends by ending the design episode.

```
IncrementalLearning  <procedure>

  Begin
     Ask(user,context)
     initialise(designScenario)
     currentNode(Scenario,CurrentAction)
     step(Scenario,CurrentAction)
  end

  step(Scenario,CurrentAction)  <procedure>
  Begin
     predictRead(Scenario,User,CurrentAction,NextAction)
     if      NextAction = NILL
     then    readFromUser(Scenario,Action)
     else    mapUserInput(Scenario,CurrentAction)
  end

  mapUserInput(Scenario,CurrentAction)  <procedure>
```

---

[1] In the case of the next node being nil (i.e. the end of the network has been reached) the algorithm will use a rote learning mechanism to add the user's next actions to the network with a frequency factor equal to zero.

```
Begin
    displaySequence(Scenario,CurrentAction,NextActionSet)
    read(User,NextActionSet, Action)
    if      NextAction = Action
    then    currentNode(Scenario, Action)
    step(Scenario, Action)
    else    addNode(Scenario,CurrentNode, Action)
    Step(Scenario, Action)
end
```

Figure 6: **Pseudo code for Design Conseqence Prediction**

The stored design process model is structured as a decision network, whose nodes represent design decision points. For instance, the task of selecting a clamping device for the tool adapter on the robot arm "toolRobotClamp" [9]. Each of the arcs is a causal link with associated strength. The latter represents a frequency factor, which is a statistical measure of how often a particular task has been used. Similar to the Samuel's checkers program [19] the DCP module modifies the strength of a link on the basis of it's experience[2]. The frequency factor's adjustment trains the system to prefer commonly used alternatives.

5.   DECISION CONSEQUENCE PREDICTION

Decision Consequence Prediction (DCP) can instigate a stored decision network (or part of it) to enable the user to assess the consequences of any given design decision before any commitments are made. For instance, as shown in Figure 7 [3], the selection of "adjacent" arrangement of tool adapters will require a re-design of the tool rack, and addition of lateral robot arm motions; because of the mentioned re-design requirements the "in-line" tool adapter rack arrangement is selected (Figure 9).

At each decision point the user can ask the DCP module to display an ordered collection of actions (events) associated with a considered decision (Sec. 5). From which the user could either use the predicted next node, or any other node. In addition, at each decision point the DCP enables the user to preview the consequences of a given network path, ("walk" down a particular decision path). This facility enables the user to "focus" the conceptual solution generation process, by a process of design exploration ("What if").

---

[2] Features that appear to be good predictors of overall success will have their weight increased, while those that do not will have their weights decreased.

[3] As part of the illustrative example detailed previously, we are required to design a rack for the tool adapters. The rack will be required to locate and support a tool adapter, and separate a tool adapter from the robot arm (Fig.8).

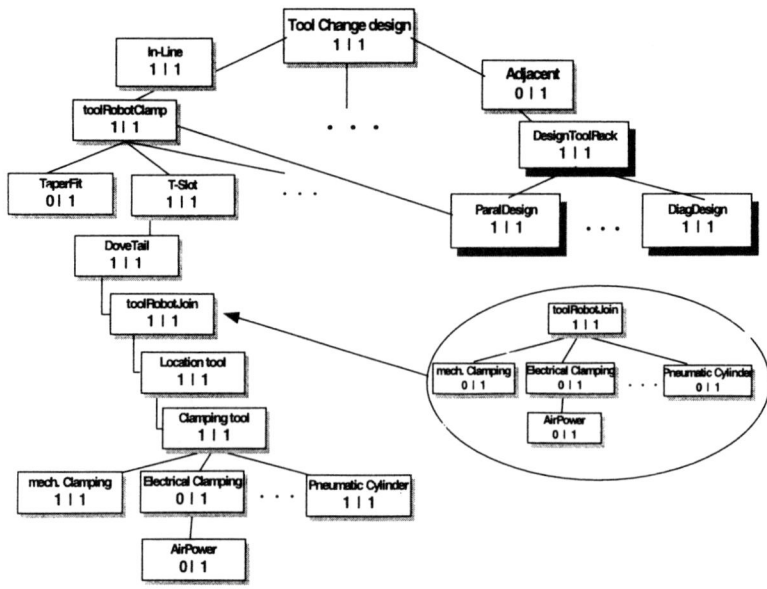

Figure 7: **Shows a simplified version of the recorder decision tree of the rapid tool change design.**

The DCP is being evaluated in the context of the concurrent conceptual design. The design of a "rapid tool change" has been used in earlier publications [9] to illustrate the use of DCP.

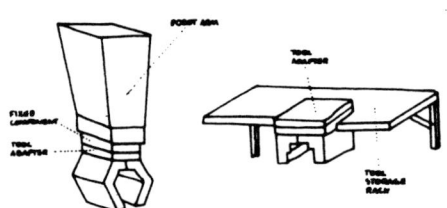

Figure 8: **Illustrates the Robot Arm & Tool Adapter Rack.**

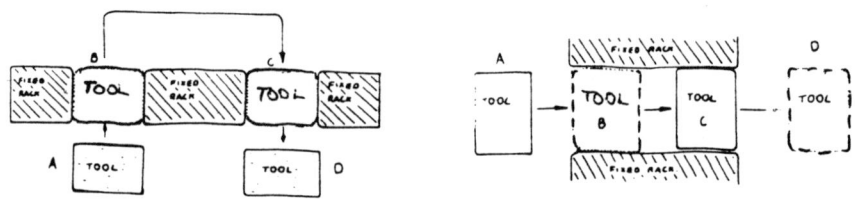

Figure 9: **Shows the in-line and adjacent Tool decomposition.**

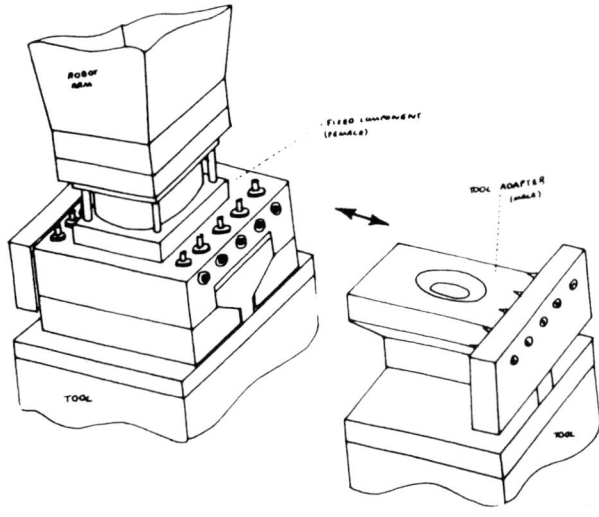

Figure 10: **Illustrates three alternatives to perform the locate function.**

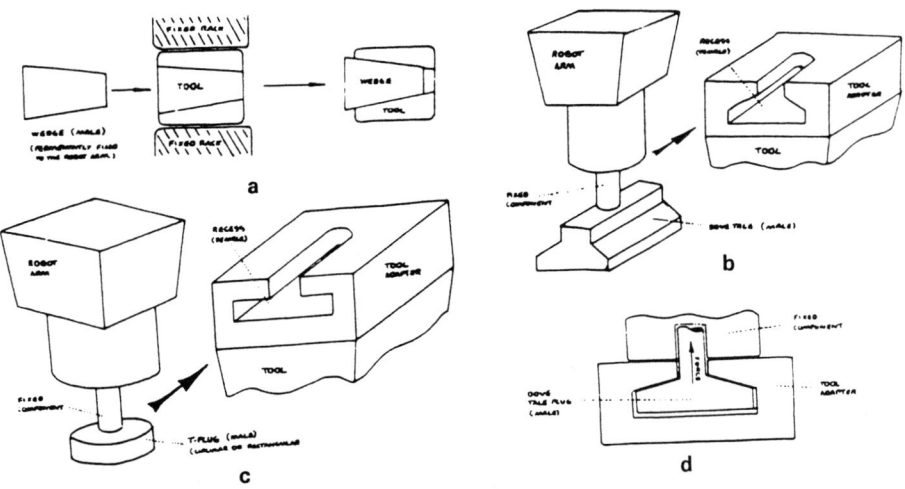

Figure 11: **The Final Design Solution.**

6. CONCLUSIONS

An increasing number of research works in conceptual design are focusing generally on the implementation of computer tools; such tools provide a wide spectrum of facilities ranging from "aide memoir" and creativity support, to geometric modeller to design decision supports [9].

It is believed that conceptual engineering design practice is best modeled as a hybrid structured (top-down) and opportunistic (bottom-up) approach. To this end, Co-Designer tool is being developed supporting step-by-step design and concept reuse method.

Furthermore, applying techniques from machine learning, such as rote learning and parameter adjustment learning, Co-Designer learns design problem-solving processes by recording the user interaction (actions, decisions, etc.), storing them as decision trees. These will be extended, improved ("fine tuned") every time the system is used. The parameter adjustment learning technique is intended to enable Co-Designer to learn by experimentation (by doing) and to give preference to commonly used design alternatives. The stored decision trees are used to provide the user (when required) with a view of the consequences of a particular decision. This facility will enable the system and the user to focus on smaller solution space.

Further development of the prototype of the Decision Consequence Prediction module will be to develop routines to assess the benefit of the addition of a new piece to an existing decision tree.

The paper has introduced the requirement of an optimisation method to enable conceptual designer to assess the cost of each alternative as early as possible in the design process.

The assessment of "all" the solutions to a particular problem is of particular importance to the conceptual design activity; where innovative solutions are often researched. In addition, the assessment processes throughout this stage are very difficult to perform, due to the lack of information which characterises this stage.

There is a pressing need for a development of a methodology to support the evaluation of conceptual designs alternatives from a multiple life-cycle facets (design, manufacturing, assembly, etc.) throughout the generation of alternative solutions. This requires access of necessary down-stream information throughout the early design stages. Inherent in this approach is the need for appropriate management of the large amount of information, and the development of an efficient method for presenting the required local design information that is required for the global optimisation.

Although design by reuse of previous work is an interesting problem-solving technique, it presents many difficult issues. The first issue is related to the documentation of the reusable work (component, concept or plan) at an appropriate level of description. The second issue is related to the formalisation of a reusable concept retrieval process. The third issue is related to the profitably and feasibility of a previous design's reuse.

This paper has presented an attempt into the adoption of the reuse philosophy into a systematic design methodology. The reuse of design concepts in particular provides an additional problem-solving tool to design problem resolution by decomposition [19, 21]. In addition, the documentation of previous cases can constitute a means for designers to learn from each other's and/or experts' solved design cases.

6. References

1. Berliner and Brimson 1988, <u>Cost Management for Todays Advanced Manufacturing: The CAM-I Conceptual Design</u>, Harvard Business School Press.
2. Pahl, G. and Beitz. P., 1984. <u>Engineering Design,</u> Springer-Verlag, Berlin, 1984.
3. John Sharpe & Vincent Oh, 1994, <u>Computer Aided Conceptual Design,</u> John Sharpe & Vincent Oh (Eds.), loose-leaf Co., ISBN 0 901800 37 6, 1994, pp.
4. Carrubba, F.P., "Designing People-Pleasing Products", Keynote Address, in <u>9th International Conference on Engineering Design ICED'93,</u> The Hague, Heurista, Zurich, 1993.
5. Taleb-Bendiab, A. 93, "ConceptDesigner: A Knowledge-Based System for Conceptual Engineering Design", *International Conference on Engineering Design, ICED'93*, 1993, The Hague, Holland, Heurista, pp. 1303-1311.
6. Chandra, D.N., S. Narasimhan, and K.P. Sycara 92, "Qualitative Reasoning Methods in Design, *Intelligent Design and Manufacturing*, A. Kusiak, Editor, 1992, Wiley-Interscience, New York, pp. 103-131.
7. Hundal, M.S. and L.D. Langholtz 1992, *"Conceptual Design by Computer-Aided Creation of Function Structures and Search for Solutions",* Journal of Engineering Design, 1992, Vol. 3, No. 2, pp 127-138.
8. Bayliss & A. Taleb-Bendiab, 1994, "A Global Optimisation Technique for Concurrent Conceptual Design", MMU, 1994
9. Taleb-Bendiab, A. 1994, "A Knowledge-Based Aid for Conceptual Engineering Design, <u>1994 Lancaster International Workshop on Engineering Desing,</u> J. Sharpe and V. Oh (Eds.), Lancaster EDC, pp. 189-203.
10. Goldberg, D.E. 91, "Genetic Algorithms as a Computational Theory of Conceptual Design", *International Conference on Application of Artificial Intelligence VI*, 1991, Elsevier Applied Science, pp. 3-16.
11. Serrano, D. and D. Gossard 87, *"Constraint management in conceptual design",* 1987, pp 211-224.
12. Rawlins,G.J.E., 1991. <u>Foundations of Genetic Algorithms</u>. Morgan Kaufmann, San Mateo, Calf.
13. van Laarhoven,P.J.M., Aarts, E.H.L., 1987. <u>Simulated Annealing: Theory and Applications</u>. D.Riedel, Dortrecht.

14. Sycara, K., D. Navin Chandra, R. Guttal, J. Koning and S. Narasimhan 92, "CADET: a Case-Based Synthesis Tool for Engineering Design", *International Journal of Expert Systems Research and Applications*, 1992, Vol. 4, No. 2, pp. 157-188.
15. Carbonell, J.G., R.S. Michalski, and T.M. Mitchell 83, "An Overview of Machine Learning, Machine Learning: An Artificial Intelligence Approach, R.S. Michalski, J.G. Carbonell, andT.M. Mitchell (Eds.), 1983, Tioga Publishing Company, Palo Alto, pp. 3-24.
16. Segre A., 93, "Learning how to plan", Towards Learning Robots, Walter Van de Velde (Ed.), A Bradford Book, The MIT Press, London, 1993, pp. 93-112.
17. French, 1985 Conceptual Design for Engineers
18. Taleb-Bendiab A. 92, "A Concept Reuse Approach for Engineering Desing Problem-Solving", *7th International Conference on the Application of Artificial Intelligence in Engineering*, 1992, Waterloo University, Ontario, Canada.
19. Samuel, A.L. 63, "Some Studies in Machine Learning using the Game of Checkers", in *Computers and Thought*, E.A. Feigenbaum and J. Feldman, Editor, 1963, McGraw-Hill, New York,
20. Taleb-Bendiab A, Oh. V, Sommerville. I & French. M, "Knowledge representation for Engineering Design Product Improvement". AIENG 92.
21. Maher, M.L. and D.M. Zhang 91, *"CADSYN: Using Case and Decomposition Knowledge for Design Synthesis"*, J. Gero (Ed.), 1991, Edinburgh, Butterworth Heinemann, pp 137-150.

THE APPLICATION OF GENETIC ALGORITHMS TO CONCEPTUAL DESIGN

M G Hudson & I C Parmee

1. INTRODUCTION

Design is not a simple hierarchical process where the designer is presented with a set of requirements and works steadily through a decomposition strategy moving from abstract concepts to the final concrete product. The design problem is ill-defined and changes as the designer explores it through solutions and partial solutions. Designers use a 'solution focused' strategy. Feasible solutions are posed to probe the 'instability of the problem' and the 'limitations of the way the problem is framed'. It is common for architects, in professional practice, to first simplify the problem so they can generate a rough solution. This solution is then used to develop understanding of the problem which leads to a gradual refinement of that solution[9]. Neither is design simply a matter of iteration around an essentially hierarchical process. Information gained during the design process can prompt the designer to transfer the design effort to higher levels, or to a location remote in the hierarchy. Design is essentially a heterarchical, possibly chaotic, process. The heterarchical nature of design is even more apparent in team design and enshrined in the philosophy of concurrent design. Given the perceived nature of the design process, adaptive computing techniques with their property of emergent behaviour present an attractive paradigm for conceptual design. An overview of the evolutionary design capabilities is presented in Appendix A for those who are unfamiliar with the techniques.

2. RELATED WORK

Genetic Algorithms and other adaptive search techniques have a long history of being applied to engineering design, most notably optimisation problems. Rechenburg's early work at the Technical University of Berlin demonstrated in a practical manner the powerful design processing capabilities of evolution-based strategies[16]. More recently French (1988) has highlighted the similarities between evolutionary processes and engineering design[3], whilst Parmee (1993, 1994) identifies strengths and weaknesses of the integration of evolutionary principles through practical application and the development of complementary techniques[10,11]. Given the attractiveness of the evolution metaphor to design, and the successful application of genetic algorithms to engineering optimisation problems, it is perhaps surprising to discover how little work has been done on integrating these techniques with conceptual design. However, during the 1990s adaptive search techniques have begun to be applied to conceptual design problems.

Goldberg (1991) outlines a framework for conceptual design in terms of GA operators[6]. The theory is supported by :-
- An outline of the similarities between a designer's recombination of previous design components to form new proposals and the GA's genetic operators of selection, recombination (crossover), and mutation.
- a "mathematical" analysis of the search properties of messy GAs.

Woodbury (1993) also draws analogies between natural evolution and the creative design process[17]. The paper presents a loose framework for Genetic Design Systems (GDS) and suggests two research directions: a theoretical mathematical description of GDS, the exploration of design issues through the application of practical GDSs. Although, Woodbury's work does not explicitly address the issue of conceptual design, there is no reason why his architecture could not be applied to these problems. Grierson (1994) tackles the conceptual design problem through the integration of a GA with a neural network[8]. The GA develops designs using the usual genetic operators of reproduction, crossover and mutation. The neural network is used to evaluate the conceptual designs produced by the genetic operators, and in effect forms part of the fitness function of the GA. The designer trains the neural network by presenting design concepts and evaluations. Grierson highlights the problem of the designer having to present the system with a large number of input-output training examples. However, the approach appears to be adequate to support routine design. TRADES (TRAnsmission DESign), Pham and Yang (1993), is a prototype GA based system which evolves conceptual / preliminary designs of vehicle transmissions[15]. This is probably the most advanced GA based "conceptual" design tool. However, the nature of the problem appears to be restricted by the sole use of quantitative evaluation functions, due to the systems emphasis on preliminary rather than conceptual design.

The work described in this paper attempts to apply GA and Adaptive Search (AS) techniques developed in optimisation problems to the conceptual design task. However, the nature of the conceptual design task, differs considerably. In an optimisation task the variables are known and the GA is left to discover their optimal values. However, according to Lawson[9] a designer explores a problem in terms of solutions and partial solutions. The nature of the problem is clarified and defined in parallel with the emergence of the solution. Hence, applying GAs to conceptual design problems raises a number of issues which are usually not present in optimisation work :-

(i) Neither the structure of the final solution nor the design space to be searched is fixed.
(ii) The evaluation of concepts is not a simple quantitative comparison.
(iii) A range of good solutions is more important than a single "optimal" solution.

The research can be seen as belonging to Woodbury's application driven path. The intention is to develop modular systems and explore the utility of various concepts and features through experimentation. One of the goals is to develop practical systems, like Pham & Yang, but based on more realistic, qualitative rather than quantitative, conceptual design problems, like Grierson.

However, the modelling of conceptual design is extremely complex and no claim is being made at this stage of the research concerning the practical integration with real world problems. The intention here is to review current approaches and present the authors' intended methodologies for discussion.

3. OVERVIEW OF SYSTEM ARCHITECTURE

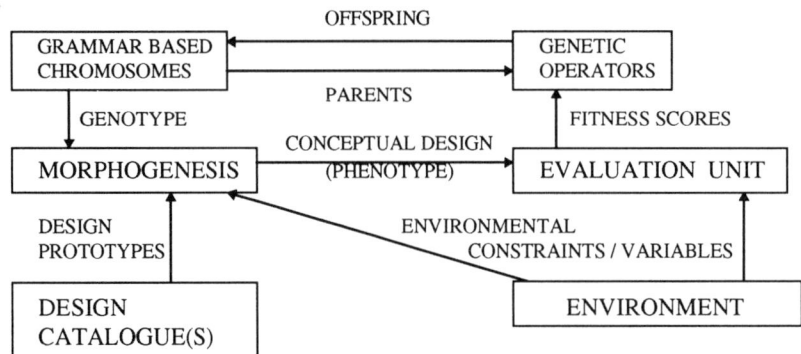

Fig 1. Architecture of the Grammar Based Chromosome GA System

An overview of the proposed system architecture is shown in figure 1. The user interface is not included, but can be regarded as an additional layer located above the generic engine. Details of the user interface are included in section 7.

An initial set of grammar based chromosomes are randomly produced from the design grammar. The growth unit takes the chromosome (the genotype) and using the design prototypes, and possibly variables in the environment generates a conceptual design, the phenotype. The phenotype consists of a set of connected design prototypes. The conceptual design is then given a fitness score by the evaluation routine, which may also utilise variables from the environment. The population is ranked according to the Pareto concept of non-dominance, and reproductive probabilities are assigned in accordance with this rank. A modified set of genetic operators create the next generation.

The designer interacts with the system initially by providing a set of functional requirements and goals, and a set of evaluation criteria. During the GA run, the designer is presented with information on the best, worst and average fitnesses, a measure of the diversity of the chromosome population, and the facility to examine any of the chromosomes of the current population. Between generations the designer can modify the functional requirements and goals, the evaluation criteria, the design grammar, or even one of the chromosomes.

4. KNOWLEDGE REPRESENTATION / DESIGN CATALOGUES

4.1 The Representation of Design Concepts / Components

The design catalogues are organised in accordance with Gero's design prototypes[4], although in a greatly reduced manner, see figure 2. The system stores frames based on function, structure

Name : Linear Pneumatic Activator
Functtion : Activator
Behavioural : Linear Motion
Structural :
    Input Port(s) :
    Output Port(s) :
    Components :
Behavioural Variables
    Boolean : Flammable = False;
    Enumerated Type : Safety Hazard = Low;
    Enumerated Type : Cost = Low;
    Qualitative : Compliance = M^+ Pressure;
    /* Qualitative equations linking input with output ports for design interpretation */

Fig 2. Part of a design prototype for a pneumatic activator

and behaviour. Components are indexed in the design catalogue(s) according to their function. The structure and behaviour slots of the components' frame are used when combined in a chromosome to prevent the generation of lethals, as shown in figure 2, and to evaluate the concept they represent to determine fitness scores.

4.2 The Representation of Designs / Chromosomes

4.2.1 Fixed Length Chromosome Strings

The first prototype uses a basic string chromosome, the length of the string being determined by the number of sub functions specified by the designer. Each sub function is allocated a bit / allele on the chromosome string. For each sub function the design catalogue is searched and a list is created of all "components" which fulfill the functional requirement. The initial population is then randomly generated from these lists. Subsequent generations are developed by the fitness evaluation and reproduction techniques of genetic algorithms, as discussed below in the section on concept generation.

    The major limitation of using fixed strings is that it prevents the addition, combination, and elimination of concepts / functional requirements. This precludes the possibility of function combination and / or seperation which is a common operation during the development of conceptual designs.

    It also forces the prototype to work on a single "design level", the classic, simple GA works on an unstructured representation. However, one of the issues we are keen to address is the application of adaptive search techniques to different levels of abstraction and hierarchy, eg component, sub-assembly, machine, mechanical system.

4.2.2 Grammar Based Chromosome Strings

The limitations of fixed length chromosome strings can be tackled in a number of ways. A grammar based GA overcomes the problem of not knowing the structure, ie number of elements, of the final conceptual design in advance. It also enables the system to handle the function / sub-function decompositon tasks. A grammar based GA enables the system to add, combine and eliminate required functions / components. It also enables the system to handle more of the conceptual design process. The first prototype provides support only in the latter stages of this process, after the completion of the functional decomposition of the design. The functional decomposition process reduces the overall search space and therefore the scope for creativity and identification of innovative solutions to the design problem, by defining a single functional structure. The second prototype, the grammar based GA, supports the functional decomposition process.

Design grammars have been successfully applied to many problems, eg Coyne (1988)[2]. One of their major advantages is the criteria of completeness. Theoretically, a design grammar is complete - it will generate all possible designs composable under the grammar. This is an important characteristic. Given that human recognition is better than recall, a system which is designed to enhance a human designer can use the completeness of the grammar to prompt the human designer with design ideas that he would otherwise not have remembered. Unfortunately, in applying design grammars to real problems, the completeness becomes a liability. The combinatorial combination of design possibilities makes an exhaustive search of the design space specified by the grammar unrealistic. By combining a design grammar with a GA, we hope to retain the "completeness" aspects of the design grammar, while using the GA to avoid the combinatorial problem by avoiding the need to conduct an exhaustive search. Fig 3, shows part of the design grammar used by the system to design robot grippers.

The initial population of chromosomes is randomly seeded from the design grammar as an embedded list, see fig 4.

In principle a context sensitive grammar could be rewritten as a context free grammar. Although, this would reduce the computational time required to generate phenotypes from the genotype, a context free grammar would have to be much larger than a context sensitive grammar to cover the same design space. In addition to being smaller and hence more readable etc, the context sensitive grammar can incorporate tests on both the contents of the design prototypes and the environment and also set actions to add / modify the contents of the environment. The test clauses also prevent the generation of lethals. The main advantages of using a context sensitive grammar over a context free grammar are :-

(i) Avoid the production of lethals
(ii) To enable the system to use "global" information and constraints stored in the environment.
(iii) Context Sensitive Grammars are more expressive than Context Free, type 2, Grammars.

```
Gripper ::= Clamping Element + Securing Type + Executive Mechanism +
            Transport Mechanism + Actuator
Clamping Element ::= Geometrical Features + Surface Properties + Construction Features +
                     Kinematics of Movement
Securing Type ::= Rigid Connection | Resilient Connection | Kinematic Pair
Executive Mechanism ::= Pushes | Levers | Resilient Rods | Cables | Yokes | Inflatable Envelopes
Transport Mechanism ::= Reducer | Lever Mechanism | Collet Mechanism | Rack Mechanism | Cam Mechanism
Reducer ::= Cylindrical | Conical | Worm | Wave | Differential
Actuator ::= Drive | Springs | Own Weight
Drive ::= Pneumatic | Hydraulic | Electric
```

Fig 3. The robot gripper design grammar

```
[Gripper
[Clamping Element
    [Geometrical Features + Surface Features + Surface Properties +
    Construction Features + Kinematics of Movement] ] +
[Securing Type
    [Rigid Connection] ] +
            .
            .
            .
[Actuator
    [Drive
        [Pneumatic]]]]
```

Fig 4. Part of the chromosome for the conceptual design of a robot gripper

4.2.2.1 Modified Genetic Operators

The "cost" of using the grammar based GA is an increase in the complexity of the GA and the genetic operators. A traditional GA could be applied to strings generated by a grammar, however the crossover operator would generate many GA legal but grammar illegal strings. Although, the grammar illegal strings could be eliminated by setting the objective function to zero for these strings, in situations where the proportion of grammar legal strings is small compared to the set of all strings in the domain, the GA would be reduced to virtually a random walk.

This problem can be avoided by the use of Antonisse, H.J. (1991)'s crossover operation[1].

(i) An arbitrary crossover point is chosen.
(ii) Split the sentence at the crossover point.
(iii) Mark the trailing edges with tags that signifies the crossover location in the derivation / parse tree. The tag consists of the nearest production whose left hand side derives both the last term of the leading sentence fragment and the first of the trailing fragment.
(iv) Find the point in the parse tree of the second string & crossover.

The modification will result in only grammar legal strings being generated.

Mutation is handled by randomly selecting a point on the chromosome and then replacing it, together with its sub-structures if appropriate, with one of the alternatives specified in the design grammar.

Although the current approach appears to be adequate for routine and innovative design, the approach cannot be easily extended to deal with creative design. The grammar based GA provides an efficient way to search the design space specified by the grammar, but the system is limited to finding solutions within this design space. In order to find creative solutions to design problems, the system needs to be able to alter the design space it is searching. Gero (1994) encodes the design grammar as part of the chromosome and therefore enables the genetic operators to alter the grammar thereby creating new design spaces to search[5]. We are currently investigating the strengths and weaknesses of this approach.

5. EVALUATION

One of the major problems in "automating" conceptual design is the problem of evaluating design alternatives. Evaluation criteria for concepts tend to be a mixture of quantitative and qualitative criteria. The fitness function must be derived from criteria which are often fuzzy, ill defined, on different levels of abstraction, expressed in a mixture of scale types and subject to frequent changes during the design process.

The evaluation criteria are derived from the requirements list. This list will already be divided into functional requirements and goals. The functional requirements should have been stated in precise quantifiable terms. In principle these are simple to integrate into the fitness function. However, although the functional requirements criteria are expressed in quantitative terms, the information stored with the design concepts tends to be expressed qualitatively. The system must therefore map between quantitative and qualitative data using a qualitative reasoning / qualitative arithmetic approach. A conceptual design either meets a functional requirement or it does not. Two approaches can be taken to conceptual designs which fail to meet functional requirements. The simplest is to set the fitness function to zero effectively excluding the genotype from the next generation. Unfortunately, this leads to the loss of potentially valuable genetic material. The second approach is to add a penalty value to the fitness score. This is a conservative strategy, ie the increase in convergence time is considered to be potentially less dangerous than the premature loss of genetic material. Goals tend to be qualitative, eg the gripper should be as fast or as cheap as possible, and usually require comparisons to be made between conceptual design alternatives.

The list of requirements, both functional requirements and goals, is presented to the designer. For each requirement the designer must indicate the scale type of the criteria, eg nominal, ordinal, interval or ratio. It is important that the system handles different scale types properly, to avoid the assignment of numeric values and the drawing of distinctions that do not exist in reality. It also simplifies the construction of the fitness function.

The result should be the production of a number of evaluation criteria. Although these criteria could be combined in a weighted function, eg cost-benefit analysis and the concept ranked on the basis of this score, human designers appear to evaluate the pros & cons across criteria. The approach adopted here is to follow this process by using Pareto Optimisation techniques. This avoids the problem of having to decide the relative importance of criteria. This eases the problem of cost assessment, one of the most important and problematic evaluation criteria. Engineering problems exist, eg building design, for which quantitative cost functions can be accurately and reliably derived from conceptual designs. Unfortunately, for other engineering design problems cost estimation may not be as reliable. Pareto Optimisation allows us to provide cost information to the designer without running the risk that the unreliability of costing calculation are distorting a weighted evaluation function. Only design concepts which fulfill the requirements are compared to avoid the inclusion of cheap, chocolate fireguards in the Pareto Optimal Set (P-set). Maintaining the P-set should provide the designer with an insight into the trade off issues between criteria.

Evaluation is not simply a matter of "summing" characterisitcs of components. Some of the criteria and characteristics of a design arise from the combination of components and their interactions, the whole is often more than the sum of its parts. To evaluate / interpret a conceptual design, the system needs to access both the structural and behavioural components of the design prototypes.

The exploratory nature of the conceptual design process prevents an apriori definition of a quantitative fitness function or even the determination of relevant variables to be manipulated by the system. Both the constraints of the problem requirements and the desirable criteria to be met, emerge as the designer develops insights into the problem by exploring the problem domain through potential solutions[12]. A GA cannot simply be set up and allowed to run, in the hope of developing a near optimum solution to the conceptual design problem. One option is to place the designer in the fitness function in order to evolve the design interactively. This would provide a number of benefits :-

(i) The designer is in control of the tool and can drive the GA in the exploration of the design domain, by modifying the fitness function on the basis of information learnt / knowledge acquired during the exploration of the design space.
(ii) The designer can provide some of the qualitative aspects of the evaluation of the solution which would otherwise be hard to evaluate.

However, it is very unlikely that a designer would blindly accept a proposed conceptual design. The system would need to justify the proposed solution. There is no single optimal solution to Conceptual Designs but a range of acceptable solutions. Part of the justification of a conceptual design is the knowledge of its value compared to other potential designs, which a designer gains by his or her exploration of the design space. Unless, the designer is able to use the system to explore the design space they will not be in a position to judge the quality of the design solution.

The disadvantage of placing the designer in the fitness function lies in the limitations that places on the number of designs which can feasibly be presented to a designer. A typical GA would run on a population of 50 chromosomes and take $\approx$ 200 generations to converge, a total of approximately 10,000 fitness evaluations. Clearly it is not practical to ask the designer to do this. In order to successfully integrate the designer as part of the fitness function, we need to reduce this number by the order of a couple of magnitudes. The current strategy is a compromise between the ideal and feasible. The designer is not directly part of the fitness function, but can interact with the GA between generations, when fitness functions etc can be modified.

A number of alternative approaches to this problem are possible and should be explored :-
(i) Reduce the size of the GA pool and the number of generations / evaluations.
(ii) Use clustering techniques[13] to reduce the number of evaluations the designer has to make as part of the fitness function. The GA is run for a number of generations, clustering techniques are then applied, and typical members of each cluster are shown to the designer to obtain evaluations for the subjective aspects of the problem.

6. THE GENERATION OF CONCEPTUAL DESIGNS

6.1 Constraint Violations and the Avoidance of Lethals
Constraints violations can occur from two sources :-

(i) The designer may specify constraints in the specification, eg equality / non-equality.
(ii) From the interaction of device components selected to fulfill the functional specification.

The second type of constraint is handled by the grammar and in principle cannot occur or at least are eliminated at the generation stage. The grammar will determine how the prototypes "fit together". This type of violation is avoided by adding tests to the grammar to ensure compatibility of the components. If the components are not compatible, the grammar backtracks and selects alternative components.

The first type of constraint could also be detected by the grammar, this time "testing" the environment. However, "Lethals" which result from violations of constraints specified by the designer, are not prevented at the generation stage. The system permits the generation of such "lethals", but adds / subtracts a penalty from the fitness scores. Constraint conflicts are also recorded, and displayed to the designer when / if he chooses to examine the chromosome string.

6.2 Pareto Optimisation

As explained in section 5, the fitness function determines the P-Set. The population is ranked on the basis of non domination. All non dominated points in the current generation are assigned the rank of 1. These points are then removed from the population and the next set of non dominated individuals is identified and assigned the rank of 2. This process continues until the entire population is ranked. The probability of each chromosome being selected to mate is determined by its rank. The chromosome is crossed using single point crossover.

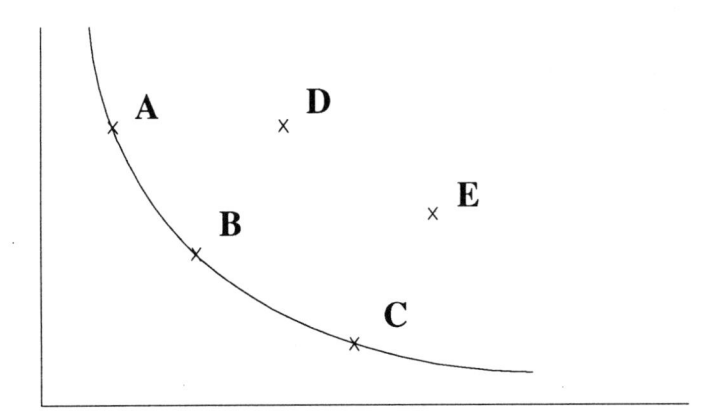

Fig 5. The Pareto Front

6.3 Maintaining Diversity

Ideally the system should provide the designer with a range of 'distinct' solutions rather than a single answer. It is therefore important to preserve diversity of the genetic pool through the encouragement of niching and speciation during a run. If one assumes the fitness landscape below, fig 6 :-

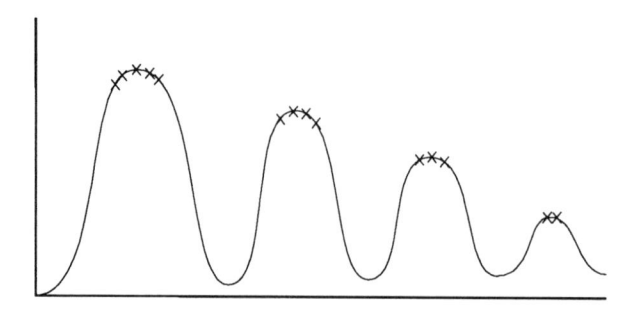

Fig 6.

it is clearly more desirable to present the designer with a collection of concepts on each of the peaks rather than a single cluster on the highest peak. There are a number of approaches outlined in the literature :-

Cavicchio (1970) - an offspring replaces the inferior parent if the offspring's fitness exceeds the parent's. Diversity is maintained because chromosomes tend to replace chromosomes similar to themselves. The approach works well in small populations.

De Jong (1975) - offspring replace existing chromosomes according to their similarity. A random subpopulation is chosen, and the chromosome in the population closest to the new offspring is replaced.

Grosso (1985) - imposes a geography on the chromosome population. Limits reproduction based on proximity of neighbours. Intermediate migration rates shown to be better than isolated subpopulation with no migration and panmictic completely mixed subpopulations.

Booker (1985) - mating templates to adaptively restrict mating between species.
These approaches are currently being investigated with a view to their performance characteristics given the nature of the conceptual design problem and the use of a grammar based GA.

Parmee (1995)[12,13] - reinforces the natural clustering tendencies of a GA by maintaining diversity in the early stages of the search and identifying clusters of high performance solutions by the introduction of a standard clustering technique.

7. HCI ISSUES

The philosophy behind the tool is one of empowerment. The intention is to enhance the design process by providing the designer with greater insight into the design space and the consequence of the trade off decision s/he must make. The nature of the interface between the GA and the designer is therefore very important.

7.1 Mode of Interaction

The designer must be allowed to work at the problem level of the design problem, not the system level of the design tool. The problem specific modules are intended to ease this interface, translating the concepts and terms which the designer uses in solving design problems in the required domain, to the system level requirements of the GA etc. The requirement of multi-criteria fitness functions to handle the often conflicting design goals and preferences, and the nature of the interface between this fitness function and the designer is being explored.

The system is basically designer driven. Although, the designer cannot interrupt the system during a generation, between generations the designer can examine any of the chromosomes, and modify the user requirements, evaluation criteria, design grammar, design prototypes or even the chromosomes.

7.2 Information Provided to the Designer

Effective control is completely dependent upon the availability and understandability of information. The designer is provided with the usual information about worst, average and best fitness, together with a measure of the diversity of the current population. Between generations the designer can also examine the chromosomes. These are presented to the designer in terms of the design prototypes, and not as a string of bits. The information a designer requires to effectively control the evolution of designs is an ongoing research topic.

7.3 Design Repository

The system is a repository of information in the form of design grammar(s) and catalogues. It is therefore desirable to enable the designer to browse, modify and extend these. Although the current research does not envisage studying the knowledge acquisition requirements of a conceptual design system, the system should, at least in theory, be extendable and maintainable by the design engineer, in part to allow customisation of the knowledge bases to reflect local technical and political preferences.

7.3.1 The Design Grammar Browser / Editor

The design grammar must be editable by the designer. Only in rare occurrences of routine design will there be only one perceived "correct" design strategy. For the tool to enhance the designer it is important for the designer to feel that the tool is an extension of his working practices and not the imposition of an alien design methodology. The designer can therefore use the design grammar browser to read the grammar and modify it to suit his/her own preferences.

7.3.2 The Conceptual Design Prototypes

The designer can browse, edit or add design prototypes to the system. Design prototypes are stored hierarchically, although no use is made of the hierarchy by the system during the generation of design concepts. The hierarchy, based on the design grammar, enables the designer to traverse the system to find design prototypes. The interface which enables the designer to add or modify a prototype is primitive. It presents a frame and expects the designer to fill the appropriate slots, it does not check the logic or consistency of the information entered. This is less than ideal, but the research does not wish to get drawn into knowledge acquisition problems. The interface as it exists has been generated to make it easier to set up the system with new problems etc.

8. FUTURE WORK

Short Term :-

(i) The addition of a thesarus to widen the initial selection of design components, both by the fixed string prototype, and by the grammar based GA system.
(ii) Implementation of other techniques for handling a variable solution size to the conceptual design task, eg messy genetic algorithms which use a variable length chromosome string[7].
(iii) Experimentation on realistic design problems to explore the strengths and weaknesses of grammar based chromosome approach.
(iv) The application of a structured genetic algorithm (sGA) to handle different levels of abstraction, and to enable the system to work on several levels of abstraction.
(v) Changing / evolving the environmental conditions to favour designs in particular regions of the search space.

Long Term :-

Evolution can be very inefficient. Large fitness functions could make the computational costs of a pure Darwinian Evolution process infeasible. The complexity of fitness functions and evaluation routine is likely to increase as GAs and Adaptive Search techniques are applied to increasingly sophisticated problems. It is therefore important to investigate techniques which might enhance an evolutionary strategy.

(i) Increase the competition between designs. Designs already compete both in terms of their fitness scores, and by direct comparison with each other during goal criteria evaluation. Clustering techniques and niching also encourage competition amongst designs. However, this could be taken a stage further and allow competition for resources during the morphogenesis stage, and a subsequent lifecycle.

(ii) GAs are poor parents. Parents do not take an interest in the development of their offspring. The classless society of evolutionary paradigms could be ended, by allowing the success of the parent(s) to privilege or handicap the offspring in terms of resources needed during the morphogenesis and subsequent lifecycle stages. This would increase the use of computational resources in searching promising areas of the design space, in preference to less interesting regions.

(iii) A simple GA replaces virtually all members of the population at roughly the same time. However, individuals could be allowed to breed as soon as they reached maturity. If the time of development is measured as the number of instructions required to generate a phenotype from the chromosome, then the speed of morphogenesis would depend upon both the complexity of the grammar encoded in the chromosome and the resources allocated to it on the basis of its parents' fitness evaluations. New offspring could replace chromosomes with the lowest fitness values, or a steady population could be maintained by defining a limited life span for organisms and resources allocated as a percentage of a fixed total.

(iv) Lamarkian Evolution (1809).

Evolutionary change through acquired characteristics. Allowing characteristics acquired during the "lifetime" of a design to be passed on to its offspring. In nature this does not appear to happen, in part because of the difficulties / costs in allowing a phenotype to affect its genotype. In the artificial environment, although Lamarkian Evolution is possible it adds, in addition to the above cost of allowing the phenotype to modify the genotype, the burden of a "lifecycle" after morphogenesis.

(v) Design Husbandry

Speed up the evolution of desired characteristics by allowing the designer to breed designs in much the same way that animal husbandry has developed breeds of cows that deliver high yields of milk, or the development of hybrid strains of wheat with increased resistance to bacteria etc. An important research question is the type of information the designer would require to effectively develop strains of designs. In a grammar based GA architecture, the chromosome would on its own be insufficient. The chromosome defines a potential set of designs not a single design, there is not a simple mapping between the phenotype and the genotype. Pareto Optimisation techniques allow the designer to select designs with desired characteristics, however like animal husbandry it might be important to also provide the designer with information about the pedigree of the design.

(vi) Genetic Engineering

The current prototypes provide the designer with basic facilities for altering the genotype of designs. Genetic engineering offers the prospect of rapid improvements to designs, but also risks seriously damaging the gene pool. The nature of the information required by the designer to effectively manipulate genotypes, and through these individuals the gene pool is currently being investigated.

9. CONCLUSIONS

Although, the work is only at a preliminary stage a number of advantages and limitations of the approach are obvious. The advantages of a grammar based chromosome architecture can be summarised as :-

(i) The seperation of the genotype and the phenotype allowing environmental information to influence the development of the design.
(ii) The system is capable of generating its own function / sub-function decomposition structure.
(iii) Function / Sub-functions can be merged or split during the design process.
(iv) The designer does not have to stipulate the solution size of the conceptual design in advance.

The main disadvantages are :-

(i) Increased computational cost.
(ii) The need to construct / generate an appropriate grammar.

Experimentation is required on more realistic design problems to assess the potential benefits and limitations of a grammar based chromosome approach to the conceptual design problem. Of potential concern is the increased computational costs. If these prove to be prohibitive the research will look more closely into the issues of reducing the number of evaluations made by the designer in an animal husbandry and genetic engineering approach.

Despite, the above reservations, there are a range of design tasks, within which the above approach can be successfully applied. In terms of Gear's classification of design - routine, innovative, and creative: it is clear that the system will support the first two categories. It will blur the distinction between routine and innovative design, as all routine design problems will be open to the possibility of innovative solutions. The position as regards creative design is not clear, and further work is required to clarify the situation.

10. REFERENCES

1. Antonisse H.J. (1991), "A Grammar Based Genetic Algorithm", In Rawlins G.J.E. (Ed) "Foundations of Genetic Algorithms", Morgan Kaufmann Publishers

2. Coyne R.D. (1988), "Logic Models of Design",Pitman

3. French M.J. (1988), "Invention and Evolution: Design in Nature and Engineering", Cambridge University Press

4. Gero J.S. (1990), "Design Prototypes: A Knowledge Representation Schema for Design", AI Magazine

5. Gero J.S. (1994), "Evolutionary Learning of Novel Grammars for Design Improvement", Artificial Intelligence for Engineering Design, Analysis and Manufacture, 8, 83-94, 1994

6. Goldberg D.E. (1991), "Genetic Algorithms as a Computational Theory of Conceptual Design", In Rzevski G. & Adeyi R.A. (Eds) "Applications of Artificial Intelligence in Engineering", Elsevier Science Publishing

7. Goldberg D.E., Deb K., Karguta H. & Harik G. (1993), "Rapid, accurate optimization of difficult problems using fast messy genetic algorithms", Proc. of the Fifth International Conference on Genetic Algorithms, pp56-64, Morgan Kaufman Publishers

8. Grierson D. (1994), "Conceptual Design using Emergent Computing Techniques", NATO Advanced Research Workshop, Nafplio, Greece, August 25-27, 1994

9. Lawson B. (1990), "How Designers Think: The Design Process Demystified" Butterworth

10. Parmee I.C. (1993), "The Concrete Arch Dam, an Evolutionary Model of the Design Process", In Albrecht R.F., Reeves C.R. & Steele N.C. (Eds), "Artificial Neural Nets and Genetic Algorithms", Proc. of the International Conference in Innsbruck, Austria, pp544-551, 1993

11. Parmee I.C. (1994) "Adaptive Search Techniques for Decision Support during Preliminary Engineering Design", Procs. Information Technologies to Support Engineering Decision Making, Institute of Civil Engineers, November 1994

12. Parmee I.C. (1995), "High Level Decision Support for Engineering Design using the Genetic Algorithm and Complimentary Techniques", Procs. Applied Decision Technologies, Brunel Conf. Centre, London, April 3-5, 1995

13. Parmee I.C. (1995), "Reinforcing the Natural Clustering Characteristics of the GA", Internal Report PEDC-04-95, April 1995

14. Parmee I.C. & Denham M.J. (1994), "The Integration of Adaptive Search Techniques with Current Engineering Design Practice", Proc. Adaptive Computing in Engineering Design and Control, pp1-13, Plymouth, 1994

15. Pham D.T. and Yang Y. (1993), "A Genetic Algorithm Based Preliminary Design System", Proc. Instn Mech. Engrs, 1993, 207, 127-133

16. Rechenburg I. (1964), "Cybernetic Solution Path of an Experimental Problem", Royal Aircraft Establishment, Library Translation 1122, Farnborough

17. Woodbury R.F. (1993), "A Genetic Approach to Creative Design", In Gero J.S. & Maher M.L. (Eds), "Modeling Creativity and Knowledge-Based Creative Design", Lawrence Erlbaum Associates Inc

11. BIBLIOGRAPHY

1. Davis L. (Ed) (1991), "Handbook of Genetic Algorithms", Van Nostrand Reinhold

2. Goldberg D.E. (1989), "Genetic Algorithms in Search, Optimisation and Machine Learning",Addison Wesley

3. Holland J. (1975),"Adaptation in Natural and Artificial Systems", The University of Michigan Press

12.   Appendix A • Adaptive Search is a collective term which describes those search and optimisation techniques with operational characteristics that are heuristic in nature. Unlike numerical optimisation where the best step forward is determined mathematically from the previous solution, a stochastic approach is introduced where reasoning from past experience is combined with a degree of randomness. This allows the adaptive search process to break away from the bounds of a strictly sequential pattern thereby ensuring that widely varying areas of the search space are sampled.

Various forms of Adaptive Search exhibit some or all of the following characteristics that are of particular interest to the engineering designer.
- They represent highly non-linear search processes.
- They have the ability to combine and subsequently manipulate large numbers of variable interactive design parameters.
- Premature convergence upon local optima can be avoided.
- Widely varying areas of the design space can be sampled in a parallel manner.
- No gradient information is required.
- 'Good' design options can rapidly be identified.
- Eventual convergence upon the global optimum design solution is possible.

Adaptive Search includes such techniques as the Genetic Algorithm (GA), Simulated Annealing, and Evolution Strategy. Initial research at the Plymouth Engineering Design Centre concentrated upon the development and application of the Genetic Algorithm. The GA offers a robust non-linear search technique that is particularly suited to the solution of multi-variate design problems. The Algorithm has the ability to negotiate search spaces which are both multi-modal and discontinuous in nature. The stochastic element within the search process ensures that premature convergence upon a local optima is avoided.

The operation of the Genetic Algorithm is described in detail in the literature (Goldberg, 1990; Davis, 1991). The designer must first determine an appropriate objective function and select those variable parameters that significantly affect the overall design of an engineering system. At this stage little, if any, knowledge of the interactive nature of the chosen parameters is required. The GA software randomly generates parameter values which populate an initial generation of around fifty parameter combinations. Each combination is passed in turn to a computer model of the engineering system under design and a value for the objective function is calculated for each case. The parameter combinations are then either rejected or reproduced depending upon the relative fitness of the returned solutions. A low performance combination may be rejected whereas high performance combinations have a greater probability of single or mulitple reproduction. This procedure is known as Fitness Proportionate Reproduction (Holland, 1974) and is analogous to the selection procedure of natural systems.

The biological analogy can be continued by the representation of the real number values of the parameters in a binary form. Each parameter combination, when being processed by the Genetic Algorithm, can be represented by a binary string. Each string is considered to represent a chromosome and each digit within the string represents a gene. Alleles, which in a natural system cause variations in the characteristics of a gene, are represented by the value of each digit.

An interchange of information between the members of the reproduced population is achieved by means of a Crossover Operator which exchanges binary data between randomly selected pairs of parameter strings. This process is repeated until a new generation of parameter combinations has been produced. A Random Mutation operator then causes randomly selected binary digits to change value according to some preset Random Mutation Probability. Fitness Proportionate Reproduction ensures that each successive generation consists mainly of increasingly fit parameter combinations whilst Crossover will gradually promote localised search within seemingly fit areas of the design space. The Random Mutation Operator prevents premature convergence upon a local optima by ensuring that widely varying areas of the search space are constantly sampled. Although the operation of the GA can be improved by the inclusion of a number of othe Genetic Operators in many cases satisfactory results can be achieved from the basic three operators described above.

The reproduced binary strings, after being manipulated by the Crossover and Random Mutation Operators then represent the second generation of parameter combinations. The real values of the parameters are again processed by the design model and the cycle is repeated until a satisfactory design is generated or the global optimum solution is achieved.

[Taken From: Parmee I.C. (1994), "Adaptive Search Techniques for Decision Support during Preliminary Engineering Design", Procs. Information Technologies to Support Engineering Decision Making, Institute of Civil Engineers, November, 1994]

SOLUTION CLUSTERING WITH GENETIC ALGORITHMS AND DFA: AN EXPERIMENTAL APPROACH

S D Santillan-Gutierrez & I C Wright

1 INTRODUCTION

During the development of a product, the conceptual and embodiment stages are critical. Although there are tools for helping designers along later stages of the design process, analysis, detail and manufacture, the tools for the conceptual and embodiment stage are few and mainly in the experimental stage.

The process that generates, evaluates and selects configurations has not been modelled in a recognised general approach. This is a consequence of its relationship with the designer's process thinking and the limits for understanding and reproducing it with the current AI tools [1].

Only recently, different authors have turned their attention to the problem of developing computer aided tools for conceptual design [2][3]. Modelling the conceptual stage is difficult, because this is a continuous process of evaluation and satisfaction of different criteria, dealing with often vague and imprecise information [4] [5].

Initial attempts for developing expert systems and other traditional tools from AI have had achieved different degree of success, due to the different and complex requirements for modelling the process of generation, evaluation and selection of solutions for a problem.

Different tools and systems are subjects of current research. Among them, there are constraint satisfaction techniques [6], modelling by mathematical and graphic techniques, like bond graph [7], or Petri Nets [8], together with development of expert systems, etc.

The research performed up to date reports different problems for developing this type of computer aided tool. Some of them are:

- The lack of a consistent and recognised product model
- The difficulties in modelling functionality, human factors and marketing constraints during the conceptual stage
- The lack of a framework to model the propagation of effects during the decision making process.

The proposed approach is aimed at helping the designer during the end of the conceptual stage. It is based on the use of genetic algorithms (GA) and more traditional techniques for locating groups of promising solutions.

2. DIFFICULTIES FOR MODELLING PRODUCTS

Research to obtain a coherent product model is currently being developed by different authors. One of the shortcomings for CIM and CE is the problematic integration of databases and tools for the product cycle development, as a consequence of the lack of a product model. This situation leads to the problem of stability of the information systems, and mainly affects the concept of integration. [9][10][11][3]

There are different proposals for developing a coherent product model. The range varies from efforts to produce standard representations based on Object Oriented Programming as STEP [3][12], to different models proposed for specific research, with more limited scope.

For the purposes of this research, the knowledge involved with the product model has been divided in seven domains, shown in figure 1.

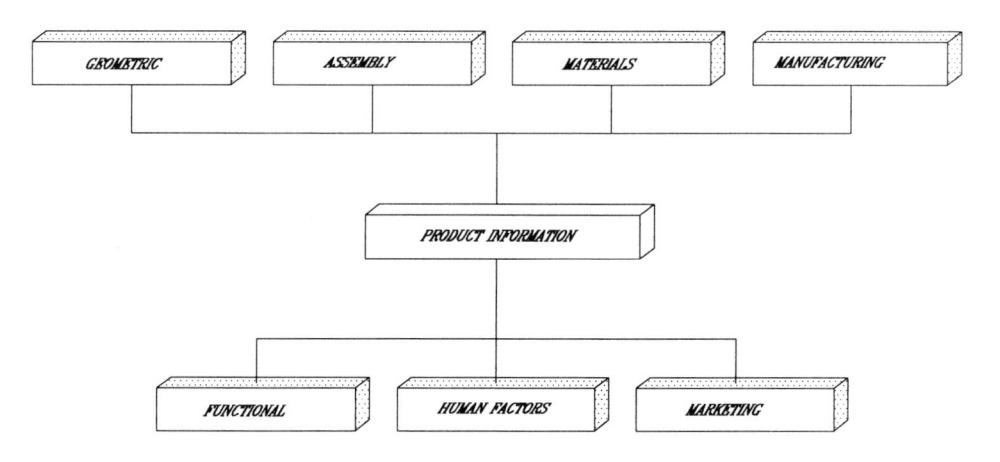

Fig. 1 Product model domains

Current research is producing systems that can handle up to four of them: Assembly, Geometry, Manufacturing and Materials. (Figure 2).

While some of the disciplines or domains involved in a product model can be stored and organised in an accessible way, the generality of their application is restricted for those disciplines that required a more specific knowledge.

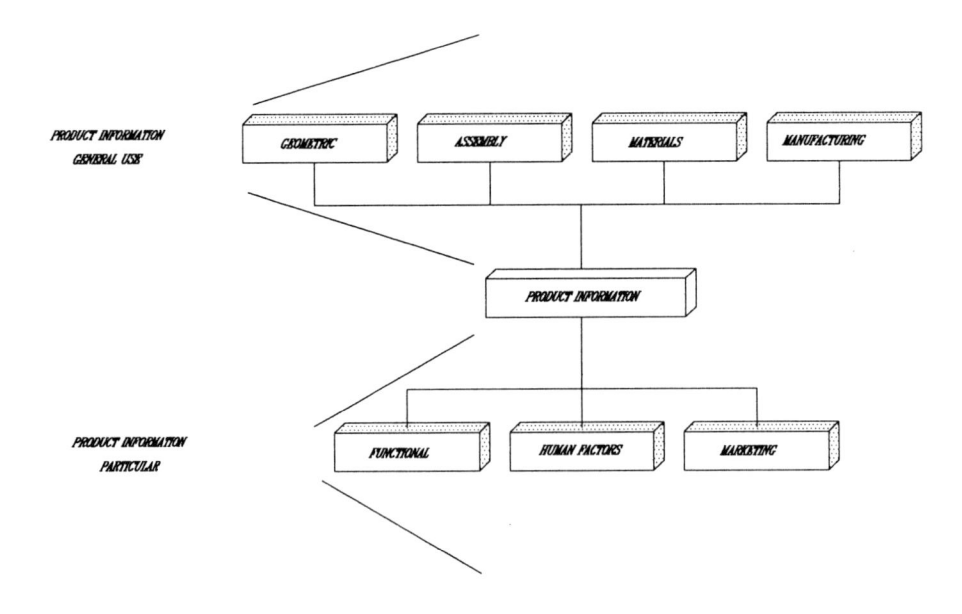

Fig.2 The domains have either particular or general attributes, features and knowledge related with the product model

The problems with the different proposed product models arise when the functional, marketing and human factors are considered, not to mention the problems for generating mathematical models to represent the behaviour of the product and system under development.

This is normal, because using the technology available for the majority of the enterprises, it is possible to define set of common parts for mechanical devices [10], common operations for assembly [13], sets of geometric features generated by the production machines and parameters to describe the behaviour of different materials according with standards and regulations.

The situation is different for modelling of functional, marketing and human factors domains. Every attribute in these domains has variation in time and space. It is very difficult to trace common patterns, attributes and standards from product families, technology involved, users or external factor involved.

While the present research aims are not focused to create a specific product model, it is important to highlight the need for a tool which can flexibly handle the different requirements of the domains .

3 DECISION MAKING

The process of decision making during design is not linear. Moreover, the designer traditionally performs tasks with different levels of complexity, analysing the problem in an iterative way with the information available, rather than following a totally systematic work.

While the designer is making decisions, he explores different regions of the design space, looking for configurations that offer the best compromise within the sets of requirements or constraints defined. This activity is not always done in a systematic way, sometimes execution of the decision making is chaotic [14].

It is difficult to define the boundaries between the conceptual and the embodiment stages. Sometimes since the definition of the Basic Operation Principles (BOP), the process of decision for embodiment starts. The selection of materials, manufacturing processes, geometry and assembly is performed with different degrees for the systems, depending on the constraints and information available.

On the other hand, modelling relations between different domains using heuristics and guidelines face the problem of growing complexity in the rules. There is not a common pattern for making decisions during the conceptual stage. The systems for decision making reported in the area are based in constraint propagation [6], constraint satisfaction [15][12], Fuzzy logic [4] , PDS Satisfaction [16], linear programming [17], etc.

Different authors report the problems of satisfying different and contradictory requirements with vague knowledge during the conceptual stage. Some of them have tried to represent the interrelation between variables or domains with lattices [18], design guidelines [2], systems theory [19], design axioms [20]. The modelling of behaviour has been attempted using Petri Nets [8], Bondgraph Theory [7], Basic Elements Theory [21], Matrix arrangements [22], while other tools are under development.

While these approaches are trying to define the decision making process in an algorithmic way, they present the following shortcomings:

- Difficulties in the definition and updating of the variables that result affected after a decision is done in one domain.
- Classification and assessment of the variables or factors to be evaluated
- Permanence of the rules for different systems or products

These problems create the need to develop a tool able to handle the information from different domains in more flexible way.

4 PROPOSED APPROACH

The proposed approach is composed of:

1. A combinatorial generation of alternatives for creating the design space
2. Use of constraints to define the different criteria for evaluating the solutions.
3. Use of the GA for exploring the space of solutions, find the regions of promising solutions
4. Redefine the constraints, according with the requirements of the different domains, tightening requirements or adopting different strategies for the search
5. Iterate steps b, c and d until a set of possible solution are defined by the convergence of the GA. The solutions should be refined using other methods.

4.1 Combinatorial Generation of Solutions

The generation of alternatives based on the combinatorial approach proposed by French [23] is a popular method. Techniques like morphological charts and configuration trees are based on the same principle.

The method is based on dividing the problem into functions, representing subsystems, according with the theory of the systems. Once the relationships between the different variables have been understood, the solutions are proposed using all the known elements that can fulfil the function. Then, the main characteristics of the elements are linked with the desired behaviour. At this stage, a process of matching the different solutions combining all the different principles is performed. This combinatorial process is developed recurrently at the level of parts, system and subsystem.

The space of solutions can be defined by a combination of n elements, with the p parameters. As the combinatorial approach produced huge amount of data, the selection of possible alternatives is performed using a process of elimination, starting with criteria like physical impossibility, increasing gradually the constraints while the process of elimination is advancing.

4.2 Constraint Satisfaction

For modelling the decision making process, the approach of constraint satisfaction is proposed. Once the constraints of a problem are defined, the solutions are chosen according with:

    a) The satisfaction of the constraint
    b) The degree of satisfaction of the constraint
    c) The degree of satisfaction of the solution referring to the set of constraints.

One of the advantages of using constraints, is the possibility of expressing the requirements from different domains in a way that involves simpler expressions than the set of rules for an expert system.

4.3 Operation and description of the GA

The GA is a simplistic way to model the Darwinian process of natural selection. Proposed first by Holland [24], they use strings of characters that represents certain characteristics of individuals. There are different sources with extended descriptions of the mechanics of the GA [25-27]. A short explanation of the technique is presented in the following paragraphs.

Generally, the representation of the attributes of an individual is made using a gene. In the case of the GA, the gene is a string of characters. Very often a binary alphabet is used, although different alphabets, and subsequently, different representations can be used.

Using the representation of the attribute, an original population is defined. The population is a set of strings, each one representing a different individual. The population can be generated randomly or created using heuristics.

The individuals are subjected to an evaluation, according to a set of criteria. These criteria are expressed in a fitness function. The qualification obtained for an individual affects the degree of adaptation to the constraints imposed.

Under the criterion of natural selection, the individuals receive a probability of reproduction The criterion is that those individual with better qualifications, are better adapted to the environmental constraints. They receive major probability of being selected for reproduction, while those individuals with lower qualifications receive a reduced probability of selection.

The "reproduction" in the GA is performed using two operations, that recombine and change the structure of the genes. These operations are called *crossover* and *mutation*.

During *crossover*, two individuals are selected, then parts of their chains are exchanged, to create two completely new individuals. See Fig 3. Meanwhile, during *mutation*, parts of the chain are randomly selected, and the values are changed.

Population generation "n"   *Population generation "n + 1"*

crossover point
XXXXXXXXX   XXXYYYYYY

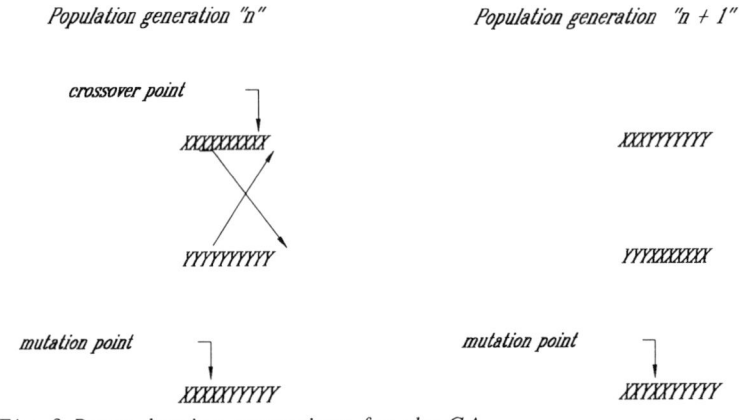

YYYYYYYYY   YYYXXXXXX

mutation point   mutation point
XXXXXYYYY   XXYXXYYYY

Fig. 3 Reproduction operations for the GA

The algorithm finishes when the population converge to a set of characteristics that match the constraints or competing groups of individuals steadily dominate the population.

Uses of the GA as a tool for exploring the space of solution to solve different problems has been reported previously. The possibility of using a tool that learns according to the performance of a population, it's capable to accept non linear relations and it is independent of the knowledge involved is very attractive for design.

The fact that the genetic algorithm seeks regions of compromise, rather than a local optimum, opens the possibility for clustering regions with the space of solutions where configurations with a good performance can be found. Due to the stochastic nature of the algorithm, there is not a guarantee that an optimal configuration will be found. Instead, sets of similar performance configurations will increasingly dominate the genetic pool, continuously spreading their presence in the population.

5 APPLICATIONS OF GA's IN ENGINEERING DESIGN

The use of GA's for conceptual design has not been reported successfully yet. Previous publications envisaged the use of the algorithm to explore the space of solutions for design [25].

However, the use of GA's for embodiment has a different degree of development, Parmee [28] used them successfully for design dams and electronic filters , Goldberg [29] reported their use for design and routing of pipelines, while Brown and Hwang [30] performed selection and matching of pulleys and belts from catalogues. Thornton [12] described a poor performance in achieving convergence for problems involving functional requirements. Other applications are related with more specific problem solving tasks, like configuration of keyboards [31].

6 FOUNDATIONS FOR THE PRESENT RESEARCH

The objective of the present work is to demonstrate the feasibility of implementing a GA capable of to exploring the space of solutions, in the boundaries between the conceptual and embodiment stages. In the present work, the search in a combinatorial space of solutions is considered similar to a search in a catalogue. The main characteristics of a search in a catalogue are:

- The configurations are defined for discrete values.
- There is a direct dependency between the code, the attributes and their effect over the object of design in the domain considered.

The code is generated according with the definition of the variables and the evaluation procedure defined for each domain. The sets of evaluation procedures represent a definition of fitness for purpose for evaluating the different configurations. The assembly domain was chosen as a departing point, due to the following factors:

- a) A solid evaluation methodology exist
- b) There are objective functions
- c) The criteria used are of general application
- d) There is consensus about its use in the research community

There are three different alternative techniques for this domain, proposed by Lucas, Hitachi and Boothroyd. From these techniques, we selected Boothroyd & Dewhurst's [32-34]. It comprises a series of criteria which define the nature of the assembly, evaluate the operations of insertion and handling parts according with their geometric requirements. The technique is documented clearly in manuals and software. Different improvements have been reported over the years. As well, it has proved to be very popular and successful in many cases [35].

One of the advantages of the technique is the code used to classify different parts. A number is assigned according with geometric characteristics of the part. The calculations of the time for inserting and handling the part are defined using this code. The codification system allows to make a comparison with a search in a catalogue.

On the other hand, the Boothroyd's technique presents different shortcomings:

- It relies on the designer. The designer is expected to solve the functional, materials and manufacturing aspects of the design before proposing the geometry for the parts.
- The "codes" for different parts have changed along the years.
- The methodology for large parts is still under development [36].

The code used in Boothroyd's method represent different geometric characteristics of the part. It consists in a two digit number, based on the main dimensions of a part. Considering this number, a standard time for handling the part is defined. A similar process is carried out to define the insertion time for the part.

The method requires the definition of a theoretical minimum number of parts for a system using the following criteria.

1. The parts should be made of different materials for functional constraints
2. The parts need to have relative movement between them
3. The needs for other assembly or disassembly operations do not allow a reduction.

CODE NUMBER 8660   CODE NUMBER 8400

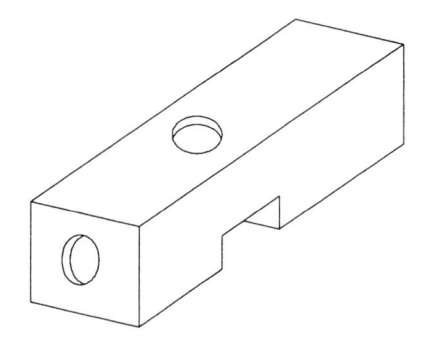

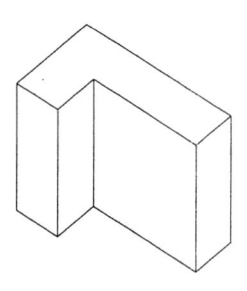

Fig. 4 Examples of Boothroyd's classification for parts (Borrowed from Boothroyd's Manual [35])

The minimum number of parts is used to define the efficiency index of an assembly. The insertion time, handling, and the efficiency are used to define a cost function to evaluate the different configurations for the system. There are two ways to map the space of solutions using DFA:
- Reduce the number of parts gradually, identifying the geometric characteristic for possible configurations.
- Minimise the fitness function (cost and assembly time) and try to find those solutions that match these major constraints.

Previous experiences, reported by Boothroyd [33, 35], indicates that in many cases, the configuration of a system with the maximum assembly efficiency for assembly violates the functional o manufacturing restrictions. For this reason, reducing gradually the number of part was considered for the research.

7 DESCRIPTION OF THE SYSTEM

The first part of the program is still under development, illustrated in figure 3. The system is based on GENESIS (Genetic Search Implementation System, developed by Geffenstrette, as public domain software). Developed primarily for function optimisation purposes, GENESIS has different features, allowing it to be adapted for the purposes of the research.

The fystem under development will complement GENESIS performing the following operations:

1. A primer consideration of production volume, number of parts, variations of the models and other factors defines the type technology suited for the sets of parts under analysis. In this stage only manual and automatic assembly are considered.
2. After this stage, the system access the user to a set of questions, following Boothroyd's method. The information provided is related with the maximum dimensions of the different parts, symmetry, present number of parts and the theoretical minimum number of parts.
3. The program uses the actual number of parts and starts the process considering this number as a departure point.
4. The system uses the maximum size to define constraints for the parts, define the efficiency of the assembly and set the parameters for evaluation using the relationships proposed by Boothroyd.
5. To create the strings of individuals, the program considers one part composed for four digits. The first two express the inserting code, the second pair represents the handling code. The system can handle assemblies up to 10 parts. The system generates a population composed for individuals proposed by the designer, and other randomly generated.
6. The population is feed into the GA , in this special case a primer evaluation of the population is performed. The strings which violate the constraints of maximum size are eliminated. New individuals are generated randomly to fill their place.
7. The fitness function is developed assigning a proportional probability to the efficiency and reduction of cost of the assembly. According with the evaluation and the probability assigned, the individuals are selected for the operation of crossover. In this stage of the research simple crossover was selected.
8. The population is sampled after a fixed number of iterations. The program will stop if one of the following conditions is present:
    - More than 90% of the individuals have a defined value of probability
    - There population 's performance does not change after a number of iterations
9. The output will be decoded and the geometric characteristics of the part for the subassemblies with better qualifications. The program decrease the number of parts and perform the steps 2-9, until the number of parts is equal to the minimum theorethical number.

10. A process for describing the different configuration obtained performed, in this case manually.

8 EXPECTED RESULTS

The codification of the program is still under development, different examples reported in the literature are under evaluation to be used as a testbed. There are different limitations expected from the output of this first stage.

One of the limitation is the absence of databases to interact with Boothroyd's method. The database should comprise parts classified as a catalogue. The interaction with databases needs the development of a product model, to define the different attributes and order the information available in the different domains. This task it is outside the scope of the present research.

The results will show that different configurations can fill the same requirements. Their full embodiment will require further work from the designer using other tools.

The behaviour of the GA is not well understood yet [26, 27, 37, 38]. Different researchers have reported great variations in the performance of the GA, according to the changes in the following parameters:

a) The number of points where the crossover is made
b) The probability of the mutation
c) The size of the sample population
d) The variation in the criteria of convergence.

This effects should be considered during the tuning of the program.

9 ENHANCEMENT OF THE RESEARCH

The system will need to be upgraded to provided a practical tool for designers. Further work is required to define a product model compatible with the use of the GA. The need to reorder the available information to create the databases requires an extensive use of Object Oriented programming techniques. The representation of different parts or elements in a way that emulate the information in catalogues is a key factor.

The search for expressing domains through the use of functions and indexes, representing general parameters of evaluation, as a part of the model is very important. Some researchers have reported evaluation procedures with this profile in different fields but further work is required.

The sets of different indexes for evaluation represent a measure of technical merit [39, 40]. The relationships to evaluate the indexes form the body of a fitness function for the configurations to be evaluated.

After this process, the GA would act as a " Design moderator", looking for those sets of solutions that share similar performances. Unlike a moderator based on sets of rules, the GA is unable to keep a record of a single configuration and their modifications. Instead it requires frequent sampling of the population to gather data about the performance of different individuals.

The tools envisaged will require extensive computing resources to translate the requirements into the different representations and handling very long strings. Nevertheless, the ability of the GA to emulate the designer's decision making process, looking for compromise rather than optimise opens new possibilities for Engineering Design.

ACKNOWLEDGEMENTS

The work outlined in this paper was supported by UNAM (National Autonomous University of Mexico) and The British Council.

REFERENCES

1.	Blessing, L. *Engineering Design and Artificial Intelligence: A promising marriage?* in *Research in Design Thinking*. 1991. Holland: Delf University Press.

2.	Hundal, M.S., *Use of functional variants in product development.* Design Theory and Methodology, 1991. **31**: p. 159-164.

3.	Krause, F.L., *Product Modelling.* Annals of the International Institution for Product Engineering Research, 1993. **42**(2): p. 695 - 699.

4.	Hamburg, I. *Using Fuzzy logic in the interactive computer support of Engineering Design.* in *ICED 93*. 1993. The Hague: Heurista.

5.	MacMahon, C.A. *Using fuzzy techniques to handle uncertainty in design information retrieval.* in *ICED 93*. 1993. The Hague: Heurista.

6.	Bascaran, E.B., *Modelling Hierarchy in Decision-Based design.* Design Theory and Methodology, 1992. **31**: p. 293-300.

7.   Bracewell. *Schemebuilder, a design aid for the conceptual stage of product design*. in *ICED 93*. 1993. The Hague: Heurista.

8.   MacMahon, C.A. *A transformation model for the integration of design computing*. in *ICED 93*. 1993. The Hague: Heurista.

9.   Erens, F. *Shortcomings of today's design frameworks*. in *ICED 93*. 1993. The Hague: Heurista.

10.   Thornton, A., *Constraint specification and satisfaction in embodiment design*. 1993, Cambridge University:

11.   Johnson, A.L. *Modelling functionality in CAD: Implications for product representation*. in *ICED 93*. 1993. The Hague: Heurista.

12.   Thornton, A. *Constraint Specification and satisfaction in Embodiment Design*. in *ICED 93*. 1993. The Hague: Heurista.

13.   Boothroyd, G., *Design for Assembly-The key to design for manufacturing*. International Journal of Advanced Manufacturing Technology, 1987. **2**(3): p. 3-11.

14.   Wallace. *Some observations in Design Thinking*. in *Research in Design Thinking*. 1991. Delft University: Delft University Press.

15.   Trousse, B. *Integration of constraint-based systems into design support tools*. in *ICED 93*. 1993. The Hague: Heurista.

16.   Hurst, K.S. *Experiences with an expert systems for concept selection*. in *ICED 93*. 1993. The Hague: Heurista.

17.   Sen, P. *Multicriteria decision support environment for engineering design*. in *ICED 93*. 1993. The Hague: Heurista.

18.   Sharpe, J. *A structured interdisciplinary approach to complex product design*. in *ICED 93*. 1993. The Hague: Heurista.

19.   Pahl, G., *Engineering Design- A systematic Approach*. ed. W. Beitz. 1988, London: The Design Council.

20.   Suh, N.P., *The principles of Design*. First ed. 1990, New York: Oxford University Press.

21.   Hubka, V., *Theory of Technical Systems*. First Edition ed. ed. W.E. Eder. 1988, New York: Springer Verlag.

22. Morrison, D., *Engineering Design- The choice of favourable systems*. 1968, London: Mc Graw Hill.

23. French, M.J., *Conceptual Design*. First ed. 1970, London: Newness and Butterworths.

24. Holland, J.H., *Adaption in Natural and Artificial systems*. First ed. 1975, Ann Arbor: The University of Michigan Press.

25. Goldberg, D.E., *Genetic Algorithms as a Computational Theory of Conceptual Design*. Artificial Intelligence in Engineering, 1991. **6**(Ch 69): p. 1-16.

26. Beasley, D., *An Overview of Genetic Algorithms: Part 2, Research topics*. University Computing, 1993. **15**(2): p. 170-181.

27. Koza, J.R., *Genetic Programming- On the programming of computers by means of natural selection*. 1st ed. 1990, Massachussets: .

28. Parmee, I.C. *Evolutionary Engineering using the genetic algorithm*. in *ICED 93*. 1993. The Hague: Heurista.

29. Goldberg, D.E., *Genetic Algorithms in search, optimization and machine learning*. First ed. 1989, New York: Addison-Wesley.

30. Brown, D.R., *Solving Fixed Configuration Problems with genetic Search*. Research in Engineering Design, 1993. **5**: p. 80-87.

31. Wong, H., *Adaptive Genetic Algorithm for optimal printed circuit board assembly planning*. Annals of CIRP, 1993. **42**: p. 17-20.

32. Boothroyd, G., *Automatic Assembly*. First Edition ed. ed. P. Corrado and L. Murch. Vol. 1. 1982, New York, USA: Marcel Dekker.

33. Boothroyd, G., *Product Design for Assembly*. First ed. ed. P. Dewhurst. 1983, Kingston, Rhode Island: Boothroyd and Dewhurst Inc.

34. Boothroyd, G., *Design for Assembly*. First Edition ed. ed. P. Dewhurst. Vol. 1. 1984, New York, USA: Marcel Dekker.

35. Boothroyd, G., *Product Design for Assembly*. Third ed. ed. D. P. 1989, Kingston, Rhode Island: Boothroyd Dewhurst, Inc.

36. Boothroyd, G., *Assembly of large products*. Annals of CIRP, 1992. **40**(1): p. 1-4.

37. KrishnaKumar, A. *Genetic Algorithms-- A robust optimization tool.* in *31st Aerospace Sciences meeting and exhibit.* 1993. Reno, Nevada: American Institute of lAeronautics and Astronautics.

38. Parmee, I.C., *Techniques to aid global search.* 1993, University of Plymouth:

39. Murdoch, T.N.S., *An approach of configuration optimization.* Journal of Engineering Design, 1992. **3**(2): p. 99-116.

40. Aguirre-Esponda, G.J., *Evaluation of Design Systems at the design stage.* 1992, Cambridge University:

HANDLING OF POSITIONAL INFORMATION IN A SYSTEM FOR SUPPORTING EARLY GEOMETRIC DESIGN

X Guan & K J MacCallum

1 INTRODUCTION

A computer aided design (CAD) system for conceptual design should support not only functional and other relevant design activities, but also geometric information processing i.e. geometric design. At the early stages of geometric design, a designer's attention and interest is mainly on exploring a variety of possible geometric configurations of a product, where *geometric configuration* refers to the total geometric structure of the product consisting of the approximate or precise geometry of components of the product and their overall spatial arrangement in forming the product [1]. Geometric configuration design is an iterative process of *development - evaluation - refinement* which leads to a suitable concept for further detail design. In this process, the designer handles various types of geometric information including shapes, sizes, locations as well as orientations of components which is known vaguely or precisely. Free-hand sketches and diagrams are frequently used during the process for effective expression, communication, and recording of both geometric and non-geometric aspects of or information about the product being designed [6, 7]. The roughness nature of such sketches and their use at early design stages reflects a desire to explore and investigate design options or concepts without committing to exactness or detail.

A CAD system for supporting early stages of design should therefore follow a principle of *minimum commitment modelling*; that is, when it is used, the utilities built in the system should not force the designer to make specific commitments earlier or more than necessary, desired or appropriate. In the case of geometric design, commitments are decisions regarding components, arrangements, shapes, sizes, positions, etc.. One requirement that can be derived from this principle for a computational support system is to assist the development of early geometric concepts using both vaguely and precisely specified geometric information. Other important requirements include support for incremental refinement of imprecise geometric concepts, support for simultaneous iterative development of multiple geometric concepts (approximate or precise), support for the handling of conflicting or inconsistent geometric information, and an adequate user system interface which facilitates, or at least does not hinder, the designer's access to and interaction with the various resources available in the system.

Although various CAD systems have been developed to support the modelling of product geometry, they often require complete, precise, or detailed specification on the geometry which is usually unavailable until the later stages of the design process. By requiring from designers commitments greater than they can make or are willing to make at the early conceptual phase, these systems are not well suited for being used in dealing with early geometric configuration problems. As a result, they are usually not used by designers in practice in concept exploration but in modelling more or less finished geometric designs at the later design stages for purposes such as visualisation and various types of analysis.

In our research, we have been following the principle of minimum commitment modelling in developing a system for supporting geometric configuration design. This paper presents one particular application of this principle in developing the system. The notion of *uncertain region* is introduced in Section 2 to describe the locations of components in a geometric configuration and, in Section 3, to facilitate the incremental reasoning about the locations of the components from both abstract spatial relationships and precise point positions. Examples from the domain of computer enclosure design modelled using the implemented prototype system are given in Section 4 to illustrate the ideas. Finally, relevant further research issues are discussed.

2 REPRESENTATION OF POSITIONAL INFORMATION

2.1 Geometric Configuration and Geometric Configuration Space

A *geometric configuration*, denoted as GC, consists of one or many *geometry* entities that are spatially related to one another. Each of the geometry entities represents a geometric model of a component of a product being designed. Each geometric configuration is constrained spatially by a *geometric configuration space*, denoted as GCS, that is defined as a 3D cubic space with a given 3D world co-ordinate system $OXYZ$ as illustrated in Figure 1. Geometric configuration space is used to provide a physical bound for geometric configuration design. It is represented by three real intervals along the three axes of the co-ordinate system:

$$GCS = (GCS_x, GCS_y, GCS_z)$$
$$GCS_x = [\underline{GCS_x}, \overline{GCS_x}]$$
$$GCS_y = [\underline{GCS_y}, \overline{GCS_y}]$$
$$GCS_z = [\underline{GCS_z}, \overline{GCS_z}]$$

where $\underline{GCS_u} < \overline{GCS_u}$ ($u \in \{x, y, z\}$). A real interval, interval for short, is a bounded set of real numbers represented by a lower and an upper bound [3].

For an object G_p placed in a geometric configuration to be completely inside the corresponding GCS, geometric configuration space imposes the following constraints on the geometry of the object:

$$D_{px} \subseteq [\underline{GCS_x} + X\_SIZE_p / 2, \overline{GCS_x} - X\_SIZE_p / 2]$$
$$D_{py} \subseteq [\underline{GCS_y} + Y\_SIZE_p / 2, \overline{GCS_y} - Y\_SIZE_p / 2]$$
$$D_{pz} \subseteq [\underline{GCS_z} + Z\_SIZE_p / 2, \overline{GCS_z} - Z\_SIZE_p / 2]$$

$\{X\_SIZE_p, Y\_SIZE_p, Z\_SIZE_p\}$ is the size of G_p along the X, Y and Z co-ordinate direction.

$D_p(D_{px}, D_{py}, D_{pz})$ is the datum point of G_p which characterises its location in a geometric configuration as described in the next section.

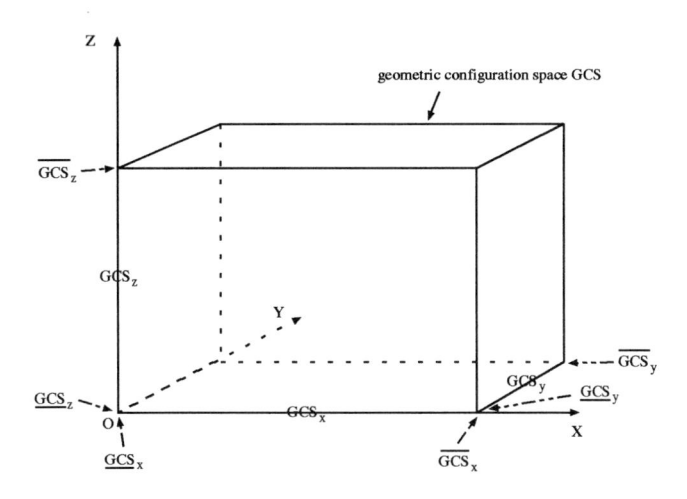

Figure 1. The concept of geometric configuration space.

2.2 Datum Point and Uncertain Region

The location of a geometry component in a geometric configuration is characterised by a *datum point*, denoted as D, on the component, e.g. its geometric centre. This datum point lies in a 3D cubic *uncertain region*, denoted as UR, which captures the approximation or uncertainty associated with the location of the component. The uncertain region is represented by three ranges (intervals) along, respectively, the X, Y, and Z axes of the world co-ordinate system associated with the corresponding geometric configuration which specify the allowable x, y, and z co-ordinates of the corresponding datum point:

$$UR = (UR_x, UR_y, UR_z)$$
$$UR_x = [\underline{UR_x}, \overline{UR_x}]$$
$$UR_y = [\underline{UR_y}, \overline{UR_y}]$$
$$UR_z = [\underline{UR_z}, \overline{UR_z}]$$

where $\underline{UR_u} \leq \overline{UR_u}$ ($u \in \{x, y, z\}$). Thus, the boundaries of the uncertain regions are represented by the bounds of the three ranges.

2.3 Spatial Relations and Point Positions

Both spatial relations and point positions may be used to specify the position of geometry components in a configuration. A spatial relation defines the location of an object in relation to another object in the same configuration. Following Retz-Schmidt [4], the object whose location is determined by the relation is called the *primary object*, and the object to which the relation is specified is called the *reference object*. Let G_p and G_r denote a primary object and a reference object in a spatial relation, respectively. Further let $\{X\_SIZE_p, Y\_SIZE_p, Z\_SIZE_p\}$ be the size of the primary object G_p along the X, Y and Z axes of the associated co-ordinate system OXYZ and $\{X\_SIZE_r, Y\_SIZE_r, Z\_SIZE_r\}$ be that of the reference object G_r. These sizes are approximated by the sizes of the orthogonal bounding boxes of the objects when deriving the location constraints on the bounds of the uncertain regions, which is considered feasible for early stages of design. Let $D_p(D_{px}, D_{py}, D_{pz})$ and $D_r(D_{rx}, D_{ry}, D_{rz})$ denote the datum points, here chosen to be the centres, of G_p and G_r, respectively. Point position and six basic spatial relations {*above, below, right, left, behind, front*} are discussed in the following.

A point position is defined by three co-ordinates in the world co-ordinate system OXYZ. Only one point position can be specified at one time to locate a geometry component in a configuration. An object $G_p{}^1$ located at a fixed point position $PP(PP_x, PP_y, PP_z)$ is defined as its datum point D_p being coincident with the point PP, i.e.:

$$D_{px} = PP_x$$
$$D_{py} = PP_y$$
$$D_{pz} = PP_z$$

In defining the spatial relations, the non-overlapping nature of objects in space is applied which requires the sizes of the involved objects to be taken into consideration. Geometric configuration space is also applied as implicit constraints. Spatial relations can be used to locate a primary object G_p in relation to a reference object G_r from the point of view of the user (the person who specifies a relation) based on a reference frame, here the co-ordinate system OXYZ. Based on this, *above*, *below*, *right*, *left*, *behind* and *front* are determined as shown in Figure 2.

> *above* Object G_p above object G_r, denoted by $above(G_p, G_r)$, constrains the location of G_p in such a way that the z co-ordinate of its datum point D_p in OXYZ is greater than that of D_r without causing any spatial interference with G_r, nor making G_p spatially outside the corresponding geometric configuration space, i.e.
>
> $$D_{pz} \subseteq [D_{rz} + (Z\_SIZE_p + Z\_SIZE_r)/2, \overline{GCS_z} - Z\_SIZE_p/2]$$

---

[1]For consistency with spatial relations, G_p is also used here and in defining geometric configuration space constraints to denote the object to be located.

below   Object G_p *below* object G_r, denoted by $below(G_p, G_r)$, constrains the location of G_p in such a way that the z co-ordinate of its datum point D_p in OXYZ is smaller than that of D_r without causing any spatial interference with G_r, nor making G_p spatially outside the corresponding geometric configuration space, i.e.

$$D_{pz} \subseteq [\underline{GCS_z} + Z\_SIZE_p / 2,\ D_{rz} - (Z\_SIZE_p + Z\_SIZE_r) / 2]$$

The other four relations *right/left* and *behind/front* can be defined similarly in the X and Y directions, respectively.

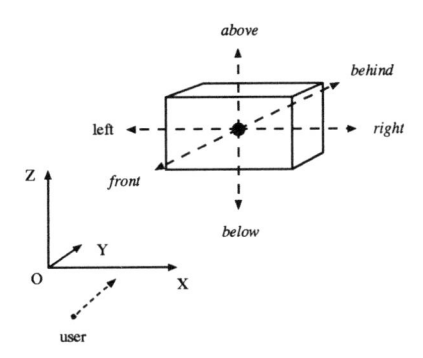

Figure 2. OXYZ as the reference frame in defining spatial relations.

As an example, Figure 3 illustrates the definition of the *above* relation. Given the *GCS*, the datum point D_r of object G_r at the position shown, as well as the shapes and sizes of the two objects, $above(G_p, G_r)$ means that G_p's datum point D_p should be situated in the cuboid region with dashed line boundaries. Note that *GCS* constraints are applied in the X and Y directions as well.

Given the chosen world co-ordinate system, the above definitions mean that *above/below*, *right/left*, and *behind/front* relations constrain only the z, x, and y co-ordinates of the datum positions of the involved objects, respectively. This facilitates the management of location constraints.

Further, for any three object G_i, G_j and G_k as well as a set of objects $\{G_i, i = 1, 2, ..., n\}$, $\omega \in \{above, below, right, left, behind, front\}$ is transitive, i.e.

$$\omega(G_1, G_j),\ \omega(G_j, G_k) \Rightarrow \omega(G_1, G_k)$$

$$\omega(G_1, G_j),\ \sum_{i=1}^{n} \omega(G_j, G_i) \Rightarrow \sum_{i=1}^{n} \omega(G_1, G_i)$$

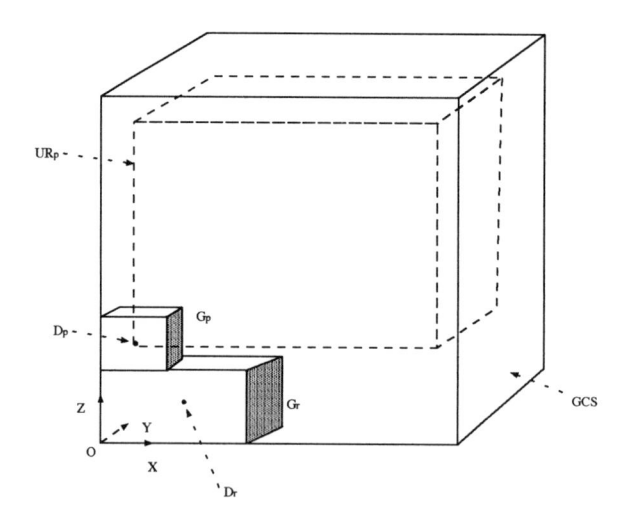

Figure 3. An example of $above(G_p, G_r)$.

and converse, i.e.

$$above(G_l, G_j) \Leftrightarrow below(G_j, G_l)$$
$$right(G_l, G_j) \Leftrightarrow left(G_j, G_l)$$
$$behind(G_l, G_j) \Leftrightarrow front(G_j, G_l)$$

3 MAINTAINING UNCERTAIN REGIONS

When being placed in a geometric configuration space, a component is given a location specification consisting of spatial relations or a point position discussed in the previous section. These relations or the point position are transformed into constraints on the bounds of the uncertain regions of the components involved. Constraints imposed by geometric configuration space on all components placed in a geometric configuration are also transformed in this way. Table 1 presents the corresponding transformations. These constraints retain the largest possible space for the uncertain regions for minimum commitment.

The constraints imposed by geometric configuration space state that, for an object placed inside a geometric configuration with a cuboid shape GCS, if the object is to be completely inside the space, its datum point, chosen as its geometric centre, should lie in a cuboid region determined by the corresponding constraints shown in Table 1. Figure 4 illustrates this through an example. It shows, in the OXZ projection, the uncertain region (dashed rectangle) of a sphere shape component G_p that belongs to a geometric configuration but has not yet been given any specific location information. Note that the component itself is displayed in a position corresponding to the minimum z co-ordinate allowed by the uncertain region.

High-Level Constraints	Constraints on Datum Point	Constraints on UR Bounds
GCS: $in(G_p, GCS)$	$D_{px} \subseteq [\underline{GCS_x} + X\_SIZE_p/2, \overline{GCS_x} - X\_SIZE_p/2]$ $D_{py} \subseteq [\underline{GCS_y} + Y\_SIZE_p/2, \overline{GCS_y} - Y\_SIZE_p/2]$ $D_{pz} \subseteq [\underline{GCS_z} + Z\_SIZE_p/2, \overline{GCS_z} - Z\_SIZE_p/2]$	$\underline{UR_{px}} = \underline{GCS_x} + X\_SIZE_p/2$ $\overline{UR_{px}} = \overline{GCS_x} - X\_SIZE_p/2$ $\underline{UR_{py}} = \underline{GCS_y} + Y\_SIZE_p/2$ $\overline{UR_{py}} = \overline{GCS_y} - Y\_SIZE_p/2$ $\underline{UR_{pz}} = \underline{GCS_z} + Z\_SIZE_p/2$ $\overline{UR_{pz}} = \overline{GCS_z} - Z\_SIZE_p/2$
Point Position: $at(G_p, PP)$	$D_{px} = PP_x$ $D_{py} = PP_y$ $D_{pz} = PP_z$	$\underline{UR_{px}} = PP_x, \overline{UR_{px}} = PP_x$ $\underline{UR_{py}} = PP_y, \overline{UR_{py}} = PP_y$ $\underline{UR_{pz}} = PP_z, \overline{UR_{pz}} = PP_z$
Spatial Relations: $above(G_p, G_r)$	$D_{pz} \subseteq [D_{rz} + (Z\_SIZE_p + Z\_SIZE_r)/2, \overline{GCS_z} - Z\_SIZE_p/2]$	$\underline{UR_{pz}} = \underline{UR_{rz}} + (Z\_SIZE_p + Z\_SIZE_r)/2$ $\overline{UR_{pz}} = \overline{GCS_z} - Z\_SIZE_p/2$ $\overline{UR_{rz}} = \overline{UR'_{rz}} - Z\_SIZE_p$
$below(G_p, G_r)$	$D_{pz} \subseteq [\underline{GCS_z} + Z\_SIZE_p/2, D_{rz} - (Z\_SIZE_p + Z\_SIZE_r)/2]$	$\underline{UR_{pz}} = \underline{GCS_z} + Z\_SIZE_p/2$ $\overline{UR_{pz}} = \overline{UR_{rz}} - (Z\_SIZE_p + Z\_SIZE_r)/2$ $\underline{UR_{rz}} = \underline{UR'_{rz}} + Z\_SIZE_p$
$right(G_p, G_r)$	$D_{px} \subseteq [D_{rx} + (X\_SIZE_p + X\_SIZE_r)/2, \overline{GCS_x} - X\_SIZE_p/2]$	$\underline{UR_{px}} = \underline{UR_{rx}} + (X\_SIZE_p + X\_SIZE_r)/2$ $\overline{UR_{px}} = \overline{GCS_x} - X\_SIZE_p/2$ $\overline{UR_{rx}} = \overline{UR'_{rx}} - X\_SIZE_p$
$left(G_p, G_r)$	$D_{px} \subseteq [\underline{GCS_x} + X\_SIZE_p/2, D_{rx} - (X\_SIZE_p + X\_SIZE_r)/2]$	$\underline{UR_{px}} = \underline{GCS_x} + X\_SIZE_p/2$ $\overline{UR_{px}} = \overline{UR_{rx}} - (X\_SIZE_p + X\_SIZE_r)/2$ $\underline{UR_{rx}} = \underline{UR'_{rx}} + X\_SIZE_p$
$behind(G_p, G_r)$	$D_{py} \subseteq [D_{ry} + (Y\_SIZE_p + Y\_SIZE_r)/2, \overline{GCS_y} - Y\_SIZE_p/2]$	$\underline{UR_{py}} = \underline{UR_{ry}} + (Y\_SIZE_p + Y\_SIZE_r)/2$ $\overline{UR_{py}} = \overline{GCS_y} - Y\_SIZE_p/2$ $\overline{UR_{ry}} = \overline{UR'_{ry}} - Y\_SIZE_p$
$front(G_p, G_r)$	$D_{py} \subseteq [\underline{GCS_y} + Y\_SIZE_p/2, D_{ry} - (Y\_SIZE_p + Y\_SIZE_r)/2]$	$\underline{UR_{py}} = \underline{GCS_y} + Y\_SIZE_p/2$ $\overline{UR_{py}} = \overline{UR_{ry}} - (Y\_SIZE_p + Y\_SIZE_r)/2$ $\underline{UR_{ry}} = \underline{UR'_{ry}} + Y\_SIZE_p$

Table 1. Transformation of spatial relations and point position as well as geometric configuration space into linear constraints on the bounds of uncertain regions. For all the cases, $\underline{UR_{pu}} \leq \overline{UR_{pu}}$ and $\underline{UR_{ru}} \leq \overline{UR_{ru}}$ ($u \in \{x, y, z\}$) must be satisfied.

The constraints imposed by spatial relations state that if an object (the primary object) is to be placed with respect to a reference object in a geometric configuration with a cuboid shape GCS, and the objects are to be completely inside the space, then the relevant constraints given in Table 1 must be satisfied. This is demonstrated by the example shown in Figure 5. Figure 5(a) shows a component G_r, which has already been placed in a configuration, and its uncertain region. Figure 5(b) illustrates the case when a new component G_p with the shape and size shown is placed *above* G_r in the same configuration. The bounds of G_p's uncertain region have been derived through the corresponding constraints shown in Table 1. Note that the upper bound of the z dimension of G_r's uncertain region has been shrunk from $\overline{UR'}_{rz}$ in (a) to $\overline{UR_{rz}}$ in (b).

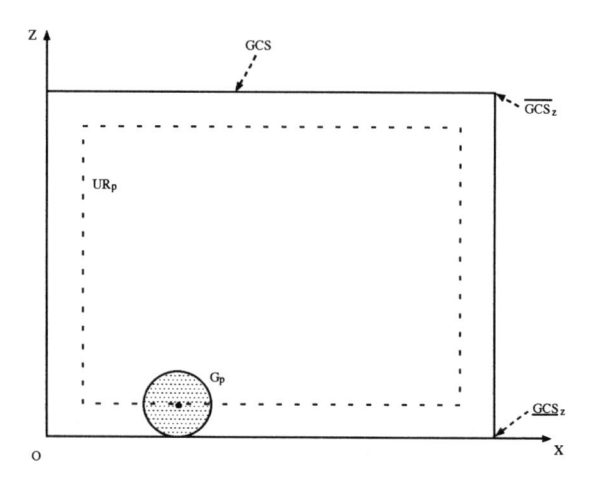

Figure 4. Implicit GCS constraints on an object placed in a configuration.

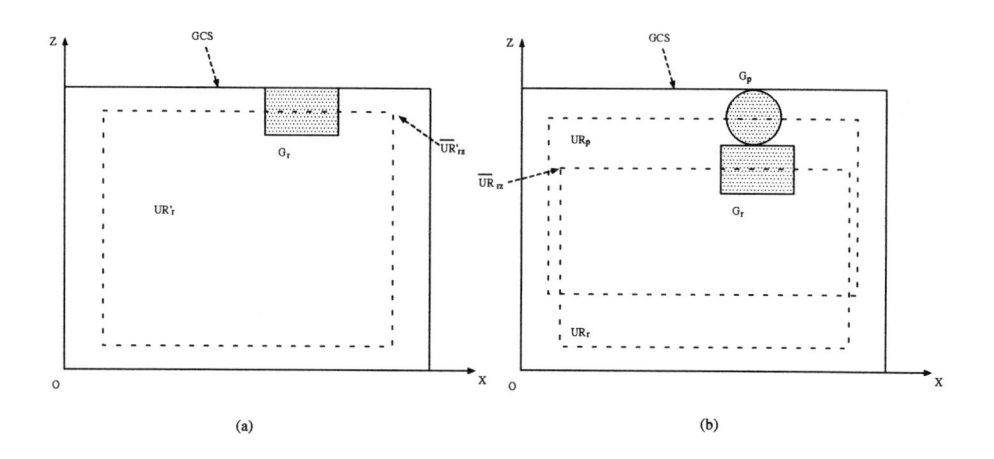

Figure 5. Location constraints from *above* relation.

Constraint based reasoning is carried out to reason and maintain uncertain regions during the modelling process. It tackles the spatial positioning problem by establishing and solving location constraints, resolving possible conflicts that arise, as well as propagating the results throughout the related configuration model. A new location specification given by a designer is checked for possible inconsistency in the specification or between the specification and existing spatial information in the configuration. If no conflict exists, spatial relations or point positions contained in the specification are then transformed into constraints on the bounds of the uncertain regions of the geometry components involved, based on Table 1. All the constraints on the bounds of the uncertain regions of all the components placed in a geometric configuration form the *location constraint model* of the configuration. This model is then solved by invoking the related constraint solving process which, if necessary, may again call the conflict handling process.

Each time when a new component is added to a configuration or the location of an existing component is modified incrementally, new constraints on the bounds of the uncertain regions of the relevant components are formed and are used to update the location constraint model of the configuration which is then solved through the constraint solving process. The location uncertain regions of the relevant components are thus established or reduced gradually. Figure 6 shows, in the OXZ projection, an example of uncertain region reduction. It illustrates that:

(a). Component G_1 placed precisely at point position P_1 in the configuration space GCS_1.

(b). Component G_2 is placed *above* G_1 in GCS_1. The uncertain region UR_2 of G_2 is shown in dashed lines.

(c). Component G_3 is placed *above* G_2 in GCS_1. This results in the shrinkage of UR_2 in the Z axis direction as shown.

(d). Instead of (c), G_3 is placed into GCS_1 on the *right* of G_2. This results in the reduction of UR_2 in the X rather than Z axis direction as shown.

(e). If G_3 is placed into GCS_1 on the *right* of and also *above* G_2, UR_2 will be reduced in both X and Z direction as shown.

4 EXAMPLES

The ideas discussed in the previous sections are being experimented in a prototype system implemented on a SunSparc platform using an object-oriented language, CLOS, a generic constraint solver, CLP(R) [2], and a solid modeller, ACIS [5]. The system implemented so far assists a designer in constructing and modifying multiple geometric configuration models incrementally by use of primitive shapes, simple size constraints, the six basic spatial relations as well as point positions, and in viewing these models visually. When using the system, modelling operations can be called directly from a Lisp Listener as Lisp functions/CLOS methods or they can be invoked via a graphical user interface. For simplicity, we will use the relevant Lisp functions and CLOS methods in the following examples.

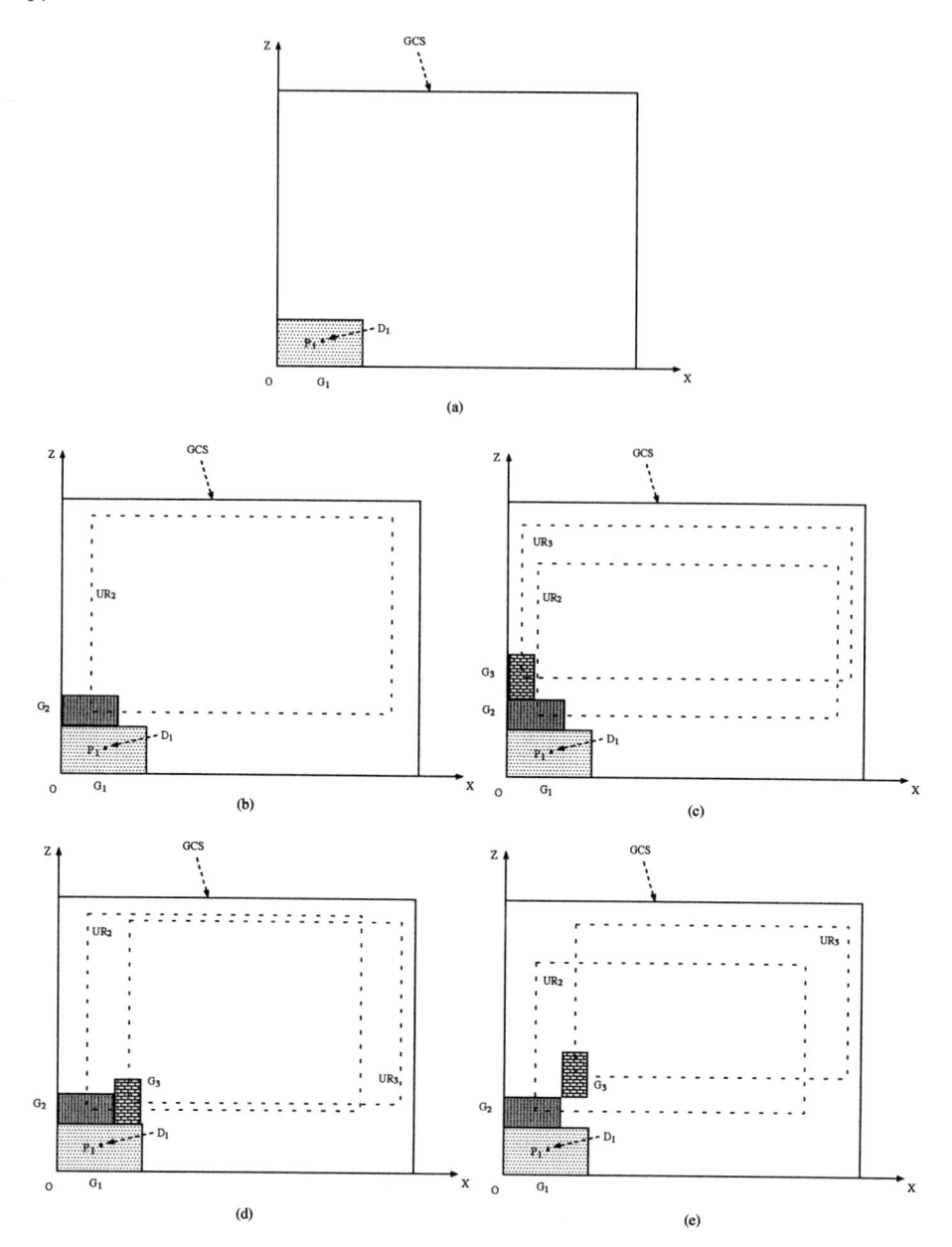

Figure 6. Gradual reduction of location uncertain regions as new components are introduced.

Suppose we would like to investigate the enclosure options for a specific computer model being designed. Using the system, an enclosure design option can be represented as a geometric configuration consisting of many geometry entities each of which represents a component or a subsystem of the computer, such as power supply unit, electric fan, hard disk drive, etc. To create a geometric configuration that represents an enclosure option, A, the following command can be issued:

>[2] (create-geometric-config :represents "Enclosure Option A")

This creates a configuration named **geometric-configuration1** and the associated *GCS* **geometric-configuration-space1** which is by default a cuboid space with a size determined by a global variable named ***gcs-size*** whose value can be set by a user. In our examples, the size of the geometric configuration space has been set to 100 units each side.

To create a geometric model of a component, say the power supply unit with cuboid shape and a size of width = 22, depth = 12.5, and height = 3, we can perform the following operation:

> (create-geometry :represents "Power Supply Unit"
            :attach-to geometric-configuration1
            :shape 'cuboid
            :size-constraints '((width = 22)
                        (depth = 12.5)
                        (height = 3)))

This generates a geometric model, **geometry1**, for the component as displayed graphically in Figure 7. The representation of the location of the component in this system is illustrated in Figure 8. As a part of **geometric-configuration1**, **geometry1** is constrained by **geometric-configuration-space1**. Therefore, although its location is not yet defined, **geometry1** (the power supply unit) has a datum position situated in **uncertain-region1** defined by three ranges, [11.0, 89.0], [6.25, 93.75], and [1.5, 98.5], derived via the method described in the previous section. The **approximate-datum-position** slot of **location1** holds the lower-left-front corner (11.0, 6.25, 1.5) of **uncertain-region1** at which the component was displayed graphically in Figure 7.

When created, the geometric model, **geometry2**, of another component, say the electric fan (with cuboid shape and size of width = 3.6, depth = 12.7 and height = 12.9), can be placed into the same configuration through the following operation:

> (specify-location geometry2 :spatial-relationships '((above geometry1)))

---

[2] > is the Lisp Listener prompt, not part of the command.

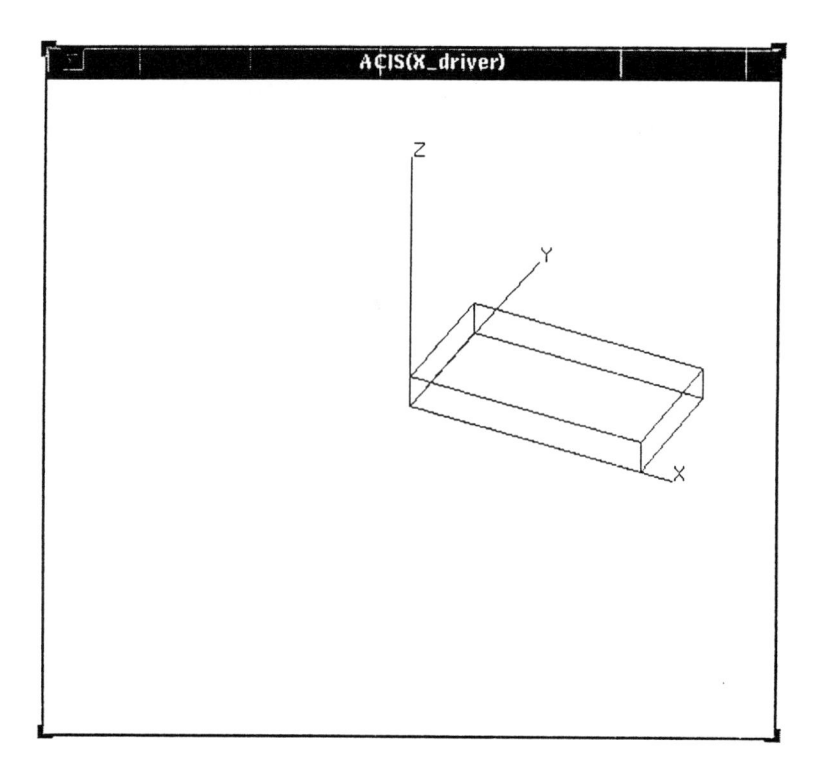

Figure 7. The geometric model of the power supply unit.

The resulting geometric configuration is displayed in Figure 9. Figure 10 shows part of the internal representation related to the location of the components. The location uncertain regions of the two components have been derived from the *above* relation imposed. As a result, the z dimension of the location uncertain region of the power supply unit, **uncertain-region1**, has been reduced from [1.5, 98.5] to [1.5, 85.6]. The other two dimensions remain unchanged.

Components can also be placed at a specified point in a configuration space. For instance, issuing the following command will establish a geometric model for the hard disk drive and position it at point (20, 15, 5) in another geometric configuration, **geometric-configuration2**:

[1:NAME] GEOMETRIC-CONFIGURATION1
[2:REPRESENTS] "Enclosure Option A"
[3:GEOMETRIES] (GEOMETRY1)
[4:SPATIAL-MAP] DAG1
[5:LOCATION-CONSTRAINT-MODEL] NIL
[6:CONSTRAINED-BY] GEOMETRIC-CONFIGURATION-SPACE1

[1:NAME] GEOMETRIC-CONFIGURATION-SPACE1
[2:CONSTRAINS] GEOMETRIC-CONFIGURATION1
[3:SHAPE] CUBOID3
[4:X-SIZE] (0.0 100.0)
[5:Y-SIZE] (0.0 100.0)
[6:Z-SIZE] (0.0 100.0)
...

[1:NAME] GEOMETRY1
[2:REPRESENTS] "Power Supply Unit"
[3:IS-PART-OF] GEOMETRIC-CONFIGURATION1
[4:SHAPE] PRIMITIVE-SHAPE1
[5:LOCATION] LOCATION1
[6:ORIENTATION] NIL

[1:NAME] LOCATION1
[2:IS-ATTACHED-TO] GEOMETRY1
[3:STATUS] UNCONSTRAINED
[4:DATUM] CENTRE
[5:DATUM-POSITION] #<Uncertain-Region 107F1DA8>
[6:CONSTRAINED-BY] NIL
[7:APPROXIMATE-DATUM-POSITION] #<Point-Position 107F2EDC>

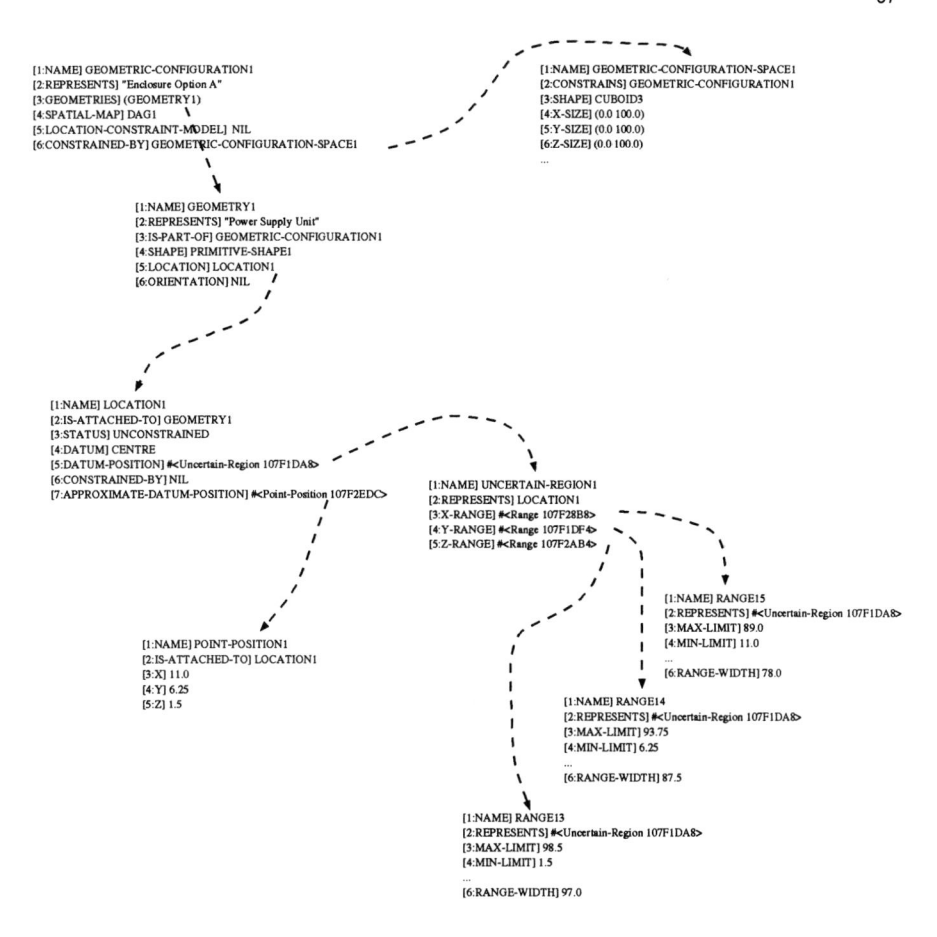

[1:NAME] UNCERTAIN-REGION1
[2:REPRESENTS] LOCATION1
[3:X-RANGE] #<Range 107F28B8>
[4:Y-RANGE] #<Range 107F1DF4>
[5:Z-RANGE] #<Range 107F2AB4>

[1:NAME] RANGE15
[2:REPRESENTS] #<Uncertain-Region 107F1DA8>
[3:MAX-LIMIT] 89.0
[4:MIN-LIMIT] 11.0
...
[6:RANGE-WIDTH] 78.0

[1:NAME] POINT-POSITION1
[2:IS-ATTACHED-TO] LOCATION1
[3:X] 11.0
[4:Y] 6.25
[5:Z] 1.5

[1:NAME] RANGE14
[2:REPRESENTS] #<Uncertain-Region 107F1DA8>
[3:MAX-LIMIT] 93.75
[4:MIN-LIMIT] 6.25
...
[6:RANGE-WIDTH] 87.5

[1:NAME] RANGE13
[2:REPRESENTS] #<Uncertain-Region 107F1DA8>
[3:MAX-LIMIT] 98.5
[4:MIN-LIMIT] 1.5
...
[6:RANGE-WIDTH] 97.0

Figure 8. Representation of the location of power supply unit.

```
> (create-geometry  :represents "Hard Disk Drive"
                    :attach-to geometric-configuration2
                    :shape 'cuboid
                    :size-constraints '((width = 15.49)
                                        (depth = 13.72)
                                        (height = 3.56))
                    :p-position '(20 15 5))
```

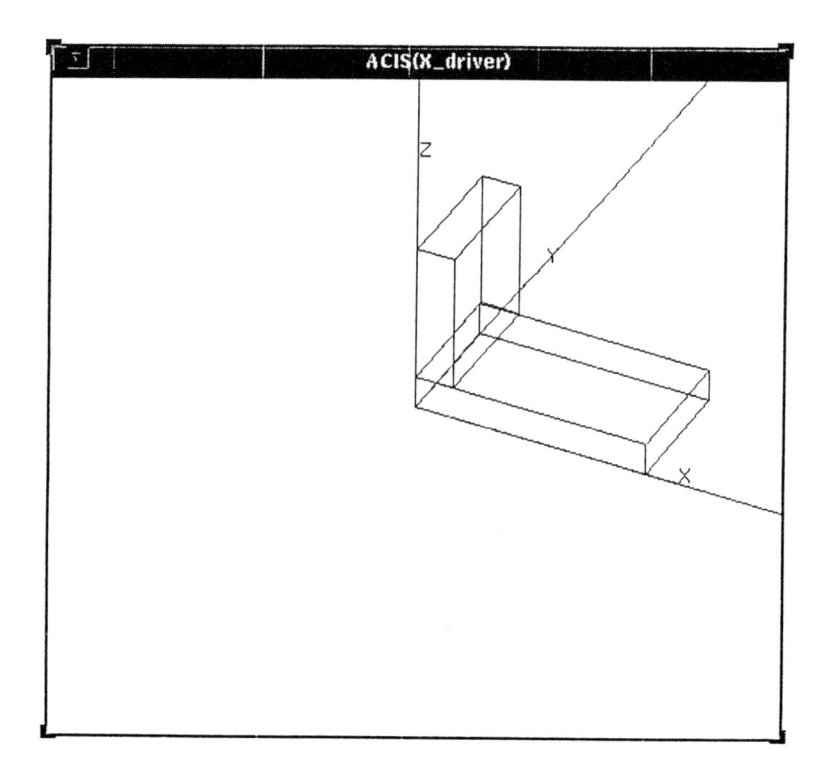

Figure 9. Electric fan is placed *above* power supply unit.

5 CONCLUSIONS

In this paper, we have presented an application of the minimum commitment modelling principle in developing a geometric configuration support system. The notion of *uncertain region* has been introduced to describe the locations of components in a geometric configuration. We have also discussed the method for reasoning about the uncertain regions from abstract spatial relations and precise point positions.

The use of uncertain region in representing and reasoning about the effect of spatial relationships on objects' positions helps a designer avoid unnecessary commitment to a fixed point position. This allows the designer to model geometric configurations in an early stage when information is still imprecise without unnecessarily reducing the corresponding solution space. It also enables a gradual refinement of approximate models when more spatial relationships are specified later to constrain the location of the components further.

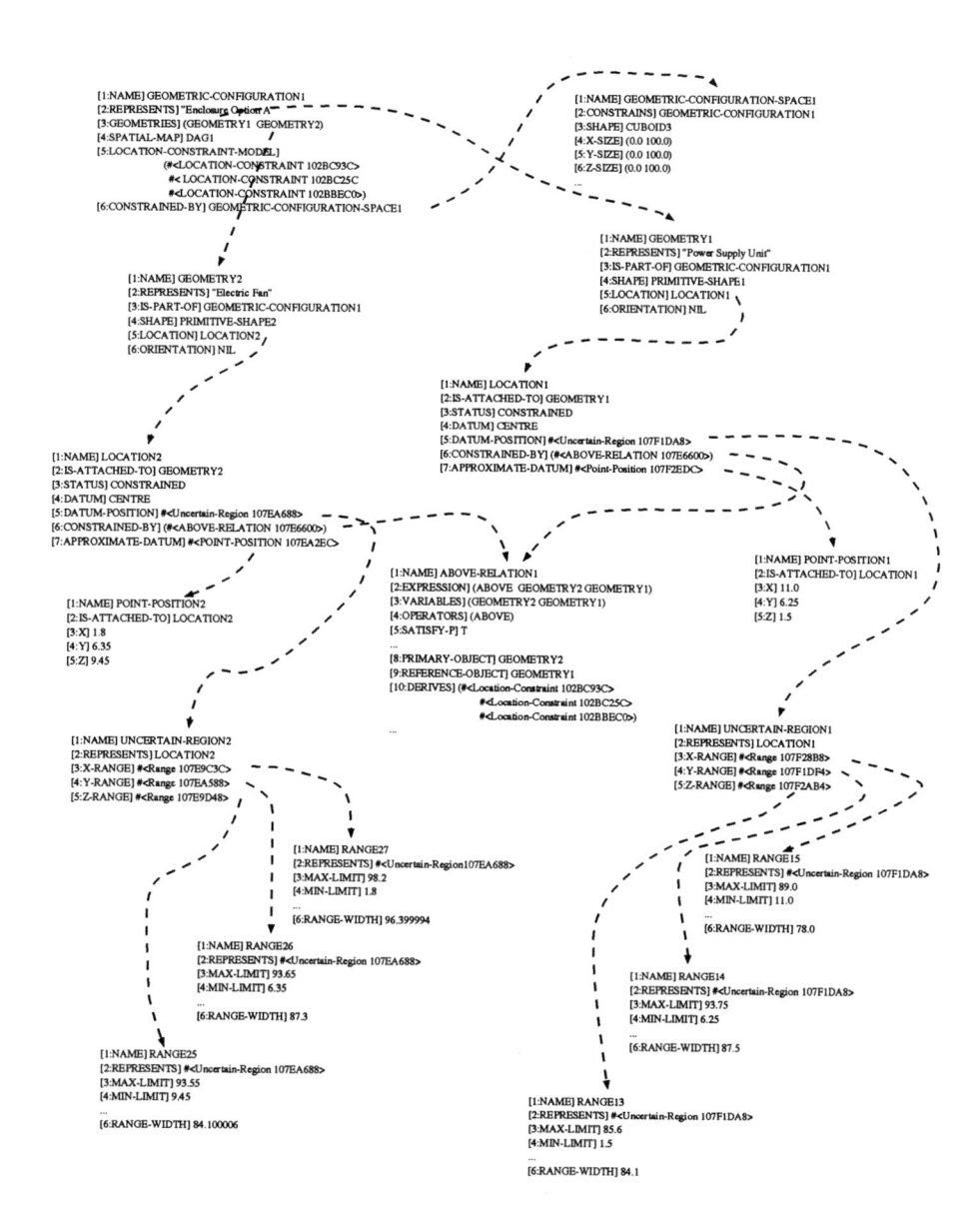

Figure 10. Representation of the location of components in **geometric-configuration1**.

As a future research issue, incorporation of more spatial relations at different levels of abstraction may be investigated to provide a better support to designers. The types of spatial relations that are useful include *distance* relations which allow a designer to specify the location of a component less precise than the point position but more than the six basic relations, *containment* relations which allow a component to contain or to be inside another component, and *adjacency* relations which allow a component to be placed in adjacent to another component.

At the moment, uncertain regions of components are reduced/refined through introducing more spatial relations or point positions. Further work may be carried out to support designers in reducing the uncertain regions by directly manipulating the bounds of these regions while still maintaining the relevant spatial relations or point positions already specified. Graphical presentation of uncertain regions is also very valuable and worth of investigation.

In this paper we have focused the discussion on handling imprecise location information. Approximation associated with the size of objects is another characteristic of early geometric configuration which, although not addressed in this paper, is also a very important part of our research.

ACKNOWLEDGEMENT

This work is supported by EPSRC, U. K.

REFERENCES

[1] Guan, X., "Computational Support for Early Geometric Design", Ph.D Thesis, University of Strathclyde, Glasgow, Scotland, United Kingdom, September 1993.

[2] Heintze, N., et al. *The CLP(R) Programmer's Manual*, version 1.1, IBM Thomas J. Watson Research Centre, USA, 1991.

[3] Moore, R. E., *Methods and Applications of Interval Analysis*, SIAM, Philadelphia, 1979.

[4] Retz-Schmidt, G., "Various views on spatial prepositions", *AI Magazine*, pp 95-105, Summer 1988.

[5] Spatial Technology Inc. *ACIS: Interface Guide*, 1992.

[6] Tovey, M., "Drawing and CAD in industrial design", *Design Studies*, Vol 10, No 1, pp 24-39, 1989.

[7] Ullman, D. G., *The Mechanical Design Process*, McGraw Hill, Inc., 1992.

AN ARCHITECTURE FOR THE INTELLIGENT SUPPORT OF KNITWEAR DESIGN

C Eckert & M Stacey

1 INTRODUCTION

Knitwear production is one of the world's largest design-driven industries. As well as being important, the knitwear design process involves a complex and interesting interaction between aesthetic and technical design, and a problematic interaction between aesthetic and technical designers.

Increasingly sophisticated CAD systems play a vital part in the design of knitted garments, but their use is usually restricted to the technical designers (knitting machine technicians), who use them to program industrial knitting machines [4]. Knitwear designers hand over their designs in the form of sketches and informal descriptions; the technicians often distort the designers' intentions out of recognition in the course of doing the detailed design. We have argued in [4] that the efficiency and the effectiveness of the knitwear design process in industry might be enhanced by giving knitwear designers more access to CAD systems and better training in how to use them. But the ability of knitwear designers to make effective use of CAD systems would also be enhanced by the development of intelligent support systems for the conceptual and aesthetic design of knitwear. In this paper we present a proposal for how to do this, in the form on an outline architecture for an intelligent support system for knitwear design.

2 KNITWEAR DESIGN

In this section we briefly describe the knitwear design process and its industrial context to motivate our proposals for ways to support it. This is a very selective summary based on the first author's extensive ethnographic study of the knitwear design process, which included taking knitwear design classes at De Montfort University, learning pattern construction and learning how to program knitting machines, as well as formal and informal interviews with knitwear designers and technicians in Britain and Germany [5]. Our proposals for the intelligent support of knitwear design are motivated both by the demands inherent in the task and by the problems arising from the organisation of knitwear design in the knitwear industry.

Knitwear design involves a combination of aesthetic and technical demands, which interlock in complex ways: small aesthetically motivated changes can have major technical consequences, and changes made to meet technical constraints unavoidably affect the appearance of a garment.

2.1 Knitwear Design in Industry

In industry the knitwear design process is shared primarily between knitwear designers (almost all young females who consider themselves middle class professionals) and knitwear technicians (almost all somewhat older males who think of themselves as skilled craftsmen) [4]. The construction of sample garments is done either by the designers or by sampling make-up staff, who construct cutting patterns from the specifications given them by the designers.

Garments are designed to fit the fashion of a particular season, and need to be produced on time, so designing in industry is done under intense time pressure. Knitwear designers are required to produce a lot of designs very quickly; only a small proportion are approved by their managers for further development. Knitwear technicians are rarely involved in the design process before they are given designs to be programmed, and they are not always treated as responsible experts whose technical judgements should be trusted. When they program designs they have very little time to consult designers and improve the designs iteratively. Very often, they make major changes for technical reasons, in order to make the garments producable at an affordable cost. These changes sometimes sabotage the designers' intentions and ruin the designs. Designers usually only get consulted at the end about whether they do or don't like the changes the technicians have made, when it's very difficult for them to object to anything the technicians have done.

2.2 Conceptual Design of Knitwear

In the design of garments, it is difficult to separate conceptual design from requirements analysis on the one hand, and detailed design on the other. Many vital decisions that would qualify as conceptual design in other industries are usually made long before anyone thinks they're designing anything [5]. These include the fundamental choices about target customers, types of garments, price points, materials to be used, and the overall images or themes to be projected, that limit the space of possibilities the designer can explore. Often quite detailed decisions are made for the designer in the specifications provided by retail chains, or are inherent in decisions to make limited changes to a previously successful garment. The range of freedom designers have varies enormously, from being asked to produce something innovative within the broad themes and looks of this year's fashion, through producing a new colour pattern for a company-standard sweater, to making the minimal number of changes to a competitor's garment needed to evade the copyright laws.

After the company has selected a set of required garments and the styles, looks or themes each should project, the knitwear designer finds and looks at a large amount of visual material related to the themes, including other companies' garments, which are used as sources of 'inspiration' [5]. Knitwear designers are trained to be able to turn anything they look at into an idea for a garment. What designers consider designing begins after this, with decisions about the overall effects to be achieved, and the balances of design elements and colours, plus the selection of motifs and stitch structures (the conceptual design of a garment). This is followed by the adaptation of these elements to produce a garment achieving the right appearance and theme projection. As with most commercial design, almost all knitwear design proceeds primarily by the selection, adaptation and combination of design elements, with ab initio design playing a more minor role.

Knitwear designers work primarily with mental images and spatial representations of designs, from the most fragmentary ideas to complete designs for garments, and evaluate them using visuospatial perceptions of balance and beauty. They use symbolic and numerical representations only to solve specific problems arising in the course of advanced detailed design [5].

The design of garment shapes is logically separate from the design of local design elements like stitch structures and motifs, and the two are usually done largely separately before the design elements are positioned on the shapes (pattern placing). However motifs and shapes constrain each other, if a good pattern placement is to be achieved - this is a distinctive characteristic of knitwear. Pattern placing often drives the revision of shapes and stitch structures. The order and priority of shape and motif design is almost always mutable: designers often want to suspend one of these design tasks to work on another.

Conceptual design blends into detailed design, as designers make progressively more detailed decisions about shapes and motifs. But the presentation of large numbers of rough sketches of garments to senior designers and managers, who select some for further work, serves to some extent as a division between conceptual and detailed design.

Detailed design may be done entirely by the designers, or entirely by the technicians, or is divided between them, depending on the structure of the company. It involves some difficult constraint satisfaction problems requiring symbolic and numerical reasoning. Pattern construction for the shapes of garment pieces involves mathematical problems in the calculation of the shapes of curves, for example sleeve crowns and sleeve insets in body blocks, and in obtaining correct and consistent sets of measurements. Designing stitch structures that can be knitted efficiently, and placing motifs on the garment so that the whole can be knitted efficiently, can involve a lot of backtracking.

2.3 The Handover from Designers to Technicians

The handover of designs from knitwear designers to knitting machine technicians is inherently problematic because of the complex interaction between aesthetic requirements and technical constraints. It is poorly managed in the knitwear industry, with the result that designers' intentions are distorted, and a large amount of effort is devoted to making infeasible designs work. The designs given to technicians to implement are frequently incoherent or technically infeasible. The technicians we've talked to estimate that only about thirty percent of the designs they're given can be produced at their intended price points. A large part of the problem is due to the lack of good ways for designers to express their ideas. Knitwear designers have no formal description language to communicate their ideas to technicians, and the very informal descriptions they use fail to meet their needs.

Knitwear designers usually hand over their designs to technicians in the form of sketches and swatches (small pieces of knitted fabric), perhaps with some measurements and verbal descriptions.

Designers' sketches often include a mixture of precise details meant to be reproduced exactly and very rough indications of unimportant or standard aspects of the design. Designers have no good standard way to indicate which aspects of their designs are meant to be taken seriously as accurate descriptions, and which not, so technicians often alter crucial aspects of the designs. Technicians have to reverse engineer swatches by working out how they are constructed; one technician at De Montfort University commented to us that his job would be made much easier if his student designers could write down their stitch structures in some formal notation, but they aren't taught one.

2.4 Technical Design of Knitwear

Knitwear designers do not consider the technical implementation of their designs to be design at all, a view we consider not just wrong but harmful [4]. But knitting machine technicians writing programs to control industrial knitting machines do a lot of detailed design; the detailed design of the appearance of a garment coincides with the conceptual planning of a program to produce it.

Almost all commercial knitwear is produced on big computer-controlled knitting machines, which are among the most complex industrial machines, some having over a hundred thousand parts. They knit by the action of cam boxes moving along two rows of needles (the front and back needle beds), which hold the top row of stitches.

The time a garment takes to knit on a commercial knitting machine is a major determinant of how much it costs. This time depends on the number of traverses of the cam boxes along the needle beds. Knitting some combinations of stitches involves traverses that move stitches between the front and back needle beds, but do not knit the stitches. (As well as changes in the relative position of the needle beds, which have no time cost.) Stitch structures involving a lot of transfer operations cost much more to knit, and this effect is magnified on the largest and most advanced machines which can knit up to four rows on a single traverse. So the detailed design of a garment can have an enormous effect on its cost.

Some combinations of stitches simply cannot be machine knitted using certain yarns. In addition to working out whether a design can be knitted fast enough to be affordable, an important part of a technician's job is working out whether designs are feasible on a given machine using a particular yarn, or whether, for instance, the tension on the yarn will be too high.

2.5 Knitwear CAD Systems

Knitting machine technicians program knitting machines using CAD systems developed by the knitting machine manufacturers, primarily Shima Seiki, Stoll and Universal. Writing these programs is the technicians' main job. Although the machine manufacturers have started targeting designers in their advertising, their CAD systems are designed to be used by technicians. Most knitwear designers get no access to the CAD systems, or only to the obsolete systems least suited to their needs. In some companies knitwear designers design directly onto CAD systems, but they only do this for the most straightforward types of designs. We discuss the technological, commercial, educational and social reasons for this sharp division in the use of knitwear CAD systems in [4].

Until the last few years, programming a knitting machine was an activity very similar to assembly language programming, involving keeping track of which stitches are where on the needle beds, and reasoning about how to minimise the number of traverses. This is still what many knitwear technicians do. The knitting machine CAD systems being produced now provide elaborate visual programming environments for designing garments by editing pictures of garment pieces, using colour codes to represent both regions in different colours and structural features like cables. When designs produced using the visual programming facilities are complete, they are compiled into lower level machine programs. Design errors resulting in a failure to generate a program are reported at this stage. Some systems allow the user to develop or edit programs using the older, lower level programming languages, to create designs that are impossible using the higher level visual languages. Producing programs using the visual programming facilities is very much faster, but the systems' users do not develop the programming skills producing simple designs that they need to develop the more elaborate designs for which the visual programming languages are inadequate. The result of this is that the knitting machine technicians are deskilled, and the range of practically feasible designs is reduced, even though the range of what is possible using industrial knitting machines has increased enormously over the last few years.

There is no universally accepted representation of stitch structures. The different machine manufacturers' visual programming languages differ. It is probably impossible to create a single notation to meet all the needs of designers and technicians, because of the number of different stitch types and the complexity of the visual effects produced by their interaction. (Nevertheless there is scope for designing and popularising better notations than are used.) Some manufacturers provide multiple representations: Universal CAD systems have a symbolic representation showing approximations to the three dimensional effects of the fabric, and a representation showing how the yarn is held on the needle beds. Stoll CAD systems have a representation using keyboard symbols for stitches.

The most advanced CAD systems use completed programs to produce simulations of the three dimensional appearance of the knitted fabric, either faked (Shima Seiki) or based on simulated knitting (Stoll). They also generate pictures of what the garment would look like on a person, by mapping an image of the fabric onto a body shape. These are very useful for marketing purposes, and for showing designers and technicians unexpected emergent properties of their designs, but their value for providing feedback for the design process is limited by the fact that they cannot be generated from incomplete designs.

3 INTELLIGENT SUPPORT FOR KNITWEAR DESIGN

The primary goal of the research reported in this paper is to find ways to support the work of knitwear designers. This means supporting the development of a design from an initial idea to a detailed specification that can be given to a knitting machine technician. Moreover, it means giving help that designers find useful, by assisting or taking over difficult tasks that are bottlenecks in the design process, and exploiting rather than supplanting designers' abilities to do what they are good at.

The work described in this paper is intended to give more control over the design process to the designers, so that they can get their own ideas realised without distortion. The intelligent support tool we envisage should enable them to make aesthetic design decisions within the technical constraints of the domain. The commercial benefit this should bring is to reduce the number of designs thrown away because they are infeasible, and the large amount of wasteful backtracking involved in making inadequately specified and over-elaborate designs work.

A crucial part of improving the knitwear design process is managing the handover of a design from the knitwear designer to the technician. In order to achieve a smooth handover, designers need to produce designs that are technically feasible at their intended price points so that the technicians do not need to modify the designs to make them work. A CAD system designed to support the designers as primary users, rather than technicians, could facilitate a smooth handover by providing first pass technical feedback to eliminate most of the infeasible or expensive designs, and by representing the designs in a format that the technicians could develop into knitting machine programs without needing to reverse engineer stitch structures.

3.1 Exploiting Human Abilities

In our view the key to developing effective CAD systems is exploiting designers' strengths while compensating for their weaknesses. Knitwear designers like to think spatially: they have strong visual imaginations, have good visual memories, and are very good at making rapid perceptual evaluations of designs. They can make perceptual judgements on aesthetic criteria and also on technical criteria when there is a straightforward relationship between the correctness of a garment and its appearance. For example, experienced designers are good at looking at garment shape curves such as necklines and seeing whether they are correct. Designers are much less good at solving the mathematical and constraint satisfaction problems involved in knitwear design, not least because these problems are often intrinsically difficult.

A CAD system for knitwear design should make as much use as possible of the designers' abilities to make perceptual evaluations. Wherever possible, it should present information about garment designs in forms that designers can understand by using their existing skills for thinking spatially about garments, and showing directly the relationship between the design and the information presented. The system should not force designers to switch to different, non-spatial ways of thinking, or require them to reason about the meaning of the system's messages, as this can easily disrupt the designers' flow of ideas.

3.2 Eliminating Bottlenecks in the Design Process

As Landauer has pointed out [11], tools intended to support human activities should be focused on the bottlenecks in the processes, where some assistance can result in major improvements in the overall performance of the activity. In knitwear design a major bottleneck is created by the mathematical problems involved in pattern construction: making sure that garment piece measurements are coherent, that the different garment pieces fit together and that the garment fits the human body in the right way. These mathematical problems are well suited to intelligent computer support: they are well-defined subtasks requiring the application of algorithms. The development of computer support tools for these mathematical tasks is made challenging by the need to develop reasoning methods for selecting the appropriate mathematical algorithms, and by the need to develop novel numerical analysis techniques for the modelling of garment shapes. Mathematical functions for shape construction are an essential part of an integrated design support environment for garment making.

Other bottlenecks are caused by the need to solve constraint satisfaction problems involving a lot of backtracking, in designing cost-effective stitch structures and finding good pattern placements. Intelligent reasoning systems can support the solution of these problems both by identifying the sources of difficulty for designers and by searching for solutions.

3.3 Intelligently Supporting Designers

When the aesthetic design of appearance affects the cost or technical feasibility of the design, and conversely technical constraints limit the search space of feasible designs, intelligent reasoning systems can support design by managing the interaction between aesthetic and technical design. There are three ways to do this; all have roles to play in supporting different aspects of knitwear design: supporting or completely automating the technical subproblems requiring mathematical reasoning or backtracking to perform constraint satisfaction; configuring CAD tools so that designers can only produce technically correct solutions; and allowing designers to produce any imaginable design, but providing feedback about technical feasibility and cost. The last approach is required when there is no simple relationship between the appearance of a design and its feasibility.

Garment shape editing is ideally suited to CAD tools configured to allow designers to create only correct designs. One can describe mathematically the subset of curves that designers consider acceptable shapes, and the constraints imposed on it by other garment pieces are mathematically simple. On the other hand, there is no simple relationship between the appearance of a stitch structure and its technical consequences. The implication of this is that designers cannot evaluate the cost and technical difficulty of knitting a design by looking at it, except in simple cases. So when knitwear designers are designing stitch structures and placing them on garment shapes, they need feedback on the technical feasibility of their designs that is generated by symbolic reasoning, either by a technician or an AI system.

The presentation of technical feedback to designers by design support systems is a problematic issue. Some AI systems generate technical criticisms of designs as the designers construct them, and present these in a window, for example Fischer's kitchen design support system [6], [7], [8]. We are deeply skeptical of the value of this approach for several reasons. Evaluations need to be both focused and timely: designers do not want criticism of designs they know are incorrect, or criticism of aspects of designs they are not currently interested in. Generating manifestly incorrect designs is often a vital step towards creating interesting correct designs. An intelligent system presenting evaluations should enable the designer to make requests for evaluations of specific aspects of a design, presented without extraneous information. Our further reason for being skeptical about extensive verbal or symbolic evaluations is that designers have to work out what the evaluations mean for their designs, which can involve complex symbolic problem solving.

The generation of evaluations by critiquing modules has an important role to play, but the evaluations should be presented on request. They should be presented graphically as far as possible, to *show* the source of the problem in the design, for example by highlighting stitch combinations that make a disproportionate contribution to the knitting time of a garment.

3.4 The Role of Automatic Design

The most direct way to exploit designers' abilities to think visually about garment designs is to show them designs. We propose using the technical evaluations generated by a design support system to guide the construction of possible completions and modifications of the partial designs created by human designers. These possible design completions could then be evaluated very quickly on aesthetic criteria by the designers - this is what they are good at - and edited further.

The system should produce all the plausible alternative design completions, and present them to the users, who can filter them very quickly to select the few they are interested in. The operation of the design completion modules could be focused by selecting completion options from a menu. For garment shape design and pattern placing this would provide the necessary degree of constraint; there would only be a small range of options required, and designers would always know which ones they wanted. An automated design module working in this way would fit neatly into the current working methods of commercial knitwear designers, who usually begin by selecting a previous design to adapt, or by generating a relatively simple or bland rough design that they then elaborate. The designer should be free to call the design completion module (as well as any other evaluation modules) at any stage in the design process. As well as assisting the development of final designs, automatic design completion would enable designers to explore the appearance and emergent properties of partial designs in context without wasting effort on parts of designs they are not currently interested in. They would be able to see if an intended arrangement of stitch structures just doesn't look right, or produces an unintended emergent pattern, or is too expensive to knit; this would be especially useful for judging the balance of small repeat patterns, which are too fiddly to visualise.

Although knitwear design is complex in other ways, it has some advantages that make automated design in knitwear relatively tractable. Knitted fabric is a discrete structure with a large grain size, so stitch structures comprise a small number of stitches in each direction. There are only a small number of different types of stitch, so knitwear lends itself to computer representation. And there are only three design problems, stitch structure design, shape design, and pattern placing, all of which only require relatively simple design descriptions. Automated knitwear design requires the implementation of reasoning engines for design construction with some design heuristics, but the evaluations used to guide search are the same as those required by designers to guide manual design. An automated design module within a design support system should not require elaborate aesthetic evaluations; the system should show the users all the plausible alternatives.

There is very little scope for the automatic completion of stitch structure designs (though automatic design of stitch structures from pictures or patterns might be a fruitful source of design ideas). The role for automatic design of stitch structures is modifying designs to produce cheaper alternatives, using evaluations of what combinations of stitches are impossible or expensive to knit. (This is what knitwear technicians currently do; putting this part of the process under the control of the designer would help them get the results they want.)

We envisage automatic design completion playing an important role in shape design and pattern placing. Garment shapes in knitwear fall into a small number of basic categories, with scope for variation in each: completing a shape design from a brief indication of how it differs from standard is a tractable problem, at least for first pass designs that can be edited. Constructing consistent correct shape descriptions without computer support is difficult and tedious, and knitwear designers frequently give technicians and make-up staff incoherent measurements; an automatic shape design module would be valuable if it could resolve inconsistent sets of measurements to present all the likely alternatives. Placing repeat patterns and cables on shapes can be automated for a limited range of situations, which could be selected from a menu [3].

The danger in the use of automatic design modules is that they might bias the design process towards producing designs very similar to those produced by the automatic design modules. Bias towards producing what is easy is a universal problem of design tools [12], observed for example by us for knitwear CAD systems and by Devane for textile CAD systems [2]. Our view is that automatic design modules, used in combination with editors that did not restrict the construction of complex designs, would exert weaker biases than those inherent in the current knitwear design process. They would allow designers to explore widely at low cost, and try out ideas they knew were unlikely to be manufacturable or affordable. The speed up provided by automatic design completion would enable designers to explore more variants before settling on one for detailed development. At the moment designers are constrained to explore only ideas they think are very likely to be feasible.

3.5 Dual Functionality: Supporting Aesthetic and Technical Design

Our aim is to support collaboration between knitwear designers and knitting machine technicians, by designing support systems to be used in different ways by users with different levels of technical knowledge, so that both knitwear designers and knitwear technicians can make appropriate use of the systems' functionality and knowledge.

Designers and technicians have different requirements for intelligent support. The designers need simple and succinct symbolic feedback about the technical implications of their designs: a limited number of simple statements about possibility, for example the number of yarn feeders a garment piece requires, and about knitting times and costs. They also need graphic illustrations of the sources of costs and problems, plus, ideally, automatic design completions. Knitting machine technicians need support that enables them to make use of their expert knowledge of knitting machines, yarns, and the local difficulties of manufacturing in their companies. This means giving them detailed information about things like tensions on yarns, in symbolic evaluations; the same calculations of yarn tensions might be used in design completion with a rule to reject any design with a tension over a fixed threshold.

The technical knowledge represented in the system should be *shared* by the different reasoning modules that need the same knowledge, so that no inconsistencies are produced when the representations of facts and inferences are edited, and the technicians who understand the idiosyncrasies of machines and yarns, and the complexities of knitting machine programs, can see directly the relationship between the inference rules in the system and the symbolic evaluations they use, and the evaluations used by the designers. This requires as clear a separation as possible between the technical knowledge, represented declaratively, and the inference engines that use it.

4 AN ARCHITECTURE

We present a proposal for an architecture for an intelligent support system for knitwear design, which is based on the considerations we have outlined above. So far very little of this architecture has been implemented. We have no intention of attempting to develop a complete CAD system for knitwear design, rather our work is complementary to the commercial development efforts of the knitting machine manufacturers. We have concentrated on the development of intelligent support modules for parts of the design process, principally the mathematical tools for modelling garment shapes, and heuristic reasoning modules for pattern placing.

For clarity we describe the architecture in the present indicative, despite its status as a proposal and a conceptual framework for the development of some of its elements in isolation.

4.1 Three Environments for Three Tasks

As we have observed, knitwear design comprises three conceptually different tasks: stitch structure design, shape design, and pattern placing. An integrated design tool for knitwear design comprises three separate design environments for these three tasks. The tool needs to allow the user to follow any path to a finished design, and do any task at any time. So each design environment allows the user to suspend jobs and work on other designs, and to switch from one environment to another.

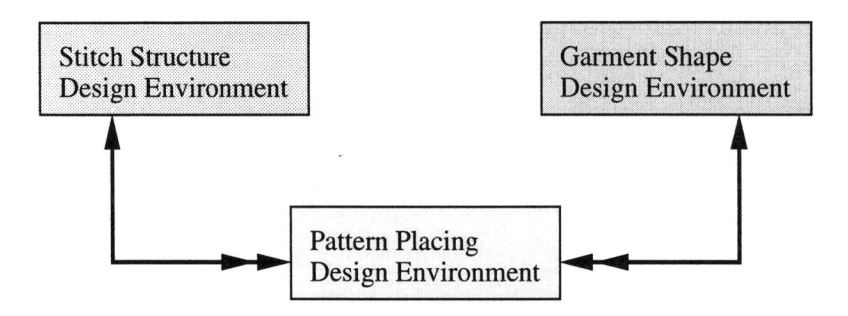

Fig 1 The three design environments

The flow of information in the architecture is shown in Figure 1; the double arrows show forward flow and single arrows show backtracking. The shape design process and the stitch structure design process are logically separate, and have no need to communicate. However the environments for stitch structure design and for shape design both feed information forward to the environment for pattern placing. Sometimes pattern placing cannot generate a satisfactory solution within the constraints of the current shape and set of motifs. In this case backtracking to the stitch structure design environment and the shape design environment is needed to make the modifications needed for successful pattern placing. If the design environments include automatic design completion modules, this backtracking process can be automatic, so that the pattern placing design environment calls the stitch structure design environment or (more likely) the shape design environment to request changes, and then presents the final result with placed patterns to the user.

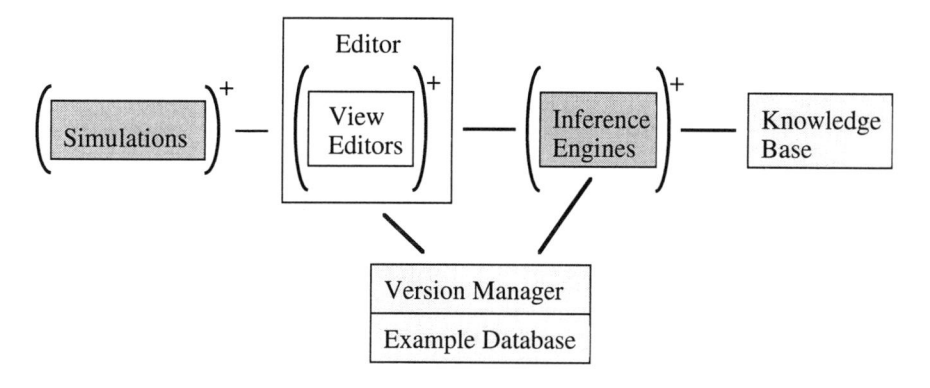

Fig 2 Outline of a design environment

4.2 Integrating Automatic and Manual Design

The manual creation and editing of designs by designers is always the central activity. The intelligent support mechanisms provide functions that are available to designers using the editors, when they choose to use them. The outline architecture we propose for a design support environment is shown in Figure 2. The editor is central. When a design can be looked at in different ways (garment shape design is a good example), the system provides different view editors acting on a shared representation of the design, with automatic translation and constraint propagation between the different views. From the editor the designers call out to intelligent reasoning modules for evaluation and design completion, and to simulation generators for showing the results of the design process (for example the appearance of a stitch structure). So far as is possible, the intelligent reasoning modules share the same declarative knowledge. The editor has access both to a database of previous designs, and to a database of other versions of the same design; the version manager keeps checkpointed partial designs and allows the designer to return to previous versions to develop them in different directions.

The automatic design completion modules we propose add alternative designs to the version manager. They can be called at any time from the design editor, to extend the existing partial design, primarily by using company-standard default values to generate complete sets of

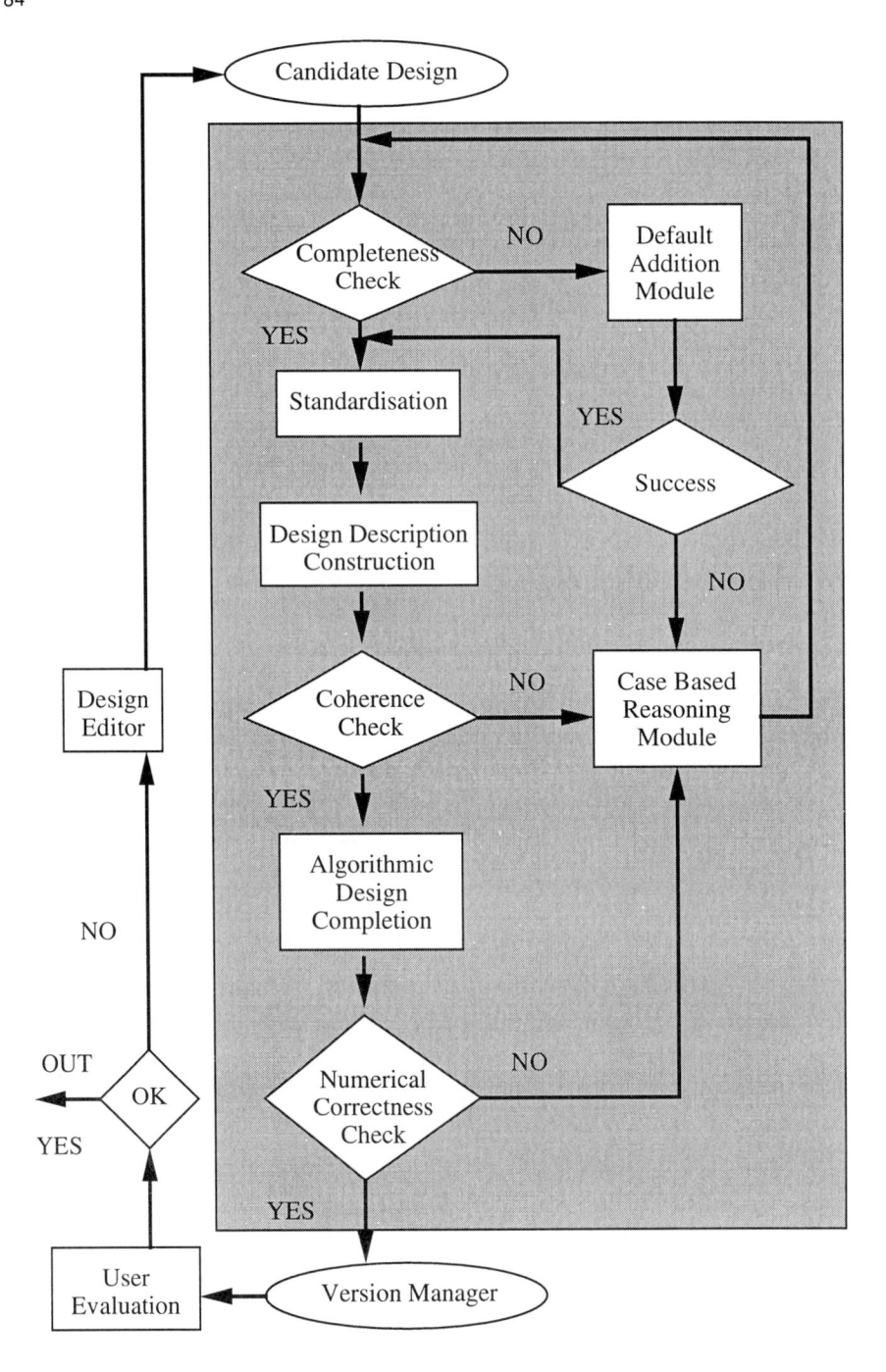

Fig 3 Automatic Design Completion

design parameters, and standard algorithms to perform the mathematical and constraint propagation tasks. Figure 3 shows a flowchart model of the automatic design completion modules for shape design and pattern placing. The input to the modules are candidate designs. We envisage that for shape design, the normal initial input from the user will be a set of measurements, or a description of variations from a previous design. For pattern placing it will be a set of motifs and a shape, plus a choice (from a menu) of how the motifs should be placed on the shape (for example, as a border, or as an all-over repeat pattern). The input is tested for completeness, before the system turns it into a standard form and constructs a description of the garment; this internal representation is required for translations between the different representations of shapes. This internal description is tested for coherence before the mathematical and constraint propagation algorithms construct a completion of the garment, by constructing shape curves or placing patterns. The choices of algorithm at this stage are inherent in the design descriptions and the designers' initial choices of what sort of completions they want. The output of this stage is subjected to a numerical correctness check.

If the partial design fails any of its checks, the automatic design completion module tries to correct it. If the input is incomplete, the system fills in gaps by adding company-standard default values for that type of garment. If the partial design is incoherent, the system uses a case-based reasoning module that finds similar (correct) designs in the database and replaces some of the attributes of the new design with values taken from the previous designs. The system should generate all the plausible alternatives; users can quickly and easily reject the ones they don't want. Completed designs can be edited, and the design completion process repeated, until the designer is entirely happy.

Figure 4 shows the architecture of the knitwear design support system. The three design environments are intended to be used both by knitwear designers and by knitting machine technicians: this entails meeting their differing requirements for intelligent support, by providing a set of evaluation modules that can provide different kinds of criticism of the design. We indicate this is figure 4 with separate boxes for the evaluation module for designers and the technical evaluation module for technicians. Each of these evaluation and design completion modules uses a common knowledge base of rules and parameters.

4.3 Designing Stitch Structures

Stitch structures are designed by drawing with a paintbox tool, using some schematic representation of stitches and stitch combinations. The stitch structure editor needs to support editing using different representations of stitch structures, but which representations are most useful is an open question. The primary representation used by the present generation of CAD systems uses colour coding both for different stitch types and for regions in different colours. From the editor the user can call simulations of the exact appearance of the stitch structure.

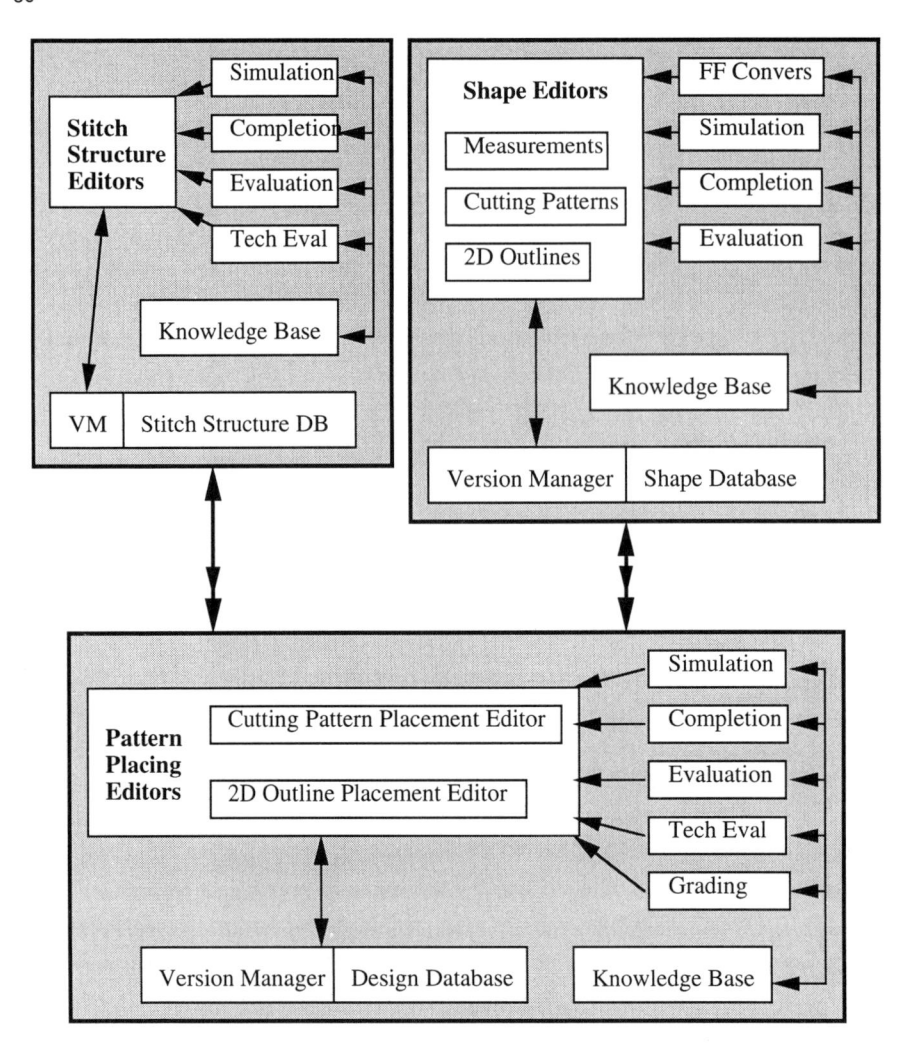

Fig 4 An architecture for the intelligent support of knitwear design

The most important function of the evaluation modules for stitch structure design is to provide feedback on the technical feasibility and cost of incomplete designs when the designers want it. In addition to symbolic descriptions, they show sources of problems and high costs graphically by highlighting areas of the editors' schematic representations of stitch structures using colour codes for transfer traverse demands and for types of technical problem. They enable the user to mark rows or regions and get symbolic and graphical feedback for those regions in isolation, at any point in the design process.

The automatic design completion module uses the evaluation mechanisms to identify combinations of stitches that create technical problems or are expensive to knit, and realigns different parts of the stitch structure so that the needle transfer operations they require do not conflict, or uses substitution heuristics to identify possible combinations of stitches that might have a similar visual effect (for example, different types of cable), and constructs designs using the alternative design elements. While its design completion heuristics could be used to complete partial designs, this is of limited value to a designer, as the search space of possible completions is too great without some definition of the goal state.

As we have observed earlier, additional technical feedback is required by technicians, who can add their own knowledge to that of the system to evaluate the design for a particular machine and yarn, and to work out how to set up knitting machines for the design. The technical evaluation module generates exact descriptions of details like yarn tensions.

The designer can search the database of patterns at any time, and incorporate any pattern into the design and edit it. The primary challenge in developing a stitch structure database is finding good methods for indexing and recognising similarities, so that it can be searched effectively. The database of patterns has a very important role to play in the design process. Books of patterns are often used as sources of inspiration by designers. Databases indexed in suitable ways would be useful for rapid selection of motifs in conceptual design, for finding alternative solutions if pattern placing fails, or for suggesting solutions for problems arising in detailed design. A great deal of designing is modification of existing designs: supporting design by editing rather than re-engineering previous stitch structures from swatches would facilitate this process without limiting it significantly. This would be especially useful for complex designs beyond the powers of straightforward automatic programming methods.

The stitch structure editor incorporates tools to support the generation of stitch structures from pictures or abstract patterns. This process is problematic because of the low resolution and limited range of colours in knitted fabric. We have explored the issue of transforming an image into a stitch structure by digitising it into large stitch-shaped regions [3]; this works reasonably well for large motifs, though for some patterns it requires a lot of manual editing to tidy up the image.

Most geometric patterns are never changed in size, because this requires an understanding of the production processes involved, and of the interactions between the various stitch types, as well as a lot of complex reasoning. However the complex reasoning can be done by an intelligent reasoning module for pattern sizing. We have implemented a heuristic reasoning system for sizing simple geometric shapes. This would serve the purpose of enhancing the power of automatic pattern placing, and increasing the range of choices open to designers without requiring them to resort to complex reasoning about a different problem that would disrupt their design thinking.

4.4 Designing Garment Shapes

As garment shapes can be described in several different ways, each of which makes different aspects of the design explicit, the garment shape module includes at least three different editors. These representations are: measurements plus a definition of the garment category (entered as text or using menus); cutting patterns for garment pieces; a two dimensional flattened outline of the garment (what the garment looks like when laid flat). The users can use any one to change a design or create a design from scratch. The system allows the user to switch rapidly between the representations, and updates all of them automatically; it should ideally display all three representations of the current design at once.

The editors for shape design using cutting patterns and two dimensional projections are essentially tailored paintbox systems. For some aspects of shape construction, the space of acceptable curves can be mathematically defined, for example armholes and sleeve crowns. The garment shape editors enable the users to explore the space of correct solutions by simple means like moving control points, and only allow the user to go beyond the bounds of the defined space of legal curves when this feature is turned off. The mathematical definitions of the spaces of acceptable curves can be edited to accommodate differences in fashion or in the individual style of a designer or a company.

Knitwear designers presently find it very difficult to communicate which parts of their designs are approximate and provisional, and which serious and exact. The paintbox editors allow the users to specify the degree of precision of different parts of their designs, and display the differences graphically using a colour code, superimposed alternatives, or gradations of fuzziness.

In addition, it would be very helpful to designers to allow them to specify designs using descriptions of the ways they differ from previous designs. The most obvious input format is typing stylised English, listing changes in no particular order. Designers will sometimes be a lot more comfortable specifying changes in a non-numerical way, at least in the earlier stages of design. Interpreting non-numerical change descriptions requires heuristic reasoning, with rules to assign numerical values to non-numerical magnitudes like "a little" that are appropriate to the context. This approach to describing shapes fits designers' standard strategies for developing garment shape designs by adapting previous designs.

We anticipate that the initial input to the shape design editors will be both incomplete and incoherent, as the information knitwear designers give to make-up staff about cutting patterns is frequently inadequate or wrong. The automatic design completion module for garment shapes uses heuristic rules and parameter calculations to add standard or company-specific default values to complete the specification. Descriptions in terms of measurements, or in terms of differences from previous garments, do not fully specify the shape. The major part of the automatic design completion module's work is using mathematical algorithms to compute shape curves to complete the internal representation, which includes both measurements and curve parameters. If the original input is incoherent this process fails or produces absurd results. The module proposes corrections to incoherent descriptions using case-based reasoning: it retrieves garments from its database with matching characteristics, and substitutes some of their parameter values or features into the current design. It presents all plausible completions to the user for selection and further editing.

Conversion between cut-and-sew shape specifications and designs for fully-fashioned garments is done by a special reasoning module. The shape design environment also provides simulations of the three dimensional appearance of the garment on a person; techniques for doing this for a standard pose have been developed by the knitting machine manufacturers.

The evaluation module checks whether the garment pieces fit together, and checks the fit to human bodies, and displays its results graphically by colour coding parts of the editors' shape representations.

4.5 Placing Patterns on Garment Shapes

The pattern placing environment provides editors for placing patterns onto shapes using the most useful representations of shapes. In the existing knitwear design process, pattern placing is usually done onto cutting patterns for garment pieces: the cutting pattern placement editor is the primary tool, but a pattern placing environment provides an editor using two dimensional projections. The patterns are placed manually by moving around outlines or schematic representations, as well as by operations for mirroring and duplicating them and for filling spaces.

The automatic design completion module can take over pattern placing at any stage and present a set of plausible completions. It can use heuristic rules to infer the users' basic approaches to pattern placing for some types of half-completed design; otherwise the users must select an approach from a menu, such as 'border' or 'diagonal fill pattern'. Finding a good pattern placement is a search task involving a lot of backtracking and iterative constraint relaxation, requiring the use of simple aesthetic heuristics as well as technical evaluations. The module can make small changes to the garment shape by calling the automatic design completion module of the shape design environment.

Grading, the task of creating variants of a design in different sizes, is done by repeating the pattern placing process for a different size of garment. We propose that the grading process should be done automatically by a grading module which calls the shape design environment with a new set of measurements which are obtained from the first version of the garment (either by multiplying by a scaling factor or using company-specific rules). The grading module uses the functionality of the automatic design completion module to fit patterns onto the new shape, using the placement parameters of the original design as a starting point. It also uses the pattern database and a module for scaling stitch structures to generate designs similar in overall appearance to the original.

Three dimensional projections of the complete garment designs onto body shapes serve to highlight potential problems like the disappearance of part of a motif into an armhole. The evaluation modules identify costs and technical difficulties caused by the juxtaposition of different motifs and stitch structures, and compute costs for producing the complete design.

5 COMMENTARY

In this paper we have presented an outline design for an intelligent system to support all aspects of knitwear design from conceptual design to programming knitting machine. Our outline architecture is intended to serve two purposes: First, to provide a focus and a plan for the development of its components - intelligent reasoning modules for supporting the work of designers, and for assisting the handover of a design from designer to technician. Second, to show how automatic design can fit into a human design process and enhance human creativity. Our own work has concentrated on building a few parts of this structure, to explore ways to provide intelligent support for designers.

5.1 Pieces in Place

The CAD systems marketed by the knitting machine manufacturers have developed rapidly in recent years. The machine builders have concentrated their efforts on developing better tools for knitwear technicians to program knitting machines, in the form of visual editors representing garment pieces made up of colour coded stitches. They don't yet have the power of the old lower level languages but rapid progress can be expected. The only forms of technical feedback available are global computations of knitting times and displays of the cam box traverses required to knit each row, which are very helpful to technicians trying to understand exactly what is going on, but hard work for designers to make use of. The machine builders are also developing impressive simulations of the appearance of the final product, which are more useful to managers and buyers than they are to designers.

Our work is intended to be complementary to the knitting machine manufacturers' activities. Our earlier work concentrated on automatic pattern placing: we have developed a system that placed motifs on garment shapes according to simple aesthetic heuristics for recognising good placements [3], and a prototype module for scaling geometric patterns according to their features, to support automatic grading. We have also developed a prototype of one part of the stitch structure design environment: a module for generating stitch structures from photographic images; for high contrast images this achieved good first-pass patterns that usually required some editing [3].

5.2 Work in Progress

Recently we have concentrated on aspects of the shape design environment. We are developing numerical methods for generating garment shape curves within the constraints imposed by the characteristics of the domain, and by the features of an individual garment. The existing knitwear and tailoring CAD systems we have seen employ splines to model garment shapes; splines have the wrong properties for modelling the spaces of curves designers consider good shapes, so we are employing Bézier curves. These methods are intended to provide the core of a support environment for shape design for knitwear and tailoring, which will embody the approach to garment shape design we have outlined here. We are also working on supporting smooth interactions between designers and technicians by analysing the properties of stitch structure representations and developing a notation that makes explicit the information technicians require to program them.

5.3 Automated Design as Design Support

We reject the widely held view (for example [7]) that automatic design and design support must be opposites, implicit in much of the work on automatic design (see for example [1], [9], [10]). Providing machine generated criticism of human generated designs is not the only way to use system-generated feedback to support human designers. Reversing this interaction, so that humans criticise, select and modify machine generated designs, can exploit the ability of AI systems to solve complex constraint satisfaction problems and do difficult mathematics, and exploit the ability of designers to do rapid and subtle perceptual evaluations of designs. In our view, automating parts of the knitwear design process is a tractable problem, and enabling designers to request design completions that they can evaluate and edit is the best way to combine the talents of designers and computers. Our work on garment shape design is intended to demonstrate the validity of this approach to design support. This approach should be valid for any design task where generating design completions subject to technical constraints is a tractable problem, and where humans can evaluate good designs perceptually.

ACKNOWLEDGEMENTS

The research reported here has been supported by SERC/ACME grant GR/J40331 and by SERC grant GR/J48689. Helen Sharp made very helpful comments on earlier drafts of this paper. Our research has benefited from conversations with the other members of the FACADE project at the Open University: Helen Sharp, George Rzevski, Marian Petre, Rodney Buckland and KK Kuan. We are very grateful to all our informants in different parts of the knitwear industry for the time and effort they put into talking to us, especially Monica Jandrisits, Annabelle Duncan and Wendy Nicholson.

REFERENCES

[1] Coyne, R.D., Rosenman, M.A., Radford, A.D., Balachandran, M.B., and Gero, J.S., *Knowledge-Based Design Systems*, Addison-Wesley, Reading, 1989.

[2] Devane, J., "Training and Education in CAD clothing/textile systems", in *CAD in clothing and textiles, a collection of expert views* (Aldrich, W., Ed.), BSP Professional, Oxford, 1992.

[3] Eckert, C.M., "A Prototype Intelligent CAD System for Knitwear Design", MSc Thesis, Department of Computing Science, University of Aberdeen, 1990.

[4] Eckert, C.M. and Stacey, M.K., "CAD Systems and the Division of Labour in Knitwear Design", in *Women, Work and Computerization: Breaking Old Boundaries - Building New Forms* (Adam, A., Emms, J., Green, E., and Owen, J., Eds.), North-Holland, Amsterdam, 1994, pp 409-422.

[5] Eckert, C.M., "The Knitwear Design Process", paper in preparation.

[6] Fischer, G., and Nakakoji, K., "Empowering designers with integrated design environments", in *Artificial Intelligence in Design '91* (Gero, J.S., Ed.), Butterworth-Heineman, Oxford, 1991, pp 191-209.

[7] Fischer, G., and Nakakoji, K., "Beyond the macho approach of artificial intelligence: empower designers - do not replace them", *Knowledge-Based Systems*, Vol 5, No 1, 1992, pp 15-30.

[8] Fischer, G., "Creativity enhancing design environments", in *Modelling Creativity and Knowledge-Based Creative Design* (Gero, J.S., and Maher, M.L., Eds.), Erlbaum, Hillsdale, 1993, pp 269-282.

[9] Gero, J.S. (Ed.), *Knowledge Engineering in Computer-Aided Design*, North-Holland, Amsterdam, 1985.

[10] Gero, J.S. (Ed.), *Expert Systems in Computer-Aided Design*, North-Holland, Amsterdam, 1987.

[11] Landauer, T.K., "Psychological research as a mother of invention", in *CHI'85: Proceedings of the ACM Conference on Human Factors in Computing Systems* (Borman, L. and Curtis, B., Eds.), ACM Press, New York, 1985, pp 44-45.

[12] Stacey, M.K., "Distorting Design: Unevenness as a Cognitive Dimension of Design Tools", Computing Department Technical Report 95/01, Open University, 1995.

REPRESENTING CONCEPTUAL DESIGN KNOWLEDGE WITH MULTI-LAYERED LOGIC

K Clibbon, E Edmonds & L Candy

1. INTRODUCTION

Any given design process begins with a problem specification, from which a set of requirements is derived. This set of requirements is not usually well defined at the beginning of the design process and does not remain static during it. For example, when designing a new car the initial requirements might be for a car that is low in price, reliable and economic. These ambiguous requirements will evolve as design decisions are made, such as to produce a new engine by modifying an existing economic low-priced one.

Conceptual design can be formalised as an activity to derive an object model whose structure meets the design requirements. An object model is the non-procedural representation of a real object (or real problem). The object model must be a faithful representation of an object possessing structural information to represent functionality; that is, attributes, properties, behaviour and relations with other objects. The approach advocated in this paper is that design starts with a hypothetical object model. If this object model does not meet the requirements, then it is modified accordingly.

From an AI perspective, conceptual design can be considered to be the process of identifying problems, specifying functionality and generating appropriate solutions; with the support of some basic building blocks. In this research the object model is a declarative representation of the real object. It is not static. Conceptual design is a dynamic activity to find this model. It is also a non-deterministic activity, in the sense that there are an infinite number of possible structures that an object model can be.

The issues discussed in this paper are illustrated with reference to the domain of car design. The knowledge used forms part of the Vehicle Packager Knowledge Support System (VPKSS) developed at the LUTCHI Research Centre. Vehicle packaging is an approach to whole vehicle design, that incorporates a representation of the key sub-systems such as engine bay, passenger cell and boot compartment. For a discussion on vehicle packaging generally and a description of the VPKSS see Candy et al. [6].

2. CONCEPTUAL DESIGN

In Engineering Design, design models are often based upon French [11] or Pahl and Beitz's [18] work. The design process is described in terms of three main stages: problem formulation or task clarification, conceptual design and detailed design. In most cases, deriving a requirements specification from the problem formulation activity is thought of as a separate stage from conceptual design. However, studies show that there is considerable overlap between these stages and indeed between all the stages of the design process.

Conceptual design has been defined as the phase "that takes the statement of the problem and generates broad solutions to it in the form of schemes" [11]. A scheme in this sense consists of design features and components; specified at an appropriate level of detail. Schemes might outline the spatial and structural relationships of the principle components that constitute the design. This definition assumes that the problem has been understood and defined and is ready to be expressed in a solution-based form. Pahl and Beitz [18] state that "conceptual design is that part of the design process in which, by the identification of the essential problems through abstraction, by the establishment of function structures and by the search for appropriate solution principles and their combination, the basic solution path is laid down through the elaboration of a solution concept.".

In many respects the traditional AI view of design is similar to the engineering design perspective, in that it is conceived of as a rational process that involves moving from one design space state to another. A basic assumption that underlies both perspectives is that a systematic design process alone can produce a truly rational approach and hence generally valid solutions. A definition of conceptual design should reflect the characteristics of the processes both at the cognitive and computational levels of design. The logic-based model discussed in this paper reflects the relationship between the representation of the design and the problem solving processes behind it.

2.1 Strategic Knowledge

Strategic knowledge is concerned with the control decisions made during the conceptual design phase, as the object model evolves. From a non-epistemological perspective knowledge is information that can be used to perform some intelligent task. Strategic knowledge is knowledge used by an agent to decide what actions to perform next, where actions have consequences that are external to the agent. Strategic knowledge is used by the conceptual designer to decide what actions to perform in a given situation, where actions are considered to have observable consequences.

Knowledge systems make control decisions at the implementation level. One such decision is conflict resolution, where the choice of which rule(s) to fire from a competing set is made. Strategic knowledge by comparison is defined at the knowledge level in terms of the effects on the observable behaviour of the agent. This occurs without reference to the internal symbolic-level organisation of the knowledge system [13]. Substantive knowledge, used to describe or model the domain of endeavour, classifies entities in the knowledge base in terms of the relations, attributes and functions between them. The procedure for deciding what to do when operating on the entities in the domain entities, is the strategy. This distinction between strategic and substantive knowledge is not the same as the difference between procedural and declarative knowledge representation, which is a symbolic-level distinction.

Substantive knowledge is represented explicitly in knowledge bases. Strategic knowledge can be represented both implicitly and explicitly. Implicit behaviour has been achieved using implementation level primitives such as numeric priorities (e.g. MYCIN's certainty factors [4]). A more explicit representation of strategic knowledge is achieved via control blocks, which are sequential procedures attached to rules (e.g. BB1's blackboard architecture [14]). A key research issue is how can strategic knowledge be represented and what kinds of strategic reasoning can be expressed during the conceptual design process.

2.1.1. Representing Strategic Knowledge. The most straightforward method of representing strategic knowledge is through the use of a procedural formulation. A procedural representation of a strategy specifies the sequence of steps that the system should take. For instance, a procedure might take the form, "do action A then B, then if condition X is true do action C else do action D."

Strategies have been encoded as procedures in programming languages, such as Lisp. In some rule-based systems task variables serve as control markers in the rule language, which direct the rule interpreter to execute modular sets of rules as subroutines [9]. S1 [10] for example, integrates procedural control in its representation language as "control blocks", which invoke rules. Another approach to procedural control is to build transition networks [12] or decision graphs that graphically specify the control flow of the program.

Although procedural formulations of control can implement a strategy, they represent it implicitly. The strategic knowledge that underlies the specific sequence of actions is hidden in the procedural representation, which specifies what to do as a procedure, but not why. Thus, if the objective is to represent the reasons underlying a strategy, an architecture capable of being driven by this knowledge is required.

Metarules guide the use of knowledge and decide what rules and methods are to be applied. There are essentially two types of metarules. Pruning metarules exclude particular rules from consideration. In terms of the goal tree, this amounts to a decision not to explore a given branch and constitutes a judgement on the overall utility of a rule, as to whether it is of any use in a specific context. The other type of metarule encodes knowledge on the relative importance of rules. At the implementation level the metarule acts to reorder rules relevant to some goal before involving them. In rule-based systems the order in which rules appear in the knowledge base influences the inferencing process. If for example one of the requirements for the design of a car was for high performance, to the detriment of passenger comfort; then the following is a simple example of a metarule that reorders:

        Metarule 001
        IF   (There are rules which are relevant to the driver)
        AND(There are rules which are relevant to rear_passenger)
        THEN(Do the former before the latter).

Metarule 001 acts as a goto, focusing inferencing on rules that relate to the driver. Metarules that reorder amount to a less drastic decision as regards the goal tree - the branches are restructured rather than pruned. Different types of meta knowledge can thus be represented, dependant on exhaustive search, reordering of goals or pruning of goals. This gives rise to a pattern of search which is not 'blind' but which is guided by heuristic knowledge. The seminal work on the explicit representation of control knowledge metarules was carried out in the MYCIN experiments [4] and by Clancy [7] [8].

The use of metarules to represent strategic knowledge is not restricted to one level of meta knowledge, but can be extended to indefinite levels. In addition, the inference engine used is still simple. Metal-level rules can select different types of inference mechanisms (such as forward- or backward- chaining) at appropriate points in the search process. The purpose of representing strategic knowledge with metarules is to tell the system which part of its substantial knowledge to apply in a given situation. In choosing among observable actions and controlling search, metarules are useful if the objective is to control the order and manner in which inferences are made.

2.2 Support for Conceptual Design

Research by the authors reported elsewhere [5], has highlighted the need for support during conceptual design. The requirements are for support during knowledge exploration and in evaluating the object model. Knowledge used in the act of designing is dynamic and in order to apply it effectively it must be evaluated during the process. At the same time however, if the design is specified to strictly, then the freedom of the model building activity is restricted and there is no room for innovation. There is clearly a trade-off between the freedom and the efficiency of design.

The designer will develop and use strategies to evaluate the object models that evolve during conceptual design. Strategic knowledge would be used by the designer to identify and formulate the design problem. Having established the requirements, the designer will require strategic support in exploration of the design, particularly as new concepts emerge. This research is concerned with logic-based support, providing tools to guide the designer through iterative model building, model analysis, model evaluation and model modification. It is hypothesised that, by having a sound theoretical logic-based foundation, support systems can check for consistency and correctness in the knowledge-base during model building.

3. A LOGIC-BASED FRAMEWORK

The logic-based framework proposed in this paper utilises layers which consist of nodes. Nodes in the object layer contain descriptions of object models. Nodes in the requirements layer consist of sets of requirements for these objects. Interaction between the two layers can be represented by links. Strategies are needed to support navigation within and between the object and requirements spaces.

3.1 The MULTIK Model

In the logic-based framework, the declarative representation of the object model is represented independently from the declarative representation of the requirements. Links between nodes in the layers represent which set of requirements is related to which object model. Static aspects of the design are concerned with the object model and requirements domain knowledge. The strategic knowledge relating to the steps to be taken during the design process are the dynamic aspects. A meta-level architecture facilitates the representation of both the static and dynamic aspects of the conceptual design process. Strategic knowledge is necessary to determine in which layer the next step of the design process is to be made. It is necessary to perform object-level reasoning, with knowledge in the object level; as well as meta-level reasoning, about the knowledge in the object level. The meta-level architecture is depicted in figure 1.

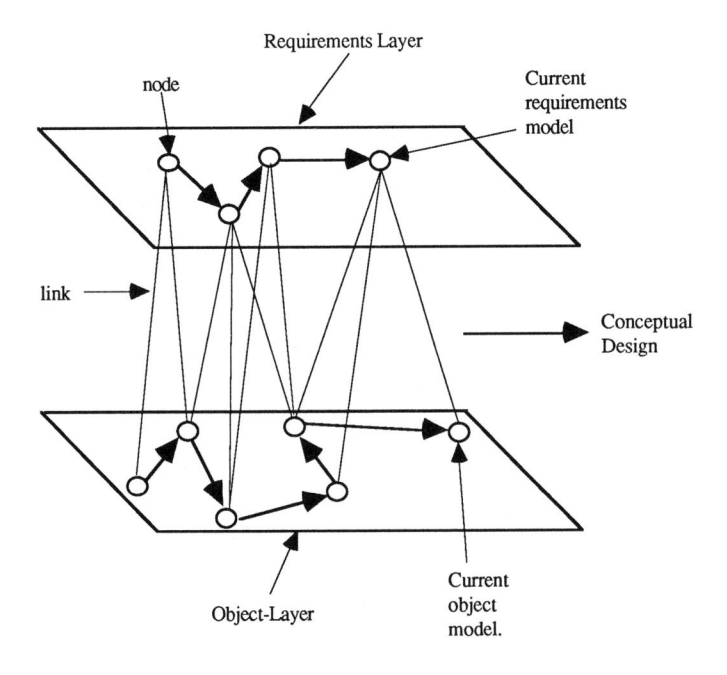

Figure 1 - A Meta-Level Architecture for Conceptual Design

3.2 Multi-Layered Logic

A multi-layered logic (MLL - [17]) knowledge representation, capable of automatic consistency checking, representing complexity and dynamic model building, has been used to model vehicle packaging knowledge. It is based on the concept of hierarchical data abstraction and is an extension to first-order logic, with data structuring capabilities. A data structure - human specified or automatically generated - is defined as a set based on ZF (Zermelo-Frankel) set theory. MLL has a set of primitive constructs including 'element-of', 'component-of', 'power-set-of', 'product-set-of', 'union-of', 'intersection-of' and 'pair-of' relations, which are used to represent the object model and its requirements. These data structures can be included as terms in predicates and are defined mathematically in terms of primitive structures according to composition rules. For example, the symbol '*' represents the power set operator, where *D means the set of subsets of D (excluding the empty set).

The syntax of MLL predicates has been expanded from First Order Logic to include a domain of each variable explicitly in the prefixes. For example, [AX/new_car](expensive X) is the same as the first-order expression $[AX]((\text{new\_car } X) \rightarrow (\text{expensive}(X)))$. Both mean new_car is expensive, but in the former expression, new_car denotes a set of new cars. To accommodate the modified syntax, an ordinary inference algorithm (resolution) has been modified in order to check for equality between data-structures - see Ohsuga [15] for details.

A MLL predicate can also include closed formulae - a formula without any free variables - as the term. For example in the following [AX/D](P,---,X,---,[AX/C](R,---,X,---)) the evaluation of the inner predicate R can be performed independently from that of the outer predicate P. The set of predicates that do not contain any predicates as terms, are located in the object-level, while those that contain object-level predicate(s) as the term(s), belong in the level above the object level - the meta-level.

4. A REALISATION OF THE MULTIK MODEL

A working prototype version of the MLL Vehicle Packaging model is now being developed in KAUS (Knowledge Acquisition and Utilisation System). KAUS has a mechanism to handle meta-level knowledge. MLL rules are indexed and partitioned into several sets (worlds). It is possible to have independent local worlds to form separate knowledge sources. KAUS utilises built-in predicates, called procedural type atoms (PTAs), to specify and change the worlds used during the inferencing process. A meta-level inferencing process can get information from the object level. Hence inferencing in one level is defined in the next highest level. For further details on MLL and KAUS refer to Ohsuga and Yumauchi [17] and Ohsuga [16].

4.1 KAUS

KAUS was developed at the University of Tokyo. It has evolved since the mid 70s into a logic programming system and can be regarded as an extension to Prolog [3]. KAUS is based on Many Sorted Logic, where sort hierarchies are described in terms of set relationships, in which each set represents the sort of a set of entities. For example, the sort 'cars' represents the set of cars, while 'sports cars' and 'family car' can be thought of as subsets of the set 'cars' - the subsorts of the sort cars. So if a Lotus Esprit is designated as a sports car and Vauxhall Cavalier as a family car, then the Lotus Esprit and Vauxhall Cavalier are members of the sorts sports car and family car respectively.

Differing from Prolog, the clausal representation of rules and facts in KAUS are arbitrary AND-OR logical formulas. A KAUS clause has a prefix in which the quantification and type declaration of variables appearing in the clause may be explicitly described. For example, a sentence "If a person Z is a parent of a male X and Y, and X is not equal to Y, then X is brother of Y." is represented in KAUS and Prolog as follows:

KAUS :[AX,Y/male][AZ/person](| (brother X Y) ~(parent X Z) ~(parent Y Z) ~($ne X Y)).
Prolog :brother(X,Y) :-male(X), male(Y), person(Z), parent(X,Z), parent(Y,Z), not(X==Y).

[AX,Y/male][AZ/person] is the prefix part of the clause. This can be read as 'for all X and Y of type male and for all Z of type person'. (| (brother X Y) ~(parent X Z) ~(parent Y Z) ~($ne X Y)) represents an OR formula which is logically equivalent to '(parent X Z) $\wedge$ (parent Y Z) $\wedge$($ne X Y) $\rightarrow$ (brother X Y)'. '|' denotes the or-connective and '~' is the logical negation symbol.

KAUS, unlike Prolog, has the expressive power to describe rules concerning sets. In KAUS the syntax of Prolog is expanded to include the domain of each variable explicitly in the prefix. Consider the following examples:

>All cars have a passenger cell.
>Sports cars and family cars are types of cars.
>Cars are a subset of vehicles.

These can be expressed in KAUS as follows:

>[AX/cars](have(X passenger_cell)).
>!ins_e cars sportsCar familyCar.
>!ins_e *vehicle cars.

In the above 'cars' refers to the set of cars and 'A' is the universal quantifier (a reserved symbol). Symbols beginning with capital letters relate to variables, and those beginning with lower-case letters represent predicates or basic objects. !ins_e defines the component_of relation. *vehicle denotes the power set of the set 'vehicle'. The first example is a predicate sentence (a rule). The second and third examples express relations between sets and objects.

In KAUS, it is possible to define type hierarchies, which can be interpreted semantically as a representation of ISA structures of objects and PART-WHOLE structures of objects. For instance, cars can be classified into sports cars and family cars and so on. In addition, cars are composed of engine bays, passenger cells and boots. The ISA and PART-WHOLE hierarchies are described in terms of set theory relationships among objects. The description of ISA and PART-WHOLE hierarchies of objects can be classified into two classes. One is concerned with the hierarchies of objects (in the object level), and another with hierarchies of objects in the meta-level. Objects in the meta-level are program clauses. Hierarchies in the meta-level describe the classification of program clauses. Figure 2 is an example description of part of an object level hierarchy for the car domain.

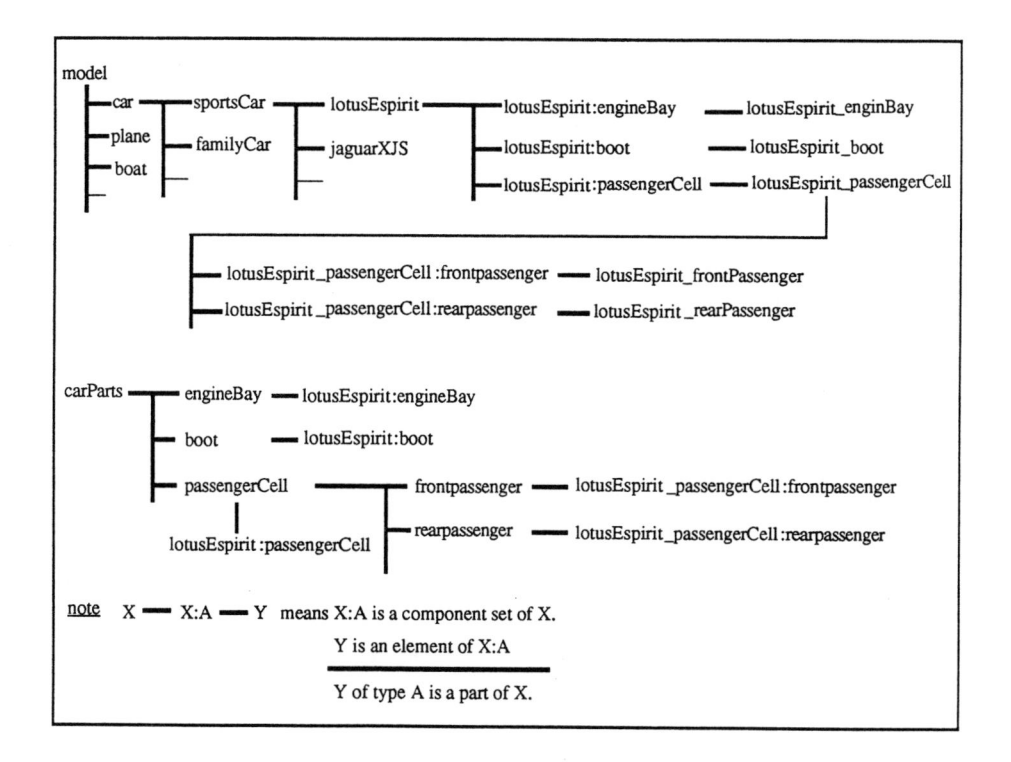

Figure 2 - An Object Level Hierarchy for the Car Domain

The above structure is described as follows:

a1).	!ins_e	*model	car, plane, boat;
a2).	!ins_e	*car	sportsCar, familyCar;
a3).	!ins_e	*carParts	enginBay, boot, passengerCell;
a4).	!ins_e	*passegerCell	frontPassenger, rearPassenger;
a5).	!ins_e	sportsCar	lotusEspirit, jaguarXJS;
a6).	!ins_e	lotusEspirit:engineBay	lotusEspirit_enginBay;
a7).	!ins_e	lotusEspirit:boot	lotusEspirit_boot;
a8).	!ins_e	lotusEspirit:passengerCell	lotusEspirit_passengerCell;
a9).	!ins_e	lotusEspirit_passengerCell:frontPassenger	lotusEspirit_frontPassenger;
a10).	!ins_e	lotusEspirit_passengerCell:rearPassenger	lotusEspirit_rearPassenger;

Note that a1 through to a4 describe hierarchies of generic objects. The statement a5 describes two instances of sportsCar and a6 to a10 describes the specific PART-WHOLE structure for the lotusEspirit. For example, in a1, car, plane and boat are described as subtypes of a supertype model. In a2, sportsCar and familyCar are defined as subtypes of a supertype cars, and so on. Figure 3 gives an example description of a hierarchy of objects in the meta-level.

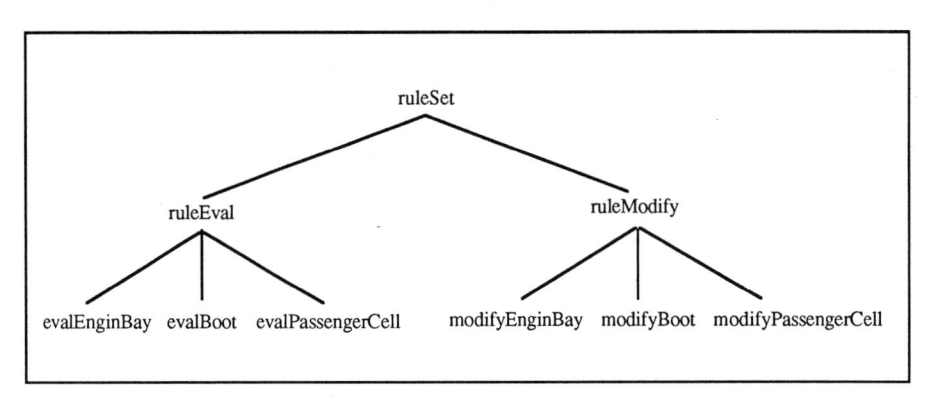

Figure 3 - A hierarchy of objects in the Meta-Level

The example above aims to classify car design rules into evaluation rules and modification rules. The evaluation and modification rules are classified into three rule sets with respect to the objects in the engine bay, boot and passenger cell. If, for example, the objects in the passenger cell were modified, then they will be so according to the associated modification rules, and evaluated according to rules on style. (i.e. rules associated with comfort requirements for the passengers).

4.1.1 Reasoning with Strategic Knowledge. Program clauses in KAUS can be classified according to their utility, availability and relevance to specific domain tasks. This results in a hierarchical categorisation of program clauses. A set of clauses which belong to a particular category in the hierarchy may be regarded as a description of a certain inference world. These inference worlds constitute strategies used by the designer during conceptual design. In terms of KAUS, strategic clauses can be built to change the current inference world. For example, with sports cars the emphasis during design is on the engine bay. Comfort in the passenger cell receives a lower priority, than when designing a family car. Most sports cars do not cater for rear passengers at all or put their comfort at a low priority. Consider the following passenger cell (see figure 4):

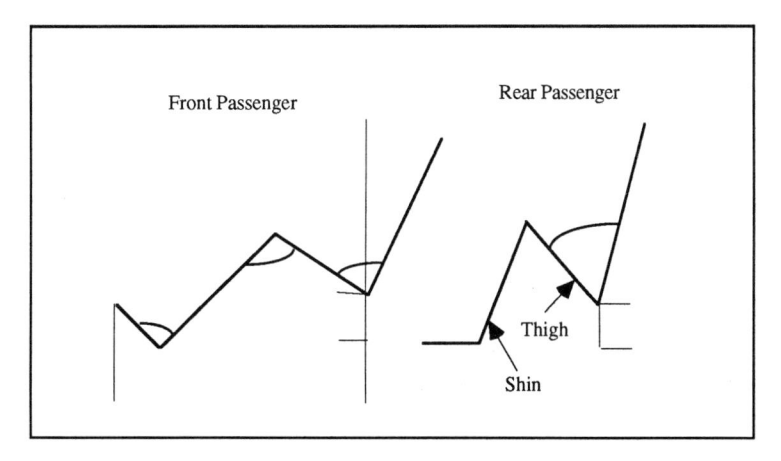

Figure 4 - A Typical Passenger Cell

The designer of a family car passenger cell will normally attempt to make the rear occupants as large as possible. The angles of the seats and the lengths of the rear occupants thigh and shin will be crucial. To illustrate how metarules can be used to represent the designers strategies for changing inference worlds, imagine that the conceptual designer is manipulating the rear passenger object by removing them or changing their shin and thigh lengths - from a set of upper and lower values. The underlying KAUS representation is as follows:

```
r1:(|(asportsCar DesignIs)    ~(rear_passenger_shinLength ShinLength),
                              ~(rear_passenger_thighlength ThighLength)).
r2:(|(afamilyCar DesignIs)    ~(numberRearSeats Number),
                              ~(rear_passenger_shinLength ShinLength),
                              ~(rear_passenger_thighlength ThighLength)).
```

f1:(numberRearSeats 2).
f2:(rear_passenger_ShinLength 470).   } rear occupants have plenty of leg room.
f3:(rear_passenger_ThighLength 447).
f4:(numberRearSeats _).
f5:(rear_passenger_ShinLength 365).   } rear occupants have limited leg room.
f6:(rear_passenger_ThighLength 344).
m1:(know numberRearSeats{\r1,\f4}).
m2:(know numberRearSeats{\r2,\f1}).
m3:(know rear_passenger_shinLength{\r1,\f5}).
m4:(know rear_passenger_shinLength{\r2,\f2}).
m5:(know rear_passenger_thighLength{\r1,\f6}).
m6:(know rear_passenger_thighLength{\r2,\f3}).
m7:([APredicate/tuple](| believe Predicate)
  ~(know Fact) ~($scopeKUs [Fact]) ~($call Predicate)).

Metarules m1 to m6 describe what is known (meta knowledge about the facts). For example, m1 describes that the numberRearSeats for a sports car can be found from f4 (fact 4). The metarule m7 is a generic rule to derive what the object model looks like, given the rules r1 and r2, facts f1 to f6, and metarules m1-m6. Metarule m7 has a conditional ($scopkeKUs [X]), which restricts the clauses available to ($call Predicate) to those known by Fact. With this setting, if the knowledge base is queried to find out what type of car (sports or family) the design tends to - i.e. (believe < X Design >)?, X will be instantiated to aSportsCar. If the design requirements were for a family car, the designer could be prompted to change the number of rear passengers and leg measurements accordingly.

In KAUS then, rules can be indexed and partitioned into several sets. A set of rules is called a 'world'. KAUS has several built in predicates, called procedural type atoms (PTAs), which specify and change the worlds used during the inferencing process. A meta-level inference process based on design strategies can get information from the object layer. Strategic control can be generalised as follows:

resolve(WFF,KB,Bool,Ans)

where WFF is a MLL clause to be solved and KB is the world or set of worlds to be used. Bool (true or false) indicates whether the WFF is inferred from KB and Ans is an answer, which amounts to bindings for the variable of the WFF.

4.2 Towards Logic-Based Support For Conceptual Design

In terms of the proposed logical framework, strategic knowledge is concerned with how best to determine the next step or action to be taken by the designer, during the design process. This might amount to a set of rules or heuristics used by the designer in certain situations. For example, if the object layer is densely populated, with object models that fit the latest set of requirements, how can the designer be supported in exploring the object space and in making decisions on which object models to include in a candidate set. Search-control knowledge could be used to constrain browsing to those areas of the object layer that are likely to contain the best solutions. This paper has concentrated on approaches to the internal knowledge-base representation of the design knowledge. The final section raises issues for computer supported conceptual design that include user interaction aspects.

5. THE WAY FORWARD - FUTURE RESEARCH

It was argued a decade ago that whilst the human expert knowledge, experience and creativity that goes into conceptual design can be aided by appropriate search algorithms, the automatic generation of complete solutions is not likely to be feasible [18]. It is still widely acknowledged that computer support for conceptual design has yet to be achieved. The product orientation that has engaged researchers in the past, and continues to do so, is directed towards creating automatically generated designs where the role of the designer is constrained in particular ways. However, more recently, there have been changes in the research agenda for computer design tools that has led to an interest in providing support for conceptual design. The arrival of knowledge-based engineering and the goal of achieving concurrency are playing a role in altering the state of what is thought to be desirable and achievable in terms of support for design as a whole.

5.1 Implications of Concurrent Engineering

The advent of concurrent engineering design, facilitated by knowledge-based techniques, has introduced new factors into the design process. The importance of multi-disciplinary team work gives rise to a need for shared understanding between the participants who may be using different languages about the same issues [1]. It also implies the use of different forms of knowledge such as the successful designer's personal strategies, vital competitive-edge corporate knowledge, as well as new legislation and economic factors.

Human involvement requires systems that enhance the designer's capability and effectiveness. In conceptual design, new ideas are generated and evaluated and support for these activities requires more interaction with the system not less. The goal of automated generation of solutions is widely recognised to be unrealistic but that has not lead to enough attention being made to the problems of human interaction with computers. This implies addressing cognitive and social issues in the system design in an explicit manner. In particular, the interaction techniques must be appropriate to the goal of amplifying the conceptual design process from the designer's perspective.

5.2 Support for Design Practice

Studies of design practice as it actually takes place, rather than an idealised notion of what it should be, provide a source of information about what contributes to successful design and what needs improvement. This view underlies the approach adopted at LUTCHI in seeking to understand what are the opportunities and constraints for computer support. If we can determine the characteristics of the cognitive processes of design, this is more valuable than building rationalist models. In particular, the importance of interactive attributes of the support system in defining requirements for computer support and the resulting system characteristics are essential. For example, if opportunistic and exploratory thinking is a vital part of conceptual design, it is important to understand how it operates and how it might be sustained and enhanced. There are implications for the type of techniques required for flexible and expressive interaction in manipulating concepts and objects.

Whilst it is one of our basic aims to take account of how design is carried out in practice rather than seeking to alter or impose a particular methodological solution, this does not mean that changes in the design process are not implicated: the new computer tools themselves will inevitably influence the design process. The difference is that we aim to take account explicitly of the human designer's needs, characteristics and working style in the system design. Thus, for example, a need to revisit the problem in the process of generating solutions, is recognised to be an important part of the clarification of the problem.

6. CONCLUSION

A logic-based framework for conceptual design has been proposed which incorporates a meta-level architecture to explicitly represent declarative design knowledge and strategies used by the designer. The KAUS logic programming language has been described with reference to some vehicle packaging knowledge. The MLL which underpins this language has been proposed as suitable formalism for representing the object model, the design requirements and the strategic knowledge used by the designer during conceptual design. It has been stated that a research framework is required to address the issue of providing computer support for design. This framework should take into account the designer-computer interaction issues, as well those issues pertaining to the underlying knowledge representation of the conceptual design process.

ACKNOWLEDGEMENTS

The work was partly funded by the Science & Engineering Research Council's Grant ref. GR/J43769. We would like to thank Professor S. Ohsuga for his help with the research.

7. REFERENCES

[1] Adzhiev, V., Beynon, M., Cartwright, A. and Yung, Y. A New Computer-Based Tool for Conceptual Design. In *Proceedings of LIWED'94*, J. Sharpe and V. Oh (Eds.) Lancaster EDC, 1994, pp 171-188.

[2] Blessing, L. *A Process-Based Approach to Computer-Supported Engineering Design.* PhD Thesis, University of Twente, Blessing, Cambridge, 1994.

[3] Bratko, I. *Prolog: Programming For Artificial Intelligence* (2nd Edition). Addison-Wesley, 1990.

[4] Buchanan, B. and Shortliffe E. *Rule-Based Expert Systems: The MYCIN Experiments of the Stanford Heuristic Programming Project.* Addison-Wesley, 1984.

[5] Candy, L. and Edmonds E. Artefacts and the Designer's Process: Implications for Computer Support to Design. *Revue Science et Techniques de la Conception*, Vol 3, Part 1, 1994, pp 11-31.

[6] Candy, L., Edmonds, E. and Patrick, D. Interactive Knowledge Support to Conceptual Design. In *Proceedings of LIWED'95*, J. Sharpe (Ed.) Lancaster EDC, 1995.

[7] Clancy, W. The Epistemology of a Rule-Based Expert System - A Framework for Explanation. *Artificial Intelligence* Vol 20, Part 3,1983, pp 215-251.

[8] Clancy, W. From GUIDON to NEOMYCIN and HERCULES in Twenty Short Lessons: ORN Final Report 1979-1985. *AI Magazine* Vol 7, Part 3, 1986, pp 40-60.

[9] Cooper, T. and Wogrin, N. *Rule-Based Programming with OPS5*. Los Altos, CA:Morgan Kaufmann, 1988.

[10] Erman, L., Scott, A. and London, P. Separating and Integrating Control in a Rule-Based Tool. *Proceedings of the IEEE Workshop on Principles of Knowledge-base Systems*, 1984, pp 37-43.

[11] French, M. *Conceptual Design for Engineers*, (2nd edition). Design Council, London, 1985.

[12] Georgeff, M. and Bonollo, U. Procedural Expert Systems. *Proceedings of the Eighth International Joint Conference on Artificial Intelligence.* 1983, pp 151-157.

[13] Gruber, T. (1989) *The Acquisition of Strategic Knowledge.* Academic Press.

[14] Hayes-Roth, B. A Blackboard Architecture for Control. *Artificial Intelligence* , Vol 26, Part 3, 1985, pp 251-321.

[15] Ohsuga S. A New method of Model Description - Use of Knowledge Base and Inference. *CAD System Framework,* K. Bo and F. Lillehagen (Eds.) North-Holland, 1983, pp 285-312.

[16] Ohsuga S. Framework of Knowledge Based Systems - Multiple Meta-Level Architecture for Representing Problems and Problem-Solving Processes. *Knowledge-Based Systems*, Vol 3, Part 4, 1990, pp 204-214.

[17] Ohsuga S. and Yamauchi H. Multi-Layer Logic - A Predicate Logic Including Data Structure as Knowledge Representation Language. *New Generation Computing*, Vol 4, 1985, Special Issue on Knowledge Representation, pp 403-439.

[18] Pahl, G. and Beitz, W. *Engineering Design*, London: The Design Council, 1984.

A SYNTHETIC REASONING METHOD BASED ON A PHYSICAL PHENOMENON KNOWLEDGE BASE

M Ishii & T Tomiyama

1 INTRODUCTION

This paper presents a reasoning method that interactively supports a designer from early synthetic phases of engineering design. We have developed a system called the Qualitative Process Abduction System (QPAS). QPAS assists the designer to model both the structure and the dynamic behavior of the design object. By employing a knowledge-based reasoning approach, QPAS can suggest appropriate structures to produce the desired dynamic behavior and execute simulation for verification. This ability to suggest structures is valuable especially when the knowledge base gets larger and when it becomes difficult to search from various engineering mechanisms for the appropriate one.

QPAS is developed as an essential tool to realize a knowledge intensive engineering framework (KIEF) that can assist engineers and designers in various engineering activities [8] [9] [10]. Knowledge intensive engineering is a new style of engineering in which engineering knowledge is used in a flexible and integrated manner and aims at generating more added-value. Knowledge intensive engineering is largely model and knowledge-based engineering activities and its key concepts are model building, simulation, model-based reasoning, model validation, and model modification.

In this paper, we focus on model building in conceptual design, since it is a crucial phase to generate the design object model to be used in the rest of the engineering activities. In conceptual design, the designer needs to select appropriate devices and formulate the structure of the design object that realizes the desired function with respect to its dynamical behavior. QPAS accepts state transitions of a physical object as required behaviors and derives structures considering physical phenomena and physical laws which are described in the knowledge base. As a result, QPAS generates a model of causal relationships among physical phenomena and parameters that explain how the derived mechanism works.

QPAS is based on two important techniques we have already developed for KIEF at the University of Tokyo.

One is the Function-Behavior-State (FBS) modeling technique [12]. In FBS modeling, the designer is supported to decompose the given function into subfunctions and then to select engineering mechanisms that realize the subfunctions. However, in the FBS modeling, the candidates of engineering mechanisms should be described as knowledge directly associated with names of function. This restricts flexibility of conceptual design. Therefore, QPAS provides a method to synthesize engineering mechanisms by tracing causality in physical behaviors described in the knowledge of physical phenomena, when candidates written in the knowledge of functions are not satisfactory.

The other is the knowledge representation technique to describe physical phenomena that occur to engineering mechanisms and physical laws that govern dynamical behaviors of the mechanisms. We have been conducting a project to build a large knowledge base to be installed in KIEF by collecting engineering knowledge including physical phenomena and physical laws [5]. We employed Qualitative Process Theory (QPT) [4] as the fundamental theory of knowledge representation and developed a qualitative reasoning system that execute behavioral simulation[6]. QPAS utilizes this knowledge base in synthetic manner to assist designers in building qualitative models.

Early work on synthetical reasoning for conceptual design includes [3] [7] [11]. Most of these authors depend on the similar ideas of confluence that regard the physical world as composed of devices and flows of energy, material, and information. Other authors describe the physical world as a set of governing equations [1]. Williams's work [14] on *Interaction-based Invention* is closely related to our work. In his method, devices are proposed from the desired interactions between quantities. Unlike our method, this method associates the interactions directly with devices, therefore this method cannot handle interactions between devices when the relationships describing these interactions are dynamic. On the contrary, in our method, interactions between quantities take place as instantiations of physical phenomena, which allows handling dynamic changes of the interaction between quantities.

In the rest of this paper, we describe our reasoning method of QPAS and show how the designer is supported in model building. Chapter 2 outlines the knowledge representation of physical phenomena and physical laws based on QPT. Chapter3 presents our reasoning method to assist the designer in synthetical composition and behavioral simulation. Chapter 4 discusses implementation of QPAS. Chapter 5 concludes this paper.

2 KNOWLEDGE REPRESENTATION OF PHYSICAL PHENOMENA

To describe the knowledge about physical phenomena, there are two levels of description that should be obtained. One is the description of physical laws governing the physical phenomenon, such as kinematic laws. For describing physical laws, we employed QPT proposed by Forbus as the base of representation. The other is the knowledge that describes the relationships between the structure of mechanical devices and the physical phenomena that should be considered on the devices. We call a piece of the latter knowledge a physical feature, for it is a cognitive feature extracted empirically by designers from the physical world. A physical feature is described as a network of physical concepts that are defined in the knowledge base.

The architecture of the knowledge base is illustrated in Figure 1. The knowledge base consists of three parts, i.e., concept base, descriptions of physical laws, and descriptions of physical features.

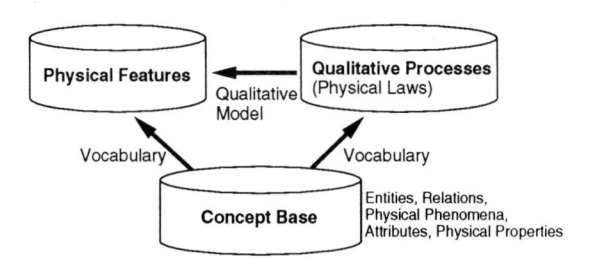

Figure 1: The Architecture of the Knowledge Base

2.1 The Concept Base

The concept base contains vocabulary of physical concepts used for engineering. The vocabulary is used to describe physical features and physical laws. Five types of concepts are defined in the knowledge base. They are;

- entities such as "shaft," "gear," and "table,"

- relations such as "on," "above," "connected," and "fixed,"

- physical phenomena such as "motion," "heat flow," "evaporation," and "friction,"

- attributes such as "position," "temperature," and "mass," and

- physical properties such as "elastic" and "magnetized."

In the concept base, the knowledge about relationships among physical concepts is also described. Physical concepts have frame-type descriptions of relationships with other physical concepts. The relationships include class hierarchies of physical concepts.

Entities are physical objects such as mechanical parts and electric devices. The concept base describes relationships between each entity and physical properties that are peculiar to the entity. Entities have a class hierarchy. For example, a "helical gear" is a subclass of a "gear." This implies that this helical gear inherits the physical properties of a gear and physical phenomena that occur on a gear pair also occur on a helical gear pair.

Relations represent topological structure among entities. By linking entities with relations, static structure of physical mechanisms is modeled.

A physical phenomenon is something that occur to entities and change the states of entities. A frame of physical phenomenon contains descriptions as follows;

- abstract class,

- pointers for linking instances of entities and show which entities the physical phenomenon occur to,

- physical properties and relations required for the entities to be linked,

- attributes to be attached to the linked entities,

- prerequisites for values of the attributes to activate the physical laws, and

- names of physical laws.

The descriptions of prerequisites for the attributes and names of physical laws are associated with descriptions of the processes in QPT, which we will present in the next section. Figure 2 describes an example of rotational transmission.

An attribute is a concept attached to entities and takes values to indicate the state of entities. Physical laws based on QPT are described as relations among these attributes.

A physical property is a concept that describes generic characteristics of entities such as "elastic" and "magnetized." A physical property is associated with a set of attributes that indicate degree of the physical property. For example, Young's modulus indicates elasticity.

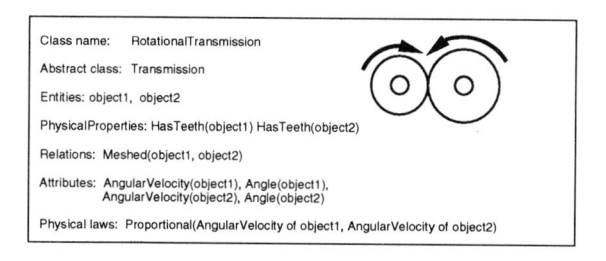

```
Class name:      RotationalTransmission
Abstract class:  Transmission
Entities: object1, object2
PhysicalProperties: HasTeeth(object1) HasTeeth(object2)
Relations: Meshed(object1, object2)
Attributes: AngularVelocity(object1), Angle(object1),
            AngularVelocity(object2), Angle(object2)
Physical laws: Proportional(AngularVelocity of object1, AngularVelocity of object2)
```

Figure 2: Description of Rotational Transmission

2.2 Qualitative Description of Physical Laws

Physical laws are relationships among attributes that consist a background theory underlying physical phenomena. For modeling physical laws, we employed QPT as the base of representation. We employed QPT because its concept of *process* matches our physical phenomena, and its qualitative relationships between attributes are more essential than precise numerical relationships for conceptual design.

Based on QPT, physical laws underlying a physical phenomenon are described as a qualitative process. A description of process contains;

- attributes required in physical laws,

- prerequisites for attributes to activate the physical laws, and

- causal relationships among attributes.

Figure 3 describes an example of a qualitative process describing "rotational transmission" which contains physical laws among angular velocity of gears.

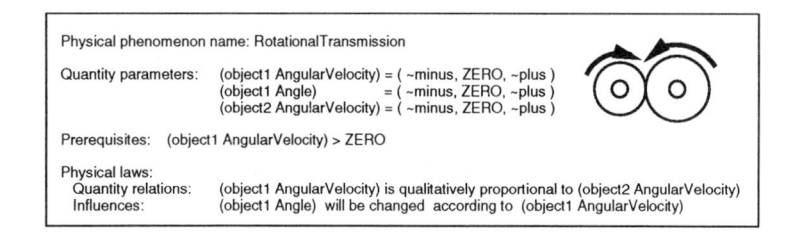

```
Physical phenomenon name: RotationalTransmission
Quantity parameters:  (object1 AngularVelocity) = ( ~minus, ZERO, ~plus )
                      (object1 Angle)           = ( ~minus, ZERO, ~plus )
                      (object2 AngularVelocity) = ( ~minus, ZERO, ~plus )
Prerequisites:   (object1 AngularVelocity) > ZERO
Physical laws:
   Quantity relations:  (object1 AngularVelocity) is qualitatively proportional to (object2 AngularVelocity)
   Influences:          (object1 Angle) will be changed according to (object1 AngularVelocity)
```

Figure 3: A Qualitative Process of Rotational Transmission

We defined two types of attributes, i.e., quantity parameters and modes. A quantity parameter is a continuous attribute assigned with qualitative values consisting of landmarks and their intervals. For example, "temperature" can take landmarks such as "melting point" and "boiling point." A mode is an attribute that describes the characteristic conditions of an entity such as "on" and "off" of a switch. Operations to a device are represented as changes of the modes of the device. For example, turning off a switch of a light is represented as a change of the modes of the light from "on" to "off." The state of an entity is decided by the values of these attributes.

Prerequisites for attributes to activate the physical laws are described as required values of the quantity parameters and modes. For example, "temperature > boiling point" is a prerequisite for the boiling phenomenon. When the state of entities satisfies the prerequisites, the physical phenomenon occurs and the physical laws associated with it become effective to cause changes of the attributes.

Physical laws are described as causal relationships among quantity parameters. Some physical laws may directly effect parameters, which is called *influence* in QPT. An influence is described such that "if a certain quantity parameter is plus, the effected quantity parameter increases." One of its examples is the relation between acceleration and velocity. Other physical laws may indirectly effect through parametric dependencies, which is called a *quantity relation*. A quantity relation is described such that "if a certain quantity parameter increases, the effected quantity parameter also increases." An example of this might be the relation between two velocities of paired gears. As a result of the effects from several physical phenomena, a quantity parameter may either increase, decrease, or stay at the same value in behavioral simulation.

2.3 Physical Features

With the physical concepts defined in Section 2.1, we can build conceptual models of mechanisms to be used for design. We call the model a *physical feature*. Figure 4 illustrates some examples of physical features.

A physical feature is represented by a network of physical phenomena, entities, and relations. Links between a physical phenomenon and entities denote that the particular physical phenomenon occurs to a set of particular entities. Links among physical phenomena represent causal dependencies, which means that activation of one of the physical phenomena causes activation of the others. Figure 5 shows representation of a physical feature of rotational transmission by two gears.

3 A REASONING METHOD FOR BUILDING QUALITATIVE MODELS

In this chapter, we propose a reasoning method of QPAS for building design object models with assistance from the knowledge base presented in the previous chapter. Our reasoning method first reasons out physical phenomena from the desired behavior of the design object referring to the physical laws, and then derives physical features that can cause the physical phenomena.

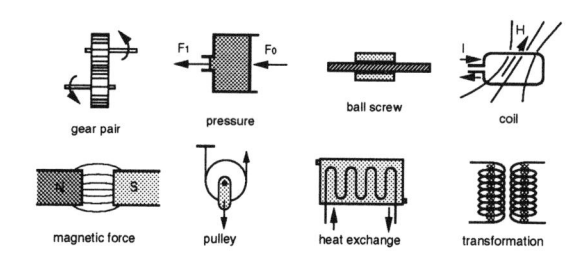

Figure 4: Examples of Physical Features

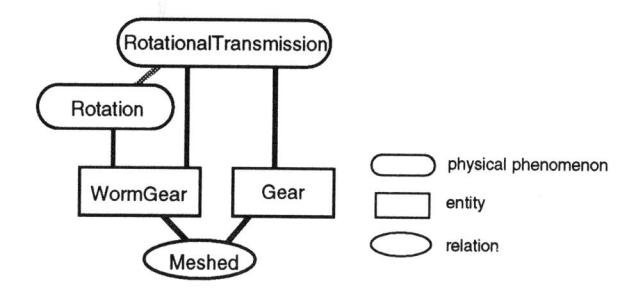

Figure 5: A Physical Feature of Gear Transmission

As a result, the reasoning system suggests to the designer devices and their mechanical relations that realize the desired behavior, and also derives a QPT model that explains how the mechanism causes the behavior.

The design flow is depicted as Figure 6. Figure 7 illustrates an example of conceptual design with QPAS, which we will explain in this chapter.

3.1 Description of Desired Behaviors

QPAS takes an input of a desired behavior represented as transitions of attributes modeled in QPT as an input for reasoning. To obtain the desired behaviors, the designer first decomposes the required function given in the beginning of conceptual design. The process is supported by a system based on the FBS modeling [13].

In the FBS modeling, a required function is decomposed into subfunctions, finally generating a function hierarchy of the design object. At the bottom of this hierarchy, concrete subfunctions appear, which can be easily interpreted as a physical behavior. In the FBS modeling, a function is expressed as a verb and a noun, e.g., "to rotate a shaft." This expression of the required function indicates that the designer knows only a part of the physical behavior

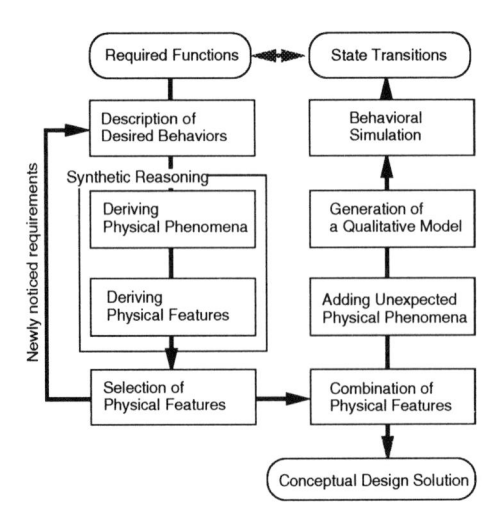

Figure 6: Design Flow in QPAS

expected in the physical world, but not the whole mechanism that causes the behavior. QPAS accepts this partially described behavior as a key to derive a mechanism.

To describe a desired behavior from one of the concrete subfunctions created in the FBS modeling, the designer selects entities and their attributes from the concept base and assigns qualitative values, i.e., landmarks and modes, which are required to describe state transitions. For example, from a function "rotate a ball screw," a ball screw, and its angular velocity and angle are selected, and landmarks such as "ZERO" and "PI" are described.

Using the description of qualitative values, the designer can represent a desired behavior as a state transition graph (STG). STG is a network of nodes that represent states of the selected entities. States are defined as a set of the qualitative value that each of the attributes takes at a certain time point. A link in STG represent changes of attributes, which contains increase or decrease of each attribute.

Figure 8 shows desired behaviors of a ball screw described as STG.

3.2 Suggestion of Physical Features

Given the desired behavior described as STG, QPAS finally derives appropriate physical features as a suggestion that can realize each specified function. The reasoning procedure to derive physical features consists of the following two steps;

Step 1 reasoning out physical phenomena, and

Step 2 reasoning out physical features.

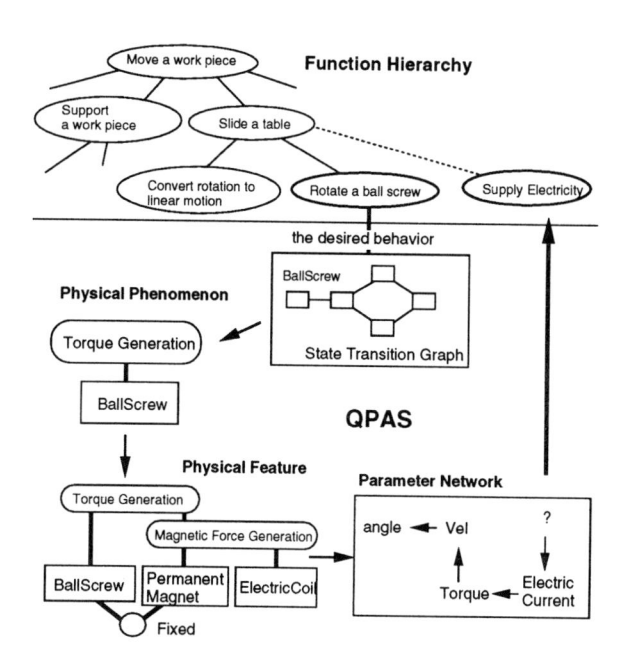

Figure 7: An Example of Design with QPAS

In the first step, QPAS searches for physical phenomena that contain the physical laws that work onto the attributes and be able to cause the transitions. To do this, QPAS extracts changes of attributes described as links in the given STG. Then, QPAS looks up qualitative descriptions of physical laws, which are influences or quantity relations in QPT, finds out the physical laws that cause each of the transitions, and derives physical phenomena that activate the physical laws. In this process, QPAS also checks prerequisites for the attributes described in physical phenomena so that qualitative values of the attributes satisfy the prerequisites during the state transitions.

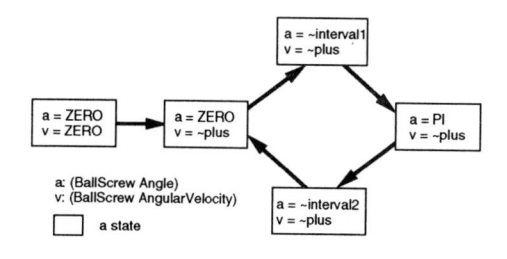

Figure 8: STG Describing Rotation of a Ball Screw

For example, from the increase of angular velocity of the ball screw, a physical phenomenon such as "torque generation" will be applied, because the physical phenomenon has influence which describes that the angular velocity can be increased when torque is positive.

In the second step, QPAS searches for physical features that can cause the physical phenomena derived in the previous step. This is done by searching for the physical features in the knowledge base that contains physical phenomena and entities, e.g., torque generation and ball screw. The class hierarchies of the physical concepts are considered during this matching, so that abstract entities in physical features matches with concrete ones. From the phenomenon "torque generation," a physical feature that models the mechanism of electromagnets used in a motor can be suggested to the designer (See Figure. 7).

The designer usually requires to achieve the function by as few devices as possible. Since a physical feature can cover more than one state transition described by a link in the given STG, QPAS tries to suggest physical features each of which can causes all the state transitions to realize a function so that it can avoid generating redundant devices. When QPAS fails to find such a physical feature, it derives physical features that realize only a part of the state transitions. In this case, the designer should select several physical features to cover all the state transitions.

In another case, there may be no physical features that cause the physical phenomena derived in the first step. In this case, the designer should select entities required to realize the physical phenomena and connect the entities by relations. In other words, the designer should build a new physical feature. QPAS assists this by suggesting entities that match with requirements of the physical phenomena. The requirements are described as physical properties and relations of entities in the frame of a physical phenomenon defined in Section 2.1. QPAS derives the entities that have the required physical properties and connects them by the required relations as suggestions to the designer. For example, "magnetic force generation" requires two entities that have "magnetized" as a physical property, so that QPAS derives "an electric coil" and "a permanent magnet" as suggestions.

3.3 Generation of Qualitative Models

The designer selects one of the suggested physical features as realization of the decomposed function. The physical features contain structure of devices and physical phenomena that cause the desired state transitions.

QPAS generates a qualitative model of physical laws that exist behind the selected physical features. Each of physical phenomena included in the physical Feature has a QPT description of physical laws. From the description, QPAS creates a causal network of quantity parameters which shows how the transitions of one parameter causes the transitions of other parameters in the mechanism of the physical feature. Figure 9 illustrates a causal network of the selected physical feature that generates torque of a ball screw by magnetic force and rotates the ball screw. The designer can understand how the mechanism works and realizes the input desired behavior by the physical feature.

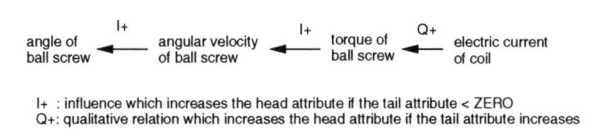

```
                I+                        I+                    Q+
angle of    ◄────────  angular velocity  ◄────── torque of  ◄──────  electric current
ball screw             of ball screw             ball screw          of coil
```

I+ : influence which increases the head attribute if the tail attribute < ZERO
Q+: qualitative relation which increases the head attribute if the tail attribute increases

Figure 9: A Causal Network of Attributes

From the causal network of parameters, the designer may notice other behaviors required to cause the first desired behavior. For example, in the mechanism of electromagnets, the network of parameters shows that electric current of the coil should be positive to cause the rotation of the ball screw. The desired behavior can be considered as a required function unnoticed when building the function hierarchy, The designer adds a function, such as "to provide electric current," to the function hierarchy. In this case, the next requirement will be something that causes the transition of the electric current from zero to plus, and this transition will be added to the inputs for QPAS.

Finally the designer selects a physical feature for each of the bottom functions in the function hierarchy and the newly noticed functions. The designer, then, combines the physical features into one network of physical concepts. Two operations are available to connect physical features. One is an operation called delegation, which unifies two entities and instantiates a new entity that has the attributes and physical properties of both of the original entities. The other operation is to connect entities with appropriate relations selected by the designer.

Once the conceptual network is generated, QPAS adds unexpected physical phenomena to the network. When building the conceptual network, only positive physical features were considered to realize the required state transitions. However, the knowledge base may contain negative physical features that can match with the conceptual network and interfere with the required state transitions. For example, there may be a physical feature that describes friction between a ball screw and a supporting object. QPAS finds such negative physical features by matching their entities and relations with the conceptual network and add the physical phenomena, such as friction, to the network.

3.4 Behavioral Simulation

When a whole network of physical concepts that represents the formulated mechanism is generated, its entire behavior can be simulated based on QPT. This reasoning is called *envisioning* in QPT. As a result of envisioning, all the state transitions of the mechanism are derived, so that the designer can check the mechanism.

For envisioning, QPAS derives the qualitative values of each attribute, which is an quantity parameter or a mode. For each quantity parameter, QPAS derives the required landmarks defined by the designer to represent the desired state transitions. Also, QPAS derives the landmarks described in the prerequisites for occurrence of physical phenomena, such as "temperature > boiling point" of boiling phenomenon. The designer should arrange order of the landmarks to create one consistent value space for each quantity parameters.

Then, QPAS generates a whole causal network of attributes from descriptions of the physical laws included in the conceptual network, as it did for the physical features in the previous section. The causal network includes all the causal dependencies among states of the attributes, occurrence of the physical phenomena, and changes of the states. We used an assumption-based truth maintenance system (ATMS) [2] to manage the causal dependencies, so that, from a state of the attributes, it can derive which physical phenomena occur and which attributes will be changed.

When the designer gives the initial state as an input, the changes of attributes caused by physical phenomena are reasoned out. The changes are increase or decrease. Then, the next states are calculated according to the changes. Repeating these, a graph of state transitions which is similar to STG is obtained.

The designer can compare the first STGs describing the required functions with the derived state transitions and check the behavior of the mechanism. When unexpected state transitions are reasoned, the designer can get their causes, since the causal dependencies of changes are managed.

4 IMPLEMENTATION OF QPAS

We developed QPAS on ObjectworksnSmalltalk. Figure 10 shows the architecture of the system. The FBS modeler is the system that support decomposition of functions.

Figure 11 shows a hardcopy of the system building a model of a mechanism that move a work piece in an NC machine as an example. The left window shows the function hierarchy and the conceptual network of the mechanism. The top right window shows the desired behaviors to rotate a ball screw. The middle right window is the causal network of quantity parameters. The bottom right window describes the attributes and its qualitative values.

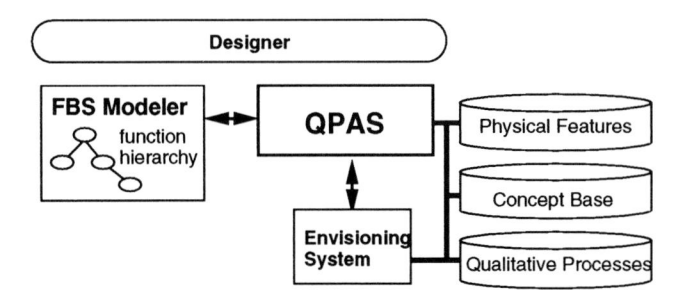

Figure 10: The Architecture of QPAS

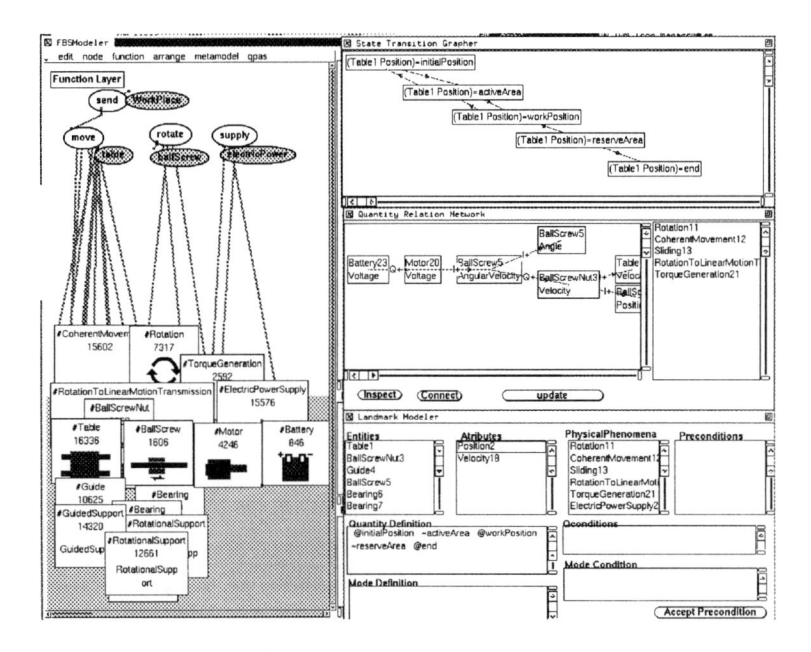

Figure 11: A Design Example of A Sending Mechanism

5 CONCLUSIONS

In this paper, we presented a reasoning method that assists the designer in building conceptual models of design objects. The reasoning method is based on the knowledge of physical phenomena and physical laws represented qualitatively. The advantage of the reasoning is that the designer can obtain not only suggestions about the mechanism but also explanations how the mechanism works. Therefore, this method is useful for the knowledge intensive engineering framework to interactively assist the designer in model building.

Future work includes usage of numerical information about desired behaviors for building numerical behavioral models, so that the designer can evaluate the design more precisely.

ACKNOWLEDGMENT

The authors would like to thank Dr. Takashi Kiriyama, who developed the qualitative reasoning system, and Dr. Yasushi Umeda, who developed the FBS modeler. We also thank to Mr. Yoshiki Shimomura, Mr. Masaharu Yoshioka and other colleagues in our group of the University of Tokyo who helped us in the discussions and implementations of the ideas presented in this paper.

REFERENCES

[1] J. Cagan and .A.M. Agogino. Innovative design of mechanical structures from first principles. *Artificial Intelligence for Engineering Design, Analysis and Manufacturing*, Vol. 1, No. 3, pp. 169–189, 1987.

[2] J. de Kleer. Problem solving with the atms. *Artificial Intelligence*, Vol. 28, pp. 197–224, 1986.

[3] M.G. Dyer, M. Flowers, and J. Hodges. Edison: An engineering design invention system operating naively. In D. Sriram and R. Addey, editors, *Proceedings of the 1st Conference of Applications of Artificial Intelligence in Engineering Problems*, pp. 327–341, 1986.

[4] K.D. Forbus. Qualitative process theory. *Artificial Intelligence*, Vol. 24, pp. 85–168, 1984.

[5] T. Kiriyama, T. Tomiyama, and H. Yoshikawa. Model generation in design. In *Fifth International Workshop on Qualitative Reasoning about Physical Systems*, pp. 93–108, 1991.

[6] T. Kiriyama, T. Tomiyama, and H. Yoshikawa. The use of qualitative physics for integrated design object modeling. In J.R. Rnderle, editor, *Design Theory and Methodology – DTM'90–*, volume DE-Vol. 27, pp. 53–60. ASME, 1991.

[7] A.J. Polovinkin. Untersuchung und entwicklung von konstruktionsmethoden. *Machinenbautechnik*, Vol. 28, No. 7, pp. 297–301, 1979.

[8] T. Tomiyama, T. Kiriyama, and Y. Umeda. Toward knowledge intensive engineering. In K. Fuchi and T. Yokoi, editors, *Knowledge Building and Knowledge Sharing*. Ohmsha, Tokyo and Osaka and Kyoto, 1994.

[9] T. Tomiyama, T. Kiriyama, and Y. Umeda. Toward knowledge intensive engineering. In *Computer Aided Conceptual Design, Proc. of the 1994 Lancaster International Workshop on CACD'94*, pp. 319–337. Lancaster University, 1994.

[10] T. Tomiyama, Y. Umeda, and T. Kiriyama. A framework for knowledge intensive engineering. In *Proceedings of the Fourth International Workshop on Computer Aided Systems Technology (CAST'94)*. University of Ottawa, Ont.,Canada, 1994.

[11] K.T. Ulrich. Computation and pre-parametric design. Technical Report AI-TR-1043, MIT AI Lab, Cambrige, MA., 1985.

[12] Y. Umeda, H. Takeda, T. Tomiyama, and H. Yoshikawa. Function, behavior, and structure. In *AIENG '90 Applications of AI in Engineering*. Computational Mechanics Publications and Springer-Verlag, 1990.

[13] Y. Umeda, T. Tomiyama, and H. Yoshikawa. A design methodology for a self-maintenance machine based on functional redundancy. In *Design Theory and Methodology (DTM '92)*. The American Society of Mechanical Engineers(ASME), 1992.

[14] B.C. Williams. Interaction-based invention: Designing novel devices from first principles. In *Proceedings of the 4th International Workshop on Qualitative Physics*, pp. 161–168, 1990.

DESIGN MODEL: TOWARDS AN INTEGRATED REPRESENTATION FOR DESIGN SEMANTICS AND SYNTAX

L B Keat, C L Tan & K Mathur

1 INTRODUCTION

Design at the early conceptual stages poses interesting, difficult and complex computational problems when it comes to building artificial intelligence or AI support systems. Design is constructive, in which the definition of objects evolve over time and re-use and re-design is standard. Information is a lot of the time incomplete and inconsistent. Such characteristics contribute to systems which are complex especially for supporting real design activities.

In the same vein, design necessarily involves the contribution of multiple design agents (structural, mechanical, architect, client, etc). Each agent of the building design process has its own design abstractions with its different features, compositions and relationships and they apply knowledge from a variety of sources (their expertise and experience, codes, user requirements, constitutive laws) to transform a design. Furthermore that transformation is applied in collaboration through the interactions of these agents. That collaboration process is the communication of information between and coordination among the agents. Given a knowledge intensive activity such as architectural design, a substantial volume of knowledge and data is generated and applied on a design but unfortunately these are not captured especially for facilitating or reducing the level of down-stream activities related to changes made to early design decisions.

Traditionally, the communication and representation of information has been carried through using specifications, plans, schedules, drawings and meetings. With CAD systems, CAD files based on various exchange standards (IGES, DXF, etc) are used however we all know that the information conveyed is restricted to just geometric and drafting data which by itself only communicates the end result of a design process and not the abstractions used in the evolution of the design and other details, such as design intent. Additionally, Eastman [1] states, " ... the representation of entities and their attributes alone is not sufficient for design; in addition to entities both functional and spatial compositions must also be represented ... ". Hence a more complete representation would consists of the entities, the functional and spatial composition. That is to say, for us to support design reasoning, the integration of knowledge-based AI support systems with CAD systems or files is crucial.

Collectively, the points made above, that design at the early stages is fluid, that the contribution and integration of the multiple agents in the design process is crucial for facilitating a more holistic approach to design right early on and that the integration with the digital media of CAD for design drawings and their subsequent reasoning, these are all important contributory factors to the development of a representation for supporting a highly dynamic design environment.

In Section 2, we will describe the features of our representation, the Design Model and the way it is structured. In Section 3, we describe our strategy for supporting an extensible representation model. In Section 4, we detail the implementation strategy which we have adopted. In Section 5, we present the system framework and we conclude in Section 6.

2   THE DESIGN OF DESIGN MODEL (DM)

The Design Model is designed to support the requirements of architectural design especially at the early conceptual stages. It elaborates and extends on knowledge representation techniques found in artificial intelligence and object-oriented concepts for a powerful and flexible data and knowledge model. The following are features most salient to our model:

i. It supports system pre-defined and user-definable and extensible simple and composite design object variables. The ability to define an extend object variables is most challenging. It questions object stability which is a most securing aspect of data storage and management. Yet, because design applications are constructive, dynamic, evolving and where re-use and re-design is standard, this feature is necessarily supported.

ii. It supports object-based and file-based integration of geometric/syntactic CAD data. Direct association between syntax and semantics optimizes semantic and syntactic data access and reduces redundancy in the information maintained. The object-based integration facilitates the increasing number of object-oriented based graphic systems. File-based integration facilitates the integration of other CAD descriptions to our representation. Interpreters would facilitate the conversion.

iii. It supports geometrical reasoning and query based on a geomerical query language (GQL) within the representation. This facilitates object discovery or emergence. It attempts to address a significant area of work centered around making CAD more responsive to design problems, their ambiguities or discontinuities.

iv. It supports the specification of prioritized multi-domain knowledge in terms of rule-sets. This facilitates the representation of design standards as a powerful design aid for facilitating continuous, real-time standards checking as the design process unfolds, domain-specific modelling knowledge and meta-knowledge that determines how and where to invoke other kinds of knowledge. This gives us the opportunity to identify and integrate in a holistic sense the different kinds of knowledge which one brings to bear in the design process especially in architectural design.

v. It supports evolving knowledge on the application domain, in terms of multi-domain knowledge sources within design objects. This facilitates knowledge refinement, the discovery of new relationships between objects, new object attributes or new inferences between object values. The knowledge-base is rule-based with quantitative and qualitative extensions that facilitate multi-domain design reasoning and understanding across both semantic and syntactic levels.

vi. It supports methods or functions for maintaining the state of semantic and syntactic data. This could involve performing procedural computations.

vii. It supports flexible user-definable semantic relationships between objects. This includes selective composition relationships and dependency relationships. Usual inheritance relationships are not sufficient to model complex artifacts.

We have as such structured the Design Model into five parts. We describe the structure and elaborate on each feature.

2.1 Semantic Aggregation (A_{sem})

The semantic aggregation encapsulates design knowledge associated with the attributes and functionalities of a design object. These functional attributes represent both simple and composite variables, their meanings contributed or volunteered during design, computationally derived from the appropriate interpretation of the object's syntax, if necessary, in combination with other functional attributes as well, or obtained by interrogation of Design Model relationships or from other Design Model instances e.g. the value to the variable "number of floors" for the object "Building" is obtained by determining the number of "Strata" objects.

2.2 Syntactic Aggregation (A_{syn})

The syntactic aggregation maintains information about variables describing the geometry, topology and appearance characteristics of a design object. One of the key ideas that we have experimented with in our research is the support of direct CAD linkages within our schema for representing design knowledge. Naturally, in integration, some assumptions have to be made concerning the type of the CAD description, in which case, an object-oriented representation of our CAD data is adopted. The integration reduces one level of association between design semantics and syntax, reduces redundancy in the information maintained and optimizes access to our graphical information as semantic and syntactic data can now be obtained with a single access to the Design Model that represents the object. This facilitates system procedures with quick access. Figure 1 illustrates the linkages where each Design

Model description represents a design object in its place within an abstraction hierarchy for representing building information.

File-based integration is more complex. Depending on the type of information represented, we can integrate to varying extends. For example, integration and interpretation of DXF files for reasoning is very difficult due to the low-level syntactic information content. This is explored further in Section 4.

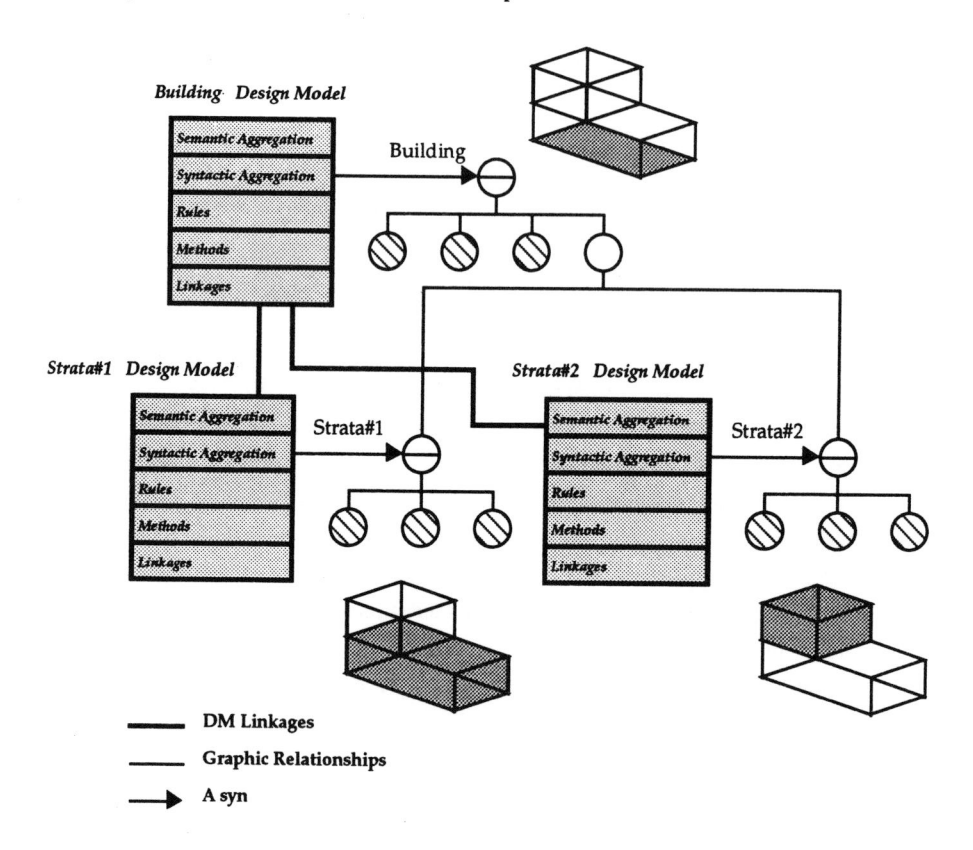

Figure 1. Design models and their syntactic linkage

2.3 Rules (R)

The concept of "Rules" here encapsulates multi-domain design knowledge about design objects in their representation themselves. Design Model addresses our representation of multi-view/domain knowledge by "re-defining" the traditional approach of separate multi-domain knowledge-bases [3], [4] as well as "refining" the granularity of attached knowledge . We do this by grouping knowledge from the multiple domains of the design process into "Rules" (Figure 2).

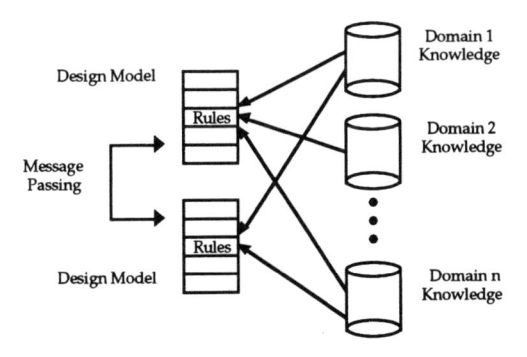

Figure 2. Structuring multi-domain knowledge into "Rules"

This means knowledge normally infered from separate knowledge-bases concerning a design object are now re-defined and refined into the representation describing the object itself and knowledge is extended from just attribute-related to object-related. "Rules" provide a new way of thinking and encapsulating architectural knowledge. It allows early identification of inter-relationships among the different, possibly competing views, contributing to the design process. And it does it in terms of design objects as it supports a more holistic approach to intelligent design assistance.

The Design Model therefore is not a mere data structure but an active entity which works to solve design problems. It has the function to maintain its state to satisfy constraints when its structure or values of its attributes are modified and to resolve conflicts resulting from conflicting demands when they arise. Because in most cases design objects do not just maintain their state in isolation, mechanisms to manage communication between design objects have to be in place. This communication is by message passing.

2.4 Methods (M).
Methods provide the techniques for maintaining the state of the semantic and syntactic aggregations and this could involve perfoming procedural calculations.

2.5 Linkages (L_{dm}).
Linkages encode information of relationships between other design objects and their knowledge.

$$DM = (A_{sem} + A_{syn} + R + M + L_{dm})$$

Whilst the adoption of object-orientation has strengths for encapsulation and re-usability of existing definitions, there are some shortcomings to the object-oriented approach in modelling the design problem. For example, they suffer from inherent limitations in modelling user-defined semantic relationships [2], [5]. In addition, concepts such as encapsulation, instantiation and method selection are not always well-suited for design. For example, as design evolves, it is usually necessary for the designer to work on the object definitions, as such encapsulation must be broken. Instances also belong to exactly one class when the designer may require the instance to belong to several points of view at the same time.

We overcome them by the design of a flexible meta-object shell whereby classes are instances of generic meta-classes and any of the set of these classes and their instances can be assembled for modelling real objects through a meta-design model shell.

3 THE META-DESIGN MODEL SHELL

The challenging issue of the Design Model is its support of evolving knowledge on the application domain. In certain applications, there is a need to integrate partial pieces of information whose interaction are not known in advance and whose structure may evolve over time. Therefore, a certain form of knowledge acquisition must be supported, necessarily with an appropriate representation model.

All the classes defined within PARADISE are instances of Meta classes. In PARADISE, there are two main meta classes. The class Meta defines the minimal attributes and behaviour of classes. This minimal list of attributes include the attributes "instances", "instance_of", "super", and "sub-classes". Further to the Meta class, there is the class Meta-DM for which the attributes "Asem", "Asyn", "Rules", "Methods" and "Linkages" are the minimal atributes.

3.1 Class Meta

The attribute "instances" gives the set of instances derived and attached to the class or meta-class considered. This attribute is system maintained and it facilitates the management of knowledge refinement and evolution. The attribute "instance of" maintains the set of classes to which the class or instance belongs. PARADISE supports multiple instantiation or multiple class membership namely that instances can belong to several classes simultaneously. This feature departs from usual object-based models where instances can belong to only one single class. The attribute "super" defines the set of super-classes of a class. It therefore supports multiple inheritance whilst the "sub-class" attribute is the inverse of "super". Sub-classes implement the inheritance relationship. A class inherits the attributes of all its super classes.

3.2 Class Meta-DM

The Meta-DM class integrates the attributes, syntactic linkage, rules, methods and relationships that represent building abstractions and their physical objects as detailed in Section 2. It is the "shell" upon which classes and instances for modelling objects in the domain of interest are assembled. The more salient classes are described in Section 4.

4 IMPLEMENTATION IN PARADISE

Everything is an object in PARADISE. In this section, we describe the more salient features of attribute, syntactic linkage and multi-domain rule integration.

4.1 Attributes

In PARADISE, attributes are C++ classes. Each attribute of a class is modelled at the base by a specific meta class called PAttribute. There are two main types of attribute classes as depicted in Figure 3. Those that derive their values by means other than through graphical derivation (PDeriveU) and those that are graphically derived (PDeriveG). For each attribute, there are slots for maintaining attribute name, attribute type, the list of rules or constraints on the attribute called the constraint list and a representation of a domain when the value is unknown.

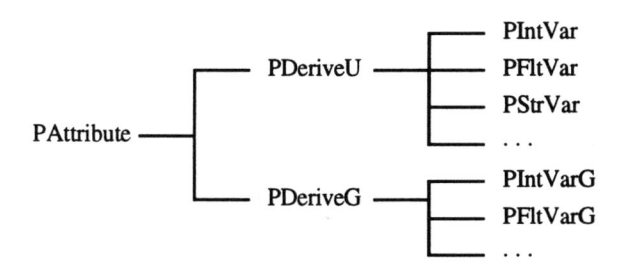

```
                              ┌── PIntVar
              ┌── PDeriveU ───┼── PFltVar
              │               ├── PStrVar
              │               └── ...
PAttribute ───┤
              │               ┌── PIntVarG
              └── PDeriveG ───┼── PFltVarG
                              └── ...
```

Figure 3. Attribute class hierarchy

Each type of variable is implemented by a class. For instance, the class for integer type attributes (PIntVar) is different from the class for floating point type attributes (PFltVar) and similarly for graphically derived attribute types (PIntVarG and PFltVarG) for which the values are obtained through geometric reasoning functions or processes that interrogates the associated graphic representation of the object through the syntactic linkage connected to the object.

4.2 Syntax

Syntactic integration at the native level, designers use PARADISE's own graphic or design system called PGS [7]. At this level, all graphics or drawings are represented as objects. This includes the representation of geometric, topological and appearance characteristics of the object. Even the management and control of user interactions with his or her drawing is object-based. As Figure 2 illustrates, graphical information can be structured based on any domain abstraction or object for which graphical as well as some non-graphical, but related information, are grouped under a control or link object (PPGSG) which forms the integration.

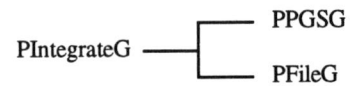

```
                          ┌─── PPGSG
         PIntegrateG  ────┤
                          └─── PFileG
```

Figure 4. Syntax class hierarchy

File-based integration is naturally more complex. Integration of this form responds to the need to communicate with other CAD drawings and to extend reasoning to such drawings. The integration here is more effective for higher-level object-based CAD formats that include feature descriptive information about their drawings from which these can be converted or decomposed into native representations for facilitating consistency and integrity checks on the drawings and their continued reasoning. For a CAD format such as DXF, in which the syntactic content is rather low-level, the integration at the file level for facilitating reasoning has proven to be very difficult.

4.3 Rules

Rules are used in PARADISE for a number of purposes. These include:

i. The computation of attribute values. Several rules can be defined for an attribute and they are ordered in order of appearance and executed in the same manner until one succeeds. A rule is executable when its antecedent arguments are instantiated and valued for which the nature of the arguments/attributes are graphic and non-graphic dependent. The execution of rules also facilitate user entry of values.

ii. The constraining of attribute value. Several constraints may be defined for an attribute and constraints may be defined as either mandatory or they are hard constraints whose compliance are strictly required or they are optional for which violation is permissible under control. Objects that exhibit such exceptions due to user manipulation are considered incomplete or inconsistent objects for which the system would support by maintaining degrees of incompleteness for every instance. These object would be monitored for managing their evolution as design unfolds further.

iii. The communication of constraints. This models the inter-communication of multiple design agents within and across design objects.

Rules or constraints in PARADISE are C++ objects. Stating a rule results in the creation of a constraint object and its posting to the constraints list of the appropriate attribute. Figure 5 depicts the class hierarchy. In PARADISE, in addition to basic constraints typified by conventional knowledge-based shells, it supports geometrical interpretation and reasoning of graphical drawings which is not supported by any knowledge or expert system shell. This process is supported by the specification of the intended reasoning using a geometrical query language (GQL), not described here, and by a geometrical reasoning and computational engine.

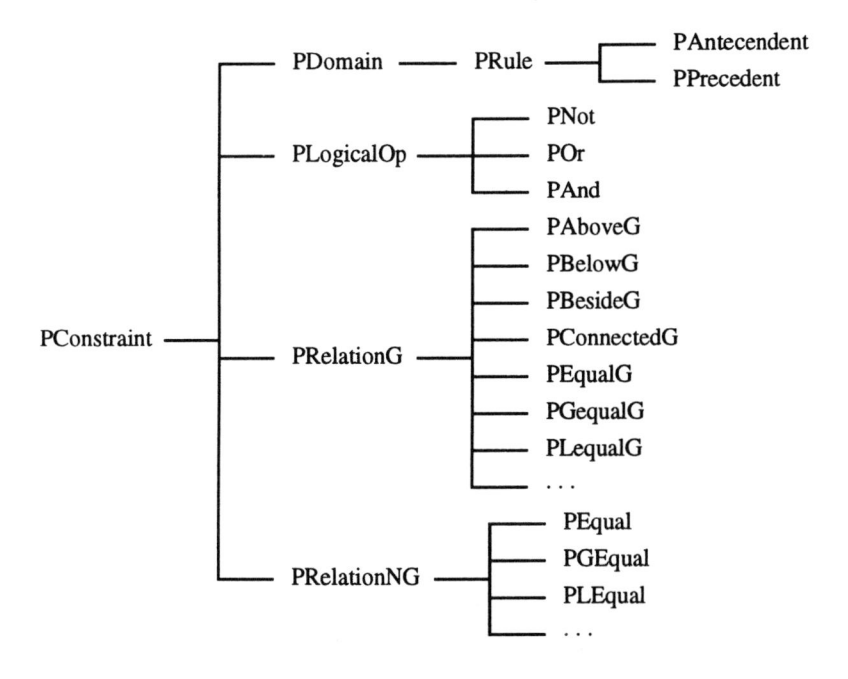

```
                           ┌─── PDomain ─── PRule ─────┬─── PAntecendent
                           │                           └─── PPrecedent
                           │
                           │                          ┌─── PNot
                           ├─── PLogicalOp ───────────┼─── POr
                           │                          └─── PAnd
                           │
                           │                          ┌─── PAboveG
                           │                          ├─── PBelowG
                           │                          ├─── PBesideG
PConstraint ───────────────┤                          ├─── PConnectedG
                           ├─── PRelationG ───────────┼─── PEqualG
                           │                          ├─── PGequalG
                           │                          ├─── PLequalG
                           │                          └─── ...
                           │
                           │                          ┌─── PEqual
                           │                          ├─── PGEqual
                           └─── PRelationNG ──────────┼─── PLEqual
                                                      └─── ...
```

Figure 5. Rules class hierarchy

5   PARADISE SYSTEM ARCHITECTURE

In this section, we describe the "managers" and "engines" which are developed to support the features and functionalities reflected by the Design Model. They comprise part of the total system that makes up our knowledge integrated design system PARADISE which is described in more detail in [6]. Figure 6 depicts the architecture.

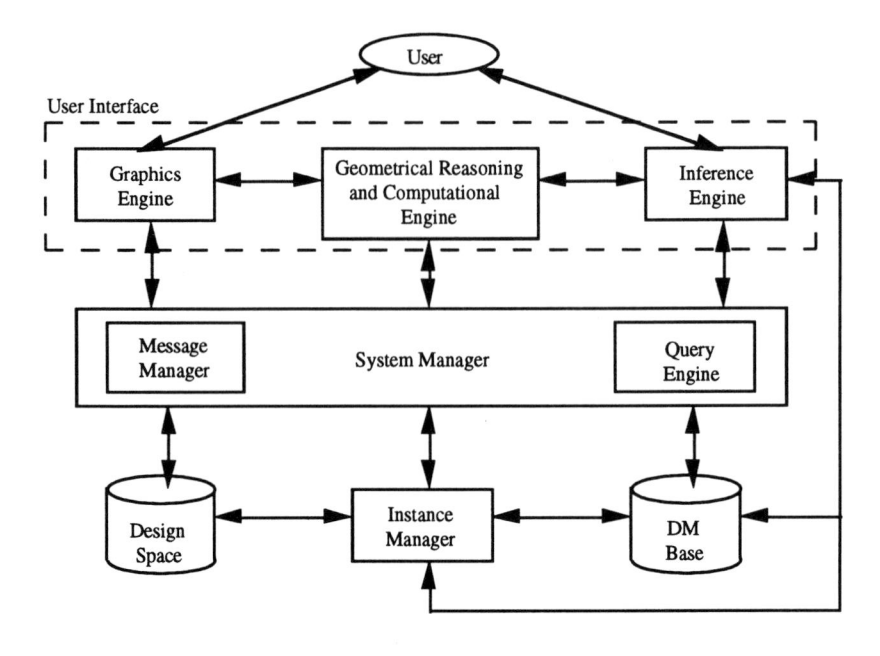

Figure 6. PARADISE system architecture

1. Graphics Engine

   It facilitates interactive composition or decomposition of three and two dimensional form or objects by providing tools that makes it easier for designing at the early stages. Tools that allow molding and shaping of form in ways very similar to shaping clay models. A major limitation with CADD systems currently is their rigidity especially with architectural design.

2. Inference Engine

   It implements two problem-solving strategies, forward chaining and backward chaining and an iterative convergence process to join opposite lines of reasoning together to yield a problem solution. Further to the inference engine, there is the Design Model Development Module that provides access to a development environment where development and maintanence tools and utilities are provided to develop, test and maintain Design Models and their associated knowledge and linkages. These tools include editors and browsers for Design Model classes and instances.

3. Geometrical Reasoning and Computational Engine

It supports the geometric reasoning processes of the system. It is the component that provides the "dual-way" communication highway between related activities involving the graphics-end of the system as well as the knowledge-end of the system. Our implementation of the geometric reasoner consists of three main components: the Design Model Event Manager (DMEM), the Graphics Event Manager (GEM) and the Multi-Domain Concept Mapper (MDCM), where DMEM and GEM represents event-triggered components for organizing and managing reasoning activities.

i. Design Model Event Manager

The Design Model Event Manager monitors user access of Design Model instances. In view of our Design Model structure, there are two main access types, namely access to entries of the semantic aggregation (A_{sem}) and access to rule (R) entries. For changes which are graphics derivable or graphics effected, they are propagated for handling by MDCM.

ii. Graphics Event Manager

The Graphics Event Manager monitors graphic events. When the designer clicks on an object and moves the object, GEM responds to the changes by invoking MDCM housed system functions for handling. That may mean having to, for example, re-compute interference checks on the object against its environment, or when an object's form is re-shaped or re-sized in response to design experimentation or exploration, re-checking the constraints relating to the semantics of the object, represented in its Design Model, to ensure satisfaction.

iii. Multi-Domain Concept Mapper

The Multi-Domain Concept Mapper provides a layer of services to both DMEM and GEM. It manages request for qualitative and quantitative information for variables whose meanings are derived from the graphical description of an object or the design space, i.e. concept mapping, as well as manage the request of the latter from multiple views, i.e. multi-domain. In a reverse of the above, it also manages the request of reasoning processes that effect change to graphical descriptions of objects.

4. Query Engine

It supports high-level Design Model query. For example, it supports query based on functional attributes. If a designer intends to address the functional performance of a particular object, it determines the object instance(s) and the variables whose parameterization affects the specific functional performance for analysis.

5. Message Manager

    It supports a message-based system for inter-object communication and notification due to rules, methods and linkages defined at the representation of design knowledge and invoked dynamically during run-time.

    Both the query engine and the message manager are components of a system manager which functions as a central controller to the various activities performed by the various components of PARADISE.

6. Instance Manager

    It support access, instantiation, initialization of object instances due to design interaction. It also supports instance management which includes modifications, additions and deletions as well as function-based index maintainence to facilitate functional query.

Additionally there is the Design Space that represents a central database for keeping record on the state of a design and the DM-base for maintaining our Design Model-base.

6 CONCLUSIONS

The development of the Design Model was motivated by the need for design systems that can assist design in which assistance meant the ability to interpret and reason about graphically presented designs, the ability to evaluate designs and suggest appropriate solutions as well as the ability to handle constraints and conflicts arising from the multiple agents of the design process. The Design Model provides the representation model that enhances the design process by providing the designer with a means for capturing and communicating a broad range of design knowledge and information derived from the multiple disciplines of the design process, as well as the means for integrating syntactic information for facilitating geometrical reasoning.

There are a number of advantages to be gained in our representation of design knowledge in support of design reasoning. Each is closely related with the concept of modularity or object-orientedness in system design. Improved performance and maintanence, easy updating and access included in a representation that we feel facilitates for an expression of design knowledge as a natural outgrowth of the design process.

At the macro level, it addresses the lack of proper collaboration between the various disciplines by the earlier detection of potential conflicts. This increases productivity by reducing development time, reducing the need for frequent engineering changes and giving better overall quality. On top of which, the system also supports continuous design verification, consistency and integrity checks, and enhanced speed in re-design when necessary.

REFERENCE

1. Eastman, C M "The Representation of Design Problems and Maintenance of Their Structure" *Artificial Intelligence and Pattern Recognition in CAD*, Latombe (ed), North-Holland Publ Company (1978)

2. Escamilla, J and Jean, P "Relationships in an Object Knowledge Representation Model" *Proc 2nd International Conference Tools for Artificial Intelligence*, Washington, USA (1990)

3. Fenves, S J "An Integrated Software Engineering Environment for Building Design and Construction", *Proceedings o the Fifth ASCE Computing in Civil Engineering* (1988)

4. Fenves, S J., Flemming, U., Hendrickson, C., Maher, M L and Schmitt, G "Integrated Software Environment for Building Design and Construction" *Computer-Aided Design* 22(1) pp 27-36

5. Giacometti, F and Chang, T C "Object-oriented Design for Modelling Parts, Assemblies and Tolerances", *Proceedings of the 2nd International Conference on TOOLS '90*, Paris France (1990)

6. Liew, B K et al "PARADISE : An Intelligent CAD System Architecture" *International Journal of Construction Information Technology* Vol 1 No 3 (1993) pp 1-24

7. Liew, B K "PGS : PARADISE Graphic Subsystem", *KIDS Research Project, Research Report R93-2*, School of Building and Estate Management and Dept. of Information Systems and Computer Science, National University of Singapore (1993)

ÉGIDE: A DESIGN SUPPORT SYSTEM FOR CONCEPTUAL CHEMICAL PROCESS DESIGN

R Bañares-Alcántara, J M P King & G H Ballinger

1 INTRODUCTION

According to Mostow [12], a comprehensive model of design should support the representation of the

1. State of design (description of the design object).

2. Goal structure (goals are prescriptions of how the descriptions of the artifact should be manipulated).

3. Design decisions (choices between alternative design paths).

4. Rationale for design decisions (justifications for goal selection).

5. Control of the design process (selection of the best goal to work on and the best plan with which to achieve it).

6. Learning in design (both, of general knowledge about the domain and of specific knowledge about the problem).

KBDS, as described in [5], can maintain design alternatives and design constraints, which are adequate for the representation of item 1 and part of item 2. We have extended such a representation with an IBIS (Issue-Based Information Systems, [13]). The result, ÉGIDE, can account for items 2 through 4, and provide a solid base for items 5 and 6.

The next section provides a brief introduction to the core of ÉGIDE. In the following section we present a brief introduction to KBDS followed by a description of the extensions to record design rationale. Finally we present a list of expected advantages from recording design rationale and future work proposed to achieve them.

2 ÉGIDE: A DESIGN SUPPORT SYSTEM

ÉGIDE is a prototype computer-based support system for integrated and cooperative chemical plant design. Its task is to assist a group of cooperating designers during the course of the design process. Design is an evolutionary process, *i.e.* it generates new design alternatives as transformations of existing alternatives. Given the complexity of chemical process design, such transformations are incremental, that is, each design step affects only a small part of the current design state. Thus, ÉGIDE is based on a representation that accounts for the evolutionary, cooperative and exploratory nature of the design process, covering design alternatives, constraints, rationale and models in an integrated manner.

Historically, ÉGIDE has been developed in two stages. The first step resulted in KBDS, a system that maintains design alternatives, models and specifications. This system was then extended to ÉGIDE, which represents design rationale. Most of the present document addresses the issues related to the maintenance and use of design rationale, which has been the main focus of industrial interest.

A design process is represented in KBDS by means of three interrelated networks that evolve through time: one for design alternatives, another one for models of these alternatives, and a third one for design specifications and constraints.

In KBDS each partial or alternative design is linked to other *design alternatives*. Furthermore, KBDS maintains a more detailed "understanding" of a design state, *i.e.* it maintains a representation of an alternative's constituent blocks, their relation to each other, and to blocks in other design alternatives (*e.g.* refinement). *Design constraints* can be used to evaluate the adequacy of a design alternative. The prediction of the behaviour of design alternatives is achieved by means of *models*.

The next section describes how ÉGIDE maintains the designers' rationale. As in the rest of ÉGIDE, the information is represented in a *prescriptive* fashion, *i.e.* in a form that is amenable to computer processing.

3 DESIGN ALTERNATIVES MAINTENANCE (KBDS)

KBDS supports features that are necessary for an effective design support system, *i.e.*:

- provides an incremental representation of the evolution of a chemical plant specification through the design process, manages the design alternatives, and allows the controlled propagation of values, Bañares-Alcántara [2];

- monitors consistency among coupled sub-designs, Bañares-Alcántara [1];

- enables the designer to navigate the different structures through the use of a GUI, Lababidi and Bañares-Alcántara [11]; and

- interfaces transparently to external application packages, Ballinger *et al.* [6], *e.g.* to synthesis and simulation packages such as CHiPS and ASPEN PLUS™.

KBDS provides a set of tools to maintain, visualise, and manipulate design alternatives and constraints. It consists of a number of interacting tools, two of the main being the *Design History* and the *Flowsheet Tool*.

The role of the *Design History* tool is to support a graphical representation of the design alternatives history by showing the relationship between its various components.

Figure 1, for example, illustrates that a particular design project starts with an alternative design, BLOCK4, which is then decomposed into a reaction and separation sections: REAC-SEC1 (hidden by its associated menu) and RSEP-SEC1. Each sub-design is then evolved through a series of modification and refinement steps, expanding both branches of the tree.

The *Flowsheet Tool* is an object-oriented flowsheet editor which enables the designer to create and edit the graphical representation of alternative designs. The most important contribution of the flowsheet editor is that process *units* are treated as objects: besides

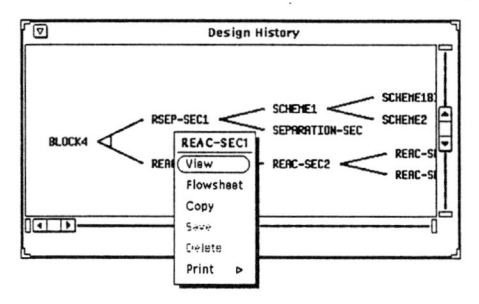

Figure 1: *Design history* tool in KBDS.

their graphical representation and connectivity (flowsheet outlook), they are also bound to the underlying object-oriented knowledge representation system, *i.e.* they are linked to various verification facilities and simulation capabilities.

Each one of the structural alternatives in the design alternatives network of Figure 1 can be expanded into one or more of what we call *operational alternatives*. Operational alternatives by our definition, share the same structure but differ among themselves in one or more design parameters, for example temperature of operation, heat transfer area, or material of construction. These differences generate diverse consequences, *e.g.* product quality, and different plants, since the processing vessels may have different sizes and forms, thus different capital costs.

One of the functions of the *Flowsheet Tool* is displaying the operational alternatives of the flowsheet. The title SCHEME4 - (1/4), in Figure 2, denotes that the alternative structure scheme4 has four operational alternatives, with the first one being currently displayed. Double clicks on an *equipment* name would invoke the "Variable Browser" (not shown), which gives the user access to the values of its constituent variables. Other alternatives of scheme4 can be checked through the three popup menus shown. This facility gives the designer the capability to browse through the existing alternatives, and select the appropriate ones for further evaluation and development.

Because design alternatives are maintained via an ATMS, de Kleer [9], value propagation across alternatives is simplified, consistency maintenance of coupled sub-designs is supported, and memory usage is economised, Bañares-Alcántara [3].

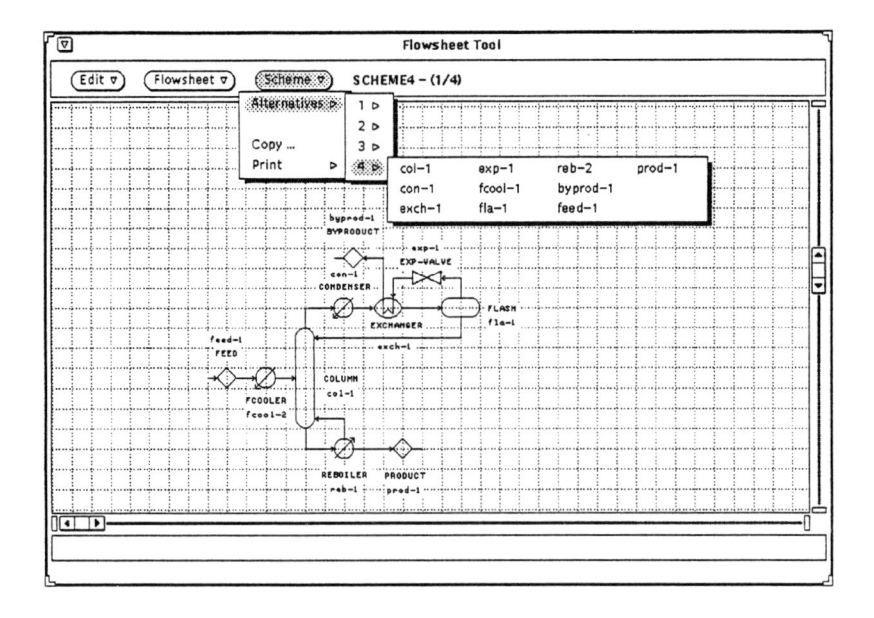

Figure 2: *Flowsheet Tool* in KBDS.

4 DESIGN RATIONALE REPRESENTATION

ÉGIDE focuses on the maintenance of the designer's rationale, for the recording and use of design deliberation, argumentation and rationale. This is done by means of IBIS networks [13], which store every decision along with its competing alternative decisions and the arguments used in the selection. A network is maintained for every problem or issue discussed; networks can be interrelated. An IBIS network consists of *issues*, *positions*, and *arguments*.

The IBIS methodology has been used to model deliberation in software development [8], but only in a declarative fashion. Extensions were required to make these ideas prescriptive and applicable to Process Engineering, for example, by connecting the IBIS network to the design alternatives history maintained by KBDS. This was achieved by means of two new objects [7]: *steps*, transformations of a design alternative suggested by a selected *position*, and *tests*, constraints placed on the design alternatives and associated to the *arguments*.

As an example, Fig. 3 shows the evolution of the design alternatives for the separation section of the HDA plant considered in Chapter 7 of [10]. This part of the design spans from the separation-block to the separation3 design alternatives.

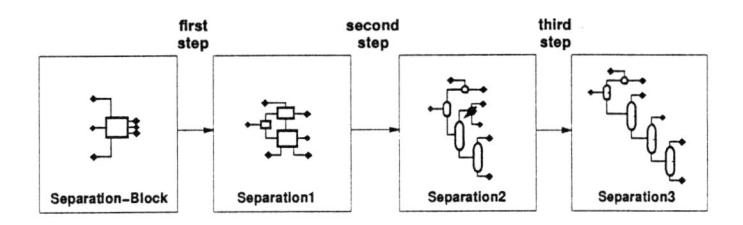

Figure 3: Design history of the HDA separation section.

The *Intent Map*, Fig. 4, shows all the IBIS networks generated during the design of the HDA plant separation section. The square at the top left-hand corner corresponds to the network shown in detail in the *Intent Tool* window, Fig. 5. The contents of

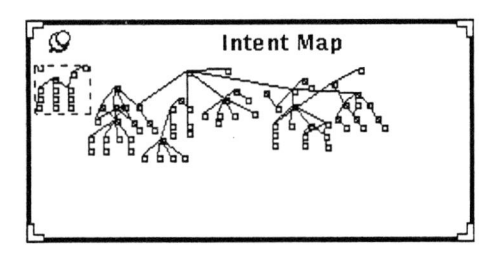

Figure 4: The *Intent Map*.

the IBIS network can be examined, manipulated and generated through the *Intent Tool*. The window shows the first (and simplest) network in the example case. It deals with the general-separation-structure? *issue* (**I**) associated with the separation-block design alternative. This *issue* raises the question of the choice of the general structure of the separation system. Also shown are three possible *positions* or design paths (**P**'s), the *arguments* or justifications related to each choice (**A**'s), and a set of *tests* that can be applied to separation-block alternative in order to evaluate the adequacy of each choice (**T**'s).

Three choices of separation structure are considered, and one is selected according to the phase of the reactor effluent stream. In Fig. 5, the *position* initial-phase-split was selected as the best choice because the reactor effluent is a VL mixture. The selection of this *position* gives rise to a *step* that transforms separation-block into separation1 (refer to Fig. 3). The other *positions* considered were to have a liquid separation system only (for liquid reactor effluents) or to partially condense the reactor effluent and then have a phase split (if the reactor effluent is a vapour).

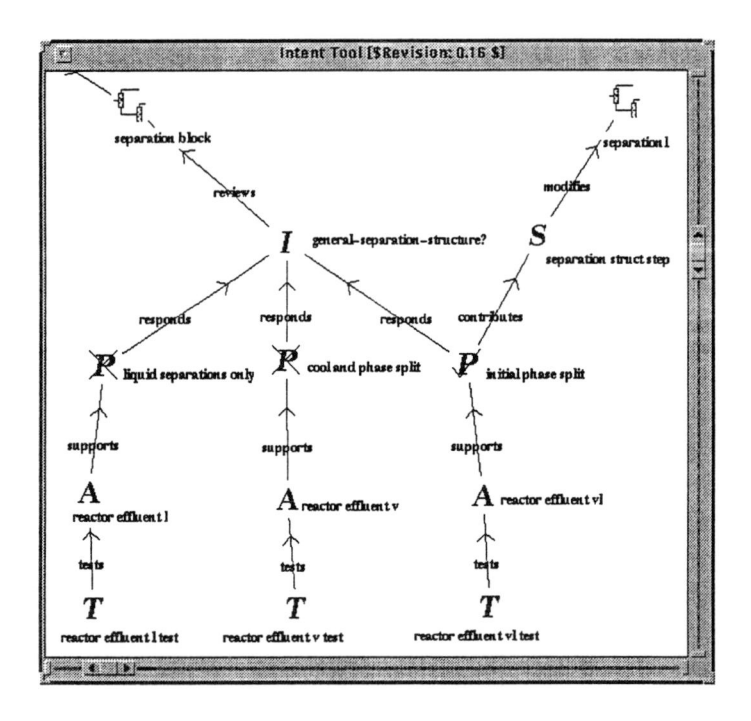

Figure 5: The *Intent Tool*.

5 USING DESIGN RATIONALE RECORDS

From the declarative point of view, an IBIS structure can be used to keep track of the issues that have been discussed, which design alternative suggested them, and, if they are resolved, which alternative they affect directly. It is also possible to find whether an issue has been resolved or not, the choices that were selected, and the reasons for their selection. Since part of the information kept in the network is in a prescriptive form, some functions have been created that rank positions responding to an issue in order of desirability and suggest the most appropriate position for a given issue. Also, parts of the design object or process can be re-used in other projects with a fuller understanding of the assumptions and implications of the action, *i.e.* re-use can be *context-sensitive*.

There are three immediate ways in which the design rationale can be used to the designer's advantage, all of them provided by ÉGIDE *Intent Tool*:

1. Storage of the design rationale in a prescriptive fashion allows the design team to identify which parts of the plant must be re-designed when there has been a change in the internal assumptions, constraints or specifications of the plant, or any external factor affecting them. This is known as *dependency-directed* backtracking.

2. An automatic evaluation of *positions* which allows *arguments* to be given a weight. Weights can be assigned their currently displayed values or be reseted to their previously stored values. Thus, the system supports "what-if" studies.

3. An automatic report generator that produces documents describing the evolution of the design alternatives and the argumentation that resulted in a given decision.

The next sections contain a detailed presentation of the dependency-directed backtracking and automatic report generation.

5.1 Dependency-directed backtracking

Design objectives, assumptions or external factors often change during the course of a design. Such changes may affect the validity of decisions previously made and thus require that the design is reviewed. If a change occurs, the *Intent Tool* allows the designer to automatically check whether all *issues* have the most promising *positions* selected and thus determine from what point in the design history the review should take place. The decisions made for each *issue* where the currently selected *position* is not the most promising *position* should be reviewed.

The *issue* general-separation-structure?, discussed in Section 4 and shown in Fig. 5, is resolved by selecting the *position* initial-phase-split, as the reactor effluent stream in separation-block is a vapour-liquid mixture.

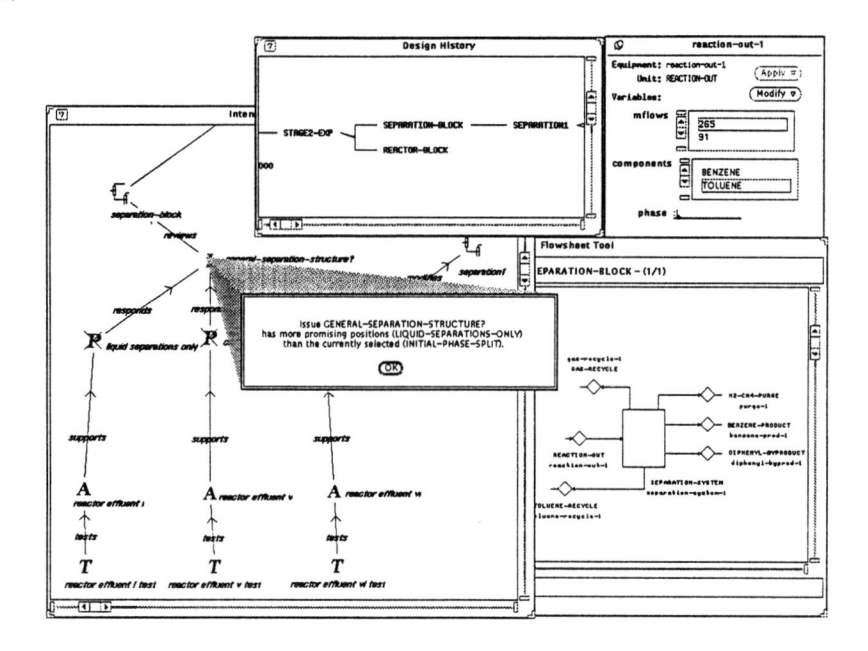

Figure 6: First scenario: A change in the phase of the reactor effluent stream (reaction-out-1) to liquid means that the most promising *position* becomes liquid-separations-only.

As a first example of a change to a previous design decision, assume that the composition of the reactor effluent has changed, due to an alteration to the reactor operating conditions. This causes the liquid-separations-only *position* to be the most promising for the general-separation-structure? *issue*. Therefore, the selection of initial-phase-split should be reviewed in light of the new reactor operation (see Fig. 6).

As a second example we use a change in the price of oil, an external factor which affects the desired design. It directly affects the *issue* treatment-of-lights which has three *positions*: recycle-light-ends (with supporting *argument* good-use-of-raw-materials), vent/flare-lights (with supporting *argument* lights-cheap-as-fuel) and use-lights--as-fuel (with supporting *arguments* lights-cheap-as -fuel and lights-are -good-fuel). If the price of oil falls from its current value of \$4.0/BTU to below \$3.6/BTU then the light ends of a separation become more valuable as a raw material than as a fuel. When this *issue* is re-examined using the *Intent Tool*, it is discovered that the *position* recycle-lights becomes more promising than the currently selected *position* (use-lights-as-fuel), since the *test* associated to the *argument* lights-are-cheap-as-fuel is no longer true.

5.2 Automatic evaluation of positions

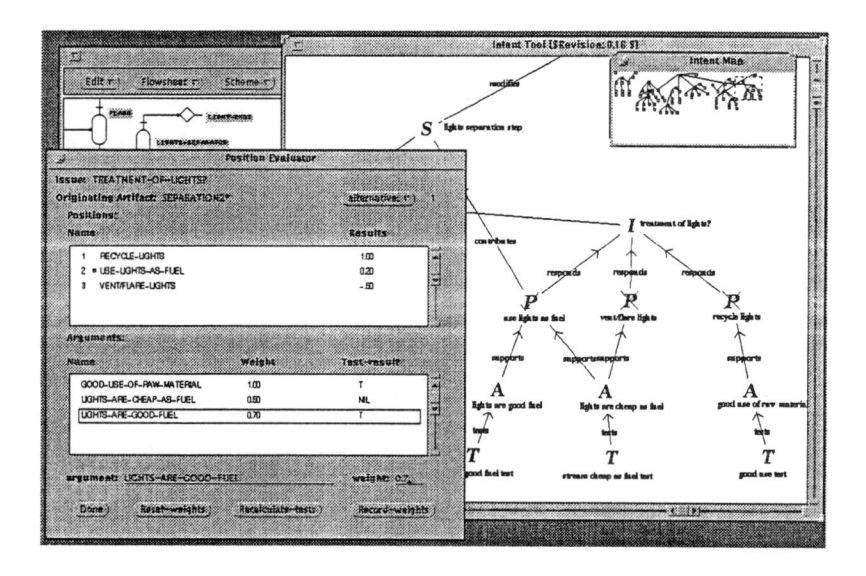

Figure 7: The Position Evaluator

When choosing from alternative *positions* it is useful to be able to rank the alternatives and weight the supporting *arguments*, so that some may be given greater importance than others. This is desirable as a designer may believe *arguments* have differing degrees of importance, for example, *arguments* relating to safety or environmental matters may be more important than those relating to product purity.

The ranking of *positions* and assignment of weights to *arguments* is carried out using the *Position Evaluator* shown in Figure 7. This window shows two lists. The first is a ranked list of *positions* which shows an overall result beside each *position* name with the currently selected *position* marked with an asterisk. This list is generated from the second list (of *arguments*) which contains, for each *argument*, its weight and the result of the associated test. The overall result for each position is the sum of the products of the weight and test result, for each supporting *argument*. The designer may edit the weight for the selected *argument* and see what effect this has on the ranking. Buttons are available so that the designer may reset the weights to the values stored in the *arguments* or record them as the new defaults.

5.3 Automatic Report Generation

Each IBIS node has a text-based description associated. This enables the automatic generation of reports. There are two types of reports that ÉGIDE can generate: *Steps Reports* which describe the design rationale for the evolution of one design alternative to another, and *Argumentation Reports* describing the deliberation resulting in a particular decision. The designer is able to select the level of detail of each report.

An extract from an automatically generated *Steps Report* can be seen in Fig. 8. It describes the evolution of `separation-block` to `separation3`, and contains information about the creator and creation time, the flowsheets for the starting and finishing design alternatives, a description of each design *step* with its selected *positions*, its supporting *arguments*, and the alternative *positions* considered.

A design *step* may be the result of the selection of various *positions* (each *position* related to a different *issue*). The *argumentation report* is centred on an *issue* and can list the *arguments* that were not included in the reasoning process but that have the potential to affect the validity of the results.

| EGIDE-Report   Date: 25/ 3/ 1994   Time: 16:03 |
| Project: YORK |
| Generated By: JOSH @ CHEMENG.ED.AC.UK |

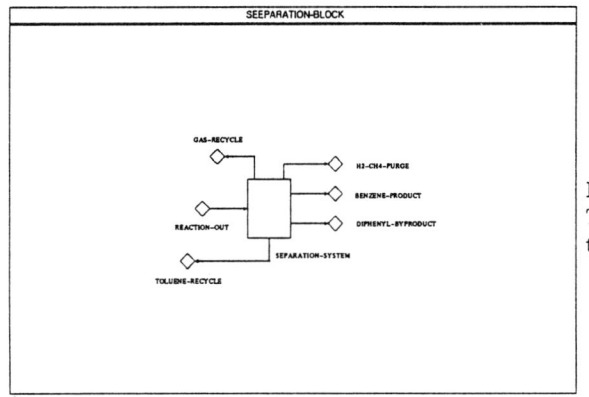

```
┌─────────────────────────────────────────┐
│            SEEPARATION-BLOCK            │
│                                         │
│                                         │
│            GAS-RECYCLE                  │
│               ◇──────┐  ┌──◇ H2-CH4-PURGE│
│                      │  │               │
│                      ┌──┴──┐ ◇ BENZENE-PRODUCT│
│         REACTION-OUT ◇     │               │
│                      │     │ ◇ DIPHENYL-BYPRODUCT│
│                      └──┬──┘               │
│                         │                  │
│                         ◇ SEPARATION-SYSTEM│
│            TOLUENE-RECYCLE              │
│                                         │
└─────────────────────────────────────────┘
```

From: separation-block
The black-box separation block after decomposition.

The evolution of **separation-block** to **separation4** involved four steps:
The **first step** modified separation-block to separation1.
It was decided

    Position: - to have a phase split as the first unit operation because

        Argument - the reactor-effluent stream contains a vapour liquid mixture.

    Alternatives considered were:

        Alternative - to have only liquid separation unit operations.

        Alternative - to cool the reactor effluent and then split the phases.

The **second step** modified **separation1** to **separation2**, this step involved four decisions:

    Position: - to remove the light ends from the benzene product stream because

        Argument - the desired benzene product purity is 99.97%.

    Alternatives considered were:

        Alternative - not to remove the light ends from the benzene product stream.

Figure 8: An extract from an example report

6 CONCLUSIONS AND FUTURE WORK

There are a number of reasons that suggest that the information supporting design decisions should be explicitly recorded. This information can be used to improve the documentation of the design process, verify the design methodology used and the design itself, and provide support for analysis and explanation of the design process. ÉGIDE is able to do this by recording the design artifact specification, the history of its evolution and the designer's rationale in a computable form.

In the future, we envisage the application of knowledge-based systems to the information recorded by ÉGIDE to achieve the following:

- Automatic generation of documents, such as summaries of achieved results, lists of pending tasks, etc.

- During retrofit, in the generation of an "artificial history" of the design process for those plants for which no computable record of their design has been kept.

- Automatic generation of design alternatives for those situations where it is possible to recognise an identical situation in the records of a previous design history.

- Case-based reasoning, drawing analogies from similar design situations in the past.

- Application of design "critics" from early stages of design, thus spreading the consideration of downstream issues, such as controllability, safety and environment, along the whole design process time span.

- Selection of competing design application packages, *e.g.* physical property prediction methods, flowsheeting packages, user interfaces.

REFERENCES

[1] Bañares-Alcántara, R., "Maintenance of consistency in designs with cooperative designers.", Presented in the *Information and Management Systems for Process Design* session, AIChE Fall Meeting. Los Angeles, USA, 17-22 November 1991. **(Conference Proceedings)**

[2] Bañares-Alcántara, R., "Representing the engineering design process: Two hypotheses.", *Computer-Aided Design (CAD)*, 23(9):595–603, November 1991. (**Journal Reference**)

[3] Bañares-Alcántara, R., "Application of an ATMS in the representation of the design process.", Proceedings of the Workshop on Application of Reason Maintenance Systems, ECAI'92, Vienna, Austria, August 1992. (**Conference Proceedings**)

[4] Bañares-Alcántara, R., "Design support systems for process engineering. I. Requirements and proposed solutions for a design process representation" Accepted in Computers & Chemical Engineering, March 1994. (**Journal Reference**)

[5] Bañares-Alcántara, R., and Lababidi, H.M.S., "Design support systems for process engineering. II. KBDS: an experimental prototype.", Accepted in Computers & Chemical Engineering, March 1994. (**Journal Reference**)

[6] Ballinger, G., Bañares-Alcántara, R.,, Costello, D., Fraga, E.S., Krabbe, J., Lababidi, H.M.S., Laing, M., McKinnel, R., Ponton, J.W., Skilling, N., and Spenceley, M., "Epee : A process engineering software environment." *Computers & Chemical Engineering*, 18(Suppl.):S283–S287, 1994. (**Journal Reference**)

[7] Ballinger, G., Bañares-Alcántara, R., and King, J.M.P., "Using an IBIS to record design rationale.", ECOSSE Technical Report 1993-17, Department of Chemical Engineering, University of Edinburgh, Edinburgh, Scotland, September 1993. (**Technical Report**)

[8] Conklin, E.J., and Yakemovic, K.C.B., "A process-oriented approach to design rationale.", *Human-Computer Interaction*, 6:357–391, 1991. (**Journal Reference**)

[9] de Kleer, J., "An assumption-based truth maintenance system.", *Artificial Intelligence*, 28:127–162, 1986. (**Journal Reference**)

[10] Douglas, J.M., *Conceptual Design of Chemical Processes*, McGraw-Hill, 1988. (**Book reference**)

[11] Lababidi, H.M.S., and Bañares-Alcántara, R.. "An integrated graphical user interface for a chemical engineering design support system." *Chemical Engineering Research and Design*, 71(A4):429–436, July 1993. (**Journal Reference**)

[12] Mostow, J., "Toward better models of the design process.", *The AI Magazine*, 6(1):44–57, 1985. **(Journal Reference)**

[13] Rittel, H.W.J., and Webber, M.M., "Dilemmas in a general theory of planning.", *Policy Sciences*, 4:155–169, 1973. **(Journal Reference)**

DEVELOPMENT OF AN INTEGRATED AI SYSTEM FOR CONCEPTUAL DESIGN SUPPORT

M X Tang

1 INTRODUCTION

Design problems are typically ill-structured: they start with a design requirement which does not contain, or imply all the criteria by which acceptable solution can be completely identified. Such a design requirement may be incomplete, inconsistent, imprecise or ambiguous. An important characteristic of design therefore involves developing a complete and consistent design requirement at the same time as developing a solution that satisfies it. This requires an integrated design environment in which domain concepts, design objects, design heuristics and design materials, are classified and structured so that their functional, structural and causal relationships can be identified and effectively reasoned about, allowing high level design tasks, such as design synthesis, design analysis, design evaluation, and design modification to be carried out whenever the design requirements are specified or changed.

Supporting design tasks at the early stage of the design process is difficult because at this stage little is known about the inconsistency of the design requirement and the structure of the design problem. The concept of the design is not formed until a designer has identified a basic knowledge structure in which information about structural, functional and causal relations of design objects becomes available and organised. The central role of an AI system supporting conceptual design is to provide solutions, through proper representations and reasoning mechanisms, for a class of design problems where the design requirement is incomplete and inconsistent, and the design problem is largely under-constrained.

Although it is possible to develop tools that automatically perform design tasks that require intensive numerical calculation, it is more useful to build design support systems that combine human and machine intelligence. Automatic design tools can only support some well defined and isolated design tasks which rely on procedural knowledge and fixed design strategies. Theses tools contribute little to decision making in the early stage of the design process which requires intensive use of both heuristic and procedural knowledge.

This paper presents an integrated AI system for conceptual design support. In this system, functional reasoning methods are used to transform an initial design requirement, using knowledge transformation rules, to an initial design solution concept structure consisting of a set of connected design elements which satisfy the initial design requirement. This knowledge transformation process is supported by a representation and reasoning scheme developed by Chakrabarti [2]. The results of functional synthesis provide a basis for embodiment design during which the physical components realising the proposed initial design solution concept structure are identified and their geometric relationships established. In this stage the values of all design variables are specified and explored using a constraint management system utilising genetic algorithms and simulated annealing techniques [4, 7]. The results of the embodiment design and kinematic analysis provide the information for evaluating the conceptual design results.

An integrated application of functional reasoning, constraint management and kinematic analysis techniques is one of the current research themes at the EDC in Cambridge University. The aim is to develop an AI-based design support system capable of deriving conceptual design solutions from functional design requirements [1]. In this paper, some of the existing systems used in this integration are reviewed. Based on this review the blackboard architecture of an integrated functional modelling system is presented. The design knowledge sources in this system are then described in the context of this blackboard architecture. Finally, some implementation issues of using a knowledge based system development tool are discussed.

2 A KNOWLEDGE-BASED APPROACH TO CONCEPTUAL DESIGN

Knowledge-based design techniques provide rich knowledge representation methods that suitably represent and manipulate different types of design objects, design heuristic knowledge, and design constraints; they provide sufficient inferencing support and control mechanisms for designers to perform design tasks, including design synthesis, evaluation and optimisation; and provide mechanisms for detecting and resolving conflicts among various design requirements and design constraints.

The following requirements are identified in developing the integrated functional modelling system:

- It should help designers to construct and extend the design knowledge base within which domain concepts, design objects, dependency information of design objects are well structured and consistently maintained.

- It should help designers to derive solutions quickly from initial, not necessarily complete and consistent, design requirements. In other words it should provide an efficient mechanism to transform an initial design requirement description to a final design specification.

- It should provide explicit explanations and justifications for any chosen aspects of the current status of a design, not only in terms of how something has been derived, but also to explain why something is not happening as expected. Locating areas of difficulty and suggesting strategies for solutions contribute to good decision making in design.

- It should allow designers to vary data, design requirements, problem solving strategies, or evaluation criteria, to obtain alternative designs. Simply speaking, a knowledge-based design support system must support multiple-context problem solving so that a design problem can be explored from many different directions.

- It should provide mechanisms for capturing and refining design knowledge so that the design knowledge base can be incrementally enlarged and enhanced. In other words, a knowledge-based design support system must have some facilities for learning.

AI methods are used in the development of the integrated functional modelling system to deal with the problems of: how to break down a design task and represent the hierarchy of components or sub-systems in a formal and natural way; how to effectively manipulate a potentially large set of design variables and constraints; how to maintain the knowledge generated during design and to detect inconsistency; and how to create and maintain multiple design solutions.

2.1 ATMB, a knowledge-based architecture for design support

Assumption-based Truth Maintained Blackboard (ATMB) is a knowledge-based design support system kernel developed in Edinburgh [5, 6]. The ATMB kernel was built on the percept that design problems at the early stage of design process are ill-structured. It supports the representation and exploration, using a number of AI-based computational techniques, of domain and design knowledge necessary for driving design solutions from an initial, not necessarily complete design requirement description.

In a knowledge-based design support system, alternative design solutions need to be explored and retained in order to compare them in contexts where design criteria, design requirements and design methods differ. In mathematical terms, features of design can be described by a set of design variables and a set of dependent design parameters. The values of design variables and dependent design parameters are determined by a set of constraints in which these variables and parameters are logically or causally related to each other. Design variables and dependent design parameters are used when it is necessary to distinguish the part of a design problem which is flexible to change (described by design variables) from other parts of the design problem (described by dependent design parameters) which are relatively dependent on design variables. It is assumed that the values of dependent design parameters can be determined by the values of the design variables through constraint propagation.

Based on this formulation, the structure of a design problem can be determined by the identification of the relationship between design variables and dependent design parameters. The constraints, in most cases, are different forms of mathematical equations, rules, or reasoning modules. These constraints typically interact in complex manners.

During design exploration, many plausible selections arise in the presence of under-constrained design variables, giving rise in turn to many plausible values of dependent design parameters. Furthermore the way in which design variables and dependent design parameters are constrained may depend on what design strategy (or method) will be employed. In other words, using a different design problem solving strategy may result in the same set of design variables and dependent design parameters being differently constrained. A design solution is a complete set of values for all the design variables and dependent parameters which jointly describe the features of the design problem and satisfy the constraints and the design requirement.

The ATMB architecture uses a frame-based representation scheme for building design knowledge bases. A design knowledge base contains design object class definitions which partially defines the spaces of possible designs for small and independent design problems. The multiple relationships and dependencies among design concepts and design objects enable them to be used together to jointly define the structure of new design problems. Such a design knowledge base provides design knowledge in a behavioural, functional and structural vocabulary. A structural relationship between design objects determines how objects are physically connected and how, for example, characteristics of a system object can be derived from its constituent component objects; a functional relationship between design objects relates objects in terms of their performances or behaviour, and their relevance to a particular design requirement and design task; a causal relationship between two design objects decides how one object might depend on another and the consequences of any change in either object.

The ATMB architecture uses a combination of computational techniques to provide general support to knowledge-based design applications. It consists of a blackboard control system, an assumption-based truth maintenance system, a design documentation system, and a graphical explanation system [6, 8]. In this architecture an assumption-based truth maintenance system (ATMS) is integrated with a blackboard control system to maintain the consistency of the design knowledge generated, and to manage the exploration of design contexts when *design requirements, initial design data* or *design methods* are changed. The ATMS-based design context management system has three primary purposes for design support: to maintain consistency in the design knowledge base; to maintain multiple design contexts; and to be used as a search control mechanism to reduce the number of combinations. It is particularly suitable for design exploration because of its incremental updating of the dependency networks and its ability to maintain a multi-contextual environment to allow different design solutions to be explored simultaneously [5].

The ATMB architecture is self-contained with a design documentation system which maintains design history records; and a graphical explanation system which provides context-dependent explanation of newly derived design knowledge or any design difficulties. The ATMB mechanism as a basic architecture for knowledge-based design applications has been used the Edinburgh Designer System (EDS) and the Castlemaine Design System [5, 6].

2.2 FUNCSION, a functional synthesis system

FUNCSION is a system developed by Chabrabarti at the EDC in Cambridge University that performs *instantaneous synthesis* of design concepts. FUNCSION was intended to support two design tasks in the early stage of the design process:

1. instantaneous synthesis of design solution concepts based on functional requirements;

2. temporal reasoning of behaviour of a design solution concept [2].

FUNCSION as a functional synthesis system is capable of deriving an initial design solution concept structure based on a design requirement description and a goal function. When the design requirement function and the goal function can be defined in advance as input and output of a transformation system, the system's roles are: to search for valid elements whose input functions match the requirement functions; and then to apply predefined transformation rules to these elements until a structure that satisfies the goal function can be derived. FUNCSION is basically a qualitative reasoning system and it has been implemented using Harlequin's KnowledgeWorks.

The important concepts in FUNCSION are *element* (predefined building blocks for design synthesis) which defines the type and constraints of a physical component, *transformation rules* which transform one function to another, *input vector* (functional requirement) and *output vector* (goal function). Both requirement function (input) and goal function (output) are represented as vectors which have *kind, orientation, sense, position* and *magnitude* as attributes. A solution concept is defined as an abstract description of a system of identifiable individual elements which can satisfy the given functional requirements [2].

Functional synthesis of design solution concept is performed first by a kind-synthesiser (reasoning on the kind attribute of input/output vectors only). This kind-synthesis process searches for a set of elements which satisfy the kind attributes of the input/output vectors. A result of this kind-synthesis is a causal network of elements in which transformation rules apply. The termination condition of the kind-synthesiser is specified by the user as the maximum number of transformations to be allowed.

The kind-synthesiser generates a list of candidates, each of which specifies a causal relation between the elements concerned. These candidates are then further specialised by checking their orientation and sense configurations using the orientation and sense transformation rules. The results of this are a number of initial design solution concept structures, each of which satisfies the kind, orientation and sense requirements.

2.3 CADET, a design embodiment tool

Computer Aided Design Embodiment Tool (CADET) is a system developed by Thornton at the EDC in Cambridge University for supporting embodiment design using systematic constraint satisfaction techniques. CADET provides facilities for: (1) constraint specification by building a product model using a library of generic functional components; and (2) constraint satisfaction by solving the design problem represented by the product model using either simulated annealing techniques or genetic algorithms [7].

A product model in CADET consists of a set of generic components connected through the relationships of either two-components interface or three-components interface. Each component has variables and pre-defined initial constraints. The initial constraints associated with any component are reasonably well understood and domain independent. A product model, once built by the user through an interactive user interface, will have a set of design variables which are constrained by those pre-defined initial constraints and any new constraints added in by the user while building up the product model. In particular, each component has a 3D geometry information associated with it. This process is referred to as constraint specification [7].

Much of the CADET's facilities deal with the problem of constraint specification during which a designer builds up a product model, using a library of components. Once a product model is built, the CADET system generates a text-based file containing design variables and constraints. This text-based file can then be used in the constraint satisfaction stage when two programs are invoked to search for a set of values of design variables that do not violate any design constraints.

During the process of constraint satisfaction, the constraint set presented by the product model is first simplified by a constraint evaluation program in order to exclude any constants and equalities, resulting in a reduced set of constraints. This reduced set of constraints is then regarded as an optimisation problem with the goal of the optimisation being defined by a constraint violation function that gives an account of how many constraints are violated. This optimisation problem is subsequently solved using either a

Genetic Algorithms program (GA) or a Simulated Annealing program (SA). Both GA and SA use the same goal function, ie., the constraint violation function in the search for a consistent set design variable values without any constraint violation. The GA program uses a two-points masked crossover strategy in its reproduction process while the SA uses a fixed strategy to change its search direction (this fixed strategy sets the values of some of coefficients used in the simulated annealing program). Both GA and SA terminate when either the goal function is satisfied to an extent, or a local minimum has been reached. This indicates that there is no guarantee that an optimised solution will be found using either GA or SA methods.

3 AN INTEGRATED ARCHITECTURE FOR FUNCTIONAL MODELLING

The ATMB architecture was used in the DtoP Project and the Castlemaine system [6]. Both FUNCSION and CADET have demonstrated the capability of providing knowledge representation and reasoning support to early stage design tasks when little is known about the basic structure of the design problem. These systems, however, do not operate in an integrated and consistent way. The need for integrating these systems arises from the fact that, with a unified knowledge representation scheme and an intelligent control system, they can form a more powerful tool for the tasks of functional synthesis of design requirement, embodiment design of product structure and kinematic analysis of product behaviour. In order to build such a system, it is necessary to abstract useful features of ATMB, FUNCSION and CADET and then to redevelop them in the form of an intelligent design support system architecture and domain independent design knowledge sources (or inference engines). The integrated functional modelling system is intended to be used by designers as a conceptual design tool in the domain of mechanical engineering.

CADET's approach to design constraint satisfaction is characterised by two features, ie., its inclusion of 3D geometry information in its generic component definitions, and its coupling of constraint satisfaction and optimisation methods (GA and SA methods) in the embodiment design process. While CADET has been developed as a self-contained design embodiment tool and has demonstrated the sufficient capability of solving a number of design problems, several problems need to be addressed in order to develop it further as a domain independent design knowledge source.

CADET successfully used genetic algorithms and simulated annealing in constraint satisfaction in a number of design examples. However, further effort needs to be made in order to establish how these methods can be used as domain independent design knowledge sources. Two important research problems need be further addressed when transferring the facilities in CADET into a general design knowledge source:

1. how to expand its product model by incorporating more complicated knowledge structures above the component level,

2. how to validate its optimisation-based (use of GA and SA in combination with constraint evaluation) approach to design constraint satisfaction.

The product model definitions in CADET need to be extended as the current ones cannot be used for functional synthesis, design embodiment and kinematic analysis. The lack of definition and classification of constraints in CADET makes it difficult to know what kind of constraints can be solved. Little heuristic design knowledge is utilised in the process of constraint satisfaction and optimisation (Both GA and SA use fixed coefficients in their algorithms). No explanation is provided when the system gets into difficulty.

One approach to enhancing CADET is to combine heuristic-based constraint management with search-based optimisation methods. This approach is more interactive than the current CADET system and therefore it is possible to utilise more design heuristic knowledge in the design embodiment stage. The automatic search methods such as simulated annealing and genetic algorithm are only used when the network of constraints has been explored and simplified by the user.

This requires the development a more powerful constraint management system capable of performing algebraic manipulation of a potentially large set of complex constraints in order to provide a good basis for using GA and SA methods in the design optimisation process. It is also important to incorporate a rule-based reasoning system into CADET so that it can reason on design heuristics or use assumptions. This rule-based system which utilises design heuristics can simplify the constraint set before the GA and SA methods are invoked.

In the integrated functional modelling system, FUNCSION system is also enhanced by enlarging its database of standard elements and transformation rules, and by integrating this database with the system's knowledge base where elements can be extended to have the links to a component knowledge base. A new interactive user interface is developed for FUNCSION that allows the functionality of the system to be fully utilised and explored by a designer. The original inference mechanism in FUNCSION is embodied with its control program. This is extended and redeveloped using a rule-based approach for representing the transformation rules.

3.1 The Blackboard architecture

The architecture of an integrated AI system identifies the necessary AI components for intelligent design support. An integrated application of FUNCSION, CADET, a kinematic analysis package [4] and AI-based methods forms an important part of a research theme at the EDC towards the development of an Integrated Design Framework (IDF) [2]. In this architecture, the ATMB kernel controls FUNCSION, CADET and a kinematic analysis package as independent design knowledge sources.

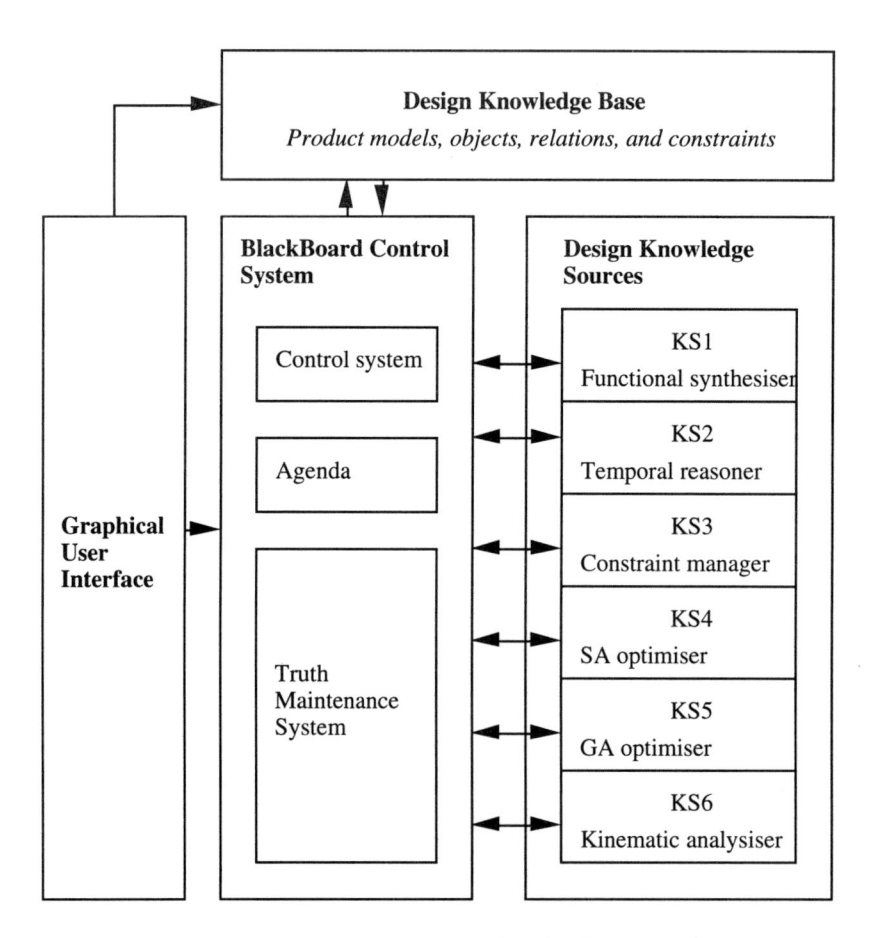

```
┌─────────────────────────────────────────────┐
│           Design Knowledge Base             │
│ Product models, objects, relations, and constraints │
└─────────────────────────────────────────────┘

BlackBoard Control System        Design Knowledge Sources

  Control system                  KS1  Functional synthesiser
  Agenda                          KS2  Temporal reasoner
  Graphical User Interface        KS3  Constraint manager
                                  KS4  SA optimiser
  Truth Maintenance System        KS5  GA optimiser
                                  KS6  Kinematic analysiser
```

Fig 1. An Integrated AI Architecture for Functional Modelling

The architecture of the integrated functional modelling system, as illustrated in Figure 1, consists of a design knowledge base containing product models which can be shared by FUNCSION, CADET and a kinematic analysis program; a blackboard upon which design knowledge sources are controlled on an opportunistic basis [3]; a number of design knowledge sources performing functional, design embodiment and kinematic analysis; and a graphical user interface capable of explaining design results, design history and producing design documents.

In this architecture, the system's design knowledge sources (KS), controlled and maintained by the ATMB kernel, interact with a designer to perform design tasks incrementally. The designer's actions such as initiating design data, specifying design constraints, and choosing design problem solving methods are transformed into the ATMB kernel as design assumptions. The system maintains the knowledge derived based on these assumptions and supports the exploration of this knowledge until a design concept solution can be fully specified in terms of functionality, physical attributes and geometric relationships between its components.

3.2 Knowledge representation

Research in EDC's product modelling team has established a hierarchical structure of generic design object definitions. In this hierarchy, design objects are defined as component, parts and artefact. An artefact assembles a number of parts and a part has a number of components. Parts and components can be connected in complex manner. The instantiation of a part of this object hierarchy creates a set of design object instances that can be explored to derive new designs [1].

In this knowledge structure, relationships between design objects are limited to simple specialisation relations called 'kindof' relations and aggregation relations called 'partof' relations. The 'kindof' relationships represent single inheritance relations so that the hierarchical structuring of objects can be defined. This hierarchical structure is achieved by packaging up small pieces of knowledge into components. Within each component, design variables and parameters are constrained through linear, non-linear equations, rules, spatial relations or tables etc. This structured object representation scheme is similar to a frame-based representation. However, it also allows spatial relationships, functional relationships and geometric relationships among design objects and variables to be explicitly represented as frames and object-oriented message passing methods [5].

The design knowledge base contains design object classes and the reasoning methods associated with each object class. For example, a component has the attributes of input/output function, type of transformation for functional synthesis purpose. The type of transformation determines how a component transfers an input to output. A component also has attributes to describe its physical and geometric features for the purpose of embodiment design and kinematic analysis.

Three kinds of reasoning modules are attached to object classes. These are: demons, constraint handlers, and forward chaining rules. Demons are functions attached to object slots. Whenever a change is made to an object instance, the demons attached to the object class are invoked to carry out inference to determine the consequence of this change within the object instance. Constraint handlers are used to propagate changes from one object instance to the others which are physically or causally related to the object instance being changes. Rules are used to perform special purposed reasoning such as orientation synthesis and constraint management.

3.3 Blackboard controlled constraint management

Constraint management is the central problem solving method in the integrated functional modelling system. It decides how the values of design variables and dependent design parameters change when assumptions are made by the designers whilst exploring an initial design solution concept structure.

In the integrated functional modelling system, embodiment design is carried out in an incremental way. The components in the design knowledge base are first selected by a designer to build up an instance of a product model. Alternatively such an instance of a product model can be derived by the functional synthesis knowledge source based on the initial user requirements. An instantiated product model only constrains the design solution space in terms of the components used, their physical connections and orientations. This needs to be further explored in the embodiment design process to determine the values of all the variables without violating any constraints. Internally, an instantiated product model is represented as a network of variables and constraints.

While genetic algorithms and simulated annealing methods can be used to carry out a systematic search once a design problem has been transferred into a network of design variables linked by constraints of different forms, it is more desirable in the integrated functional modelling system to combine systematic search methods with heuristic reasoning methods. In such a system the designer's input is monitored by the control system and the system infers as much as possible whenever the designer makes any changes to the constraint network. This process is supported using a blackboard control system.

A blackboard model represents a highly structured control scheme, and a special case of opportunistic problem solving. It has a simple architecture consisting of a number of knowledge sources, a blackboard data structure and a control system which controls knowledge sources on an opportunistic basis. The working memory of the system is viewed as a blackboard where the communication between different independent knowledge sources takes place. The blackboard provides globally available data structures for all the knowledge sources and controls them in a systematic way. Opportunistic control strategy of the blackboard system is probably most appropriate for a design support system [5, 6]. Such a control strategy contributes to design support tasks in the following ways:

- it allows a design support system to work as a self-organising system whose problem solving knowledge sources can respond dynamically and differently to design situations;

- it encourages the development of independent and self-contained design knowledge sources, thus making it easy to integrate design knowledge of different forms with the other parts of a design support system; and

- it has a control mechanism that is suitable for different knowledge representation schemes, ie, any thing that can be treated as a black box such as a rule set, an internal message passing handler, or an external software package, as long as it is self-contained.

The blackboard control system consists of an agenda and a number of knowledge sources. The agenda contains all the design variables to be determined. Whenever a component is included in the instantiated product model, the variables in that component are put by the system onto the blackboard agenda. A variable remains on the blackboard agenda until its value has been decided. Whenever the system receives an input from the user, the system verifies this input and then take the necessary actions. In order for the system to do so, user actions in this stage are classified into:

1. add a new design variable,

2. add a new constraint concerning several variables,

3. specify the value of a variable,

4. delete a variable,

5. delete a constraint

The first action only adds a new agenda item onto the blackboard agenda and triggers no system's action. The second action triggers the system to check whether any values can be derived for the variables concerned in the newly added constraint. In general adding a constraint contributes to tightening the design solution space and reducing the number of variables to be solved. The third action also triggers the system to carry out constraint propagation during which the newly added value of a variable is propagated throughout the constraint network. The deletion of a variable or a constraint also results in serious work of the system in retreating the results that have been derived based on the deleted variable or constraints. These five user actions and the relevant blackboard inferences reflect an interactive process in which the system actively supports a designer in trying to solve a constraint-based design problem.

The system also explains, at any point of the design process, how variables and constraints are related, what are the unsolved variables and remaining constraints, and what are the likely consequences if the value of one of the remaining variable is specified or changed. In this way, the system helps designers to known what has been derived, what are the remaining problems, and what to do next. Throughout this process the designers still play a decisive role in exploring the design concept. The blackboard control system actively supports this decision making by regularly checking the blackboard agenda and then taking the necessary actions.

If this process does not result in all the variable being solved, then the remaining constraint network can be transferred into a search problem. Simulated annealing methods or genetic algorithm can then be used to search the solution space by defining a utility function based on the number of constraints that are still being violated [7]. With the support of the blackboard controlled constraint management system, the search space will be more confined than it is originally defined by the instantiated product model.

3.4 Design knowledge sources

Design knowledge sources in the integrated functional modelling system are self-contained programs controlled by the blackboard control system. These design knowledge sources perform inferences and return results which are subsequently managed by the blackboard control system in a design document. An important feature of them is that they share a common data structure on the blackboard and they act opportunistically and independently. Each design knowledge source consists of: a preconditions part which decides when it should get invoked; and an action part that actually performs a specified task.

In the first version of the integrated functional modelling system, the following design knowledge sources are included.

1. a rule-based knowledge source performing functional synthesis based on Charkarabarti's method [2]. This knowledge source defines generic element of conceptual design in terms of name, input, output, and type of transformation. The type of transformation of an element determines how an element transfers an

orientation or sense input to output. In this design knowledge source, knowledge about these transformations are represented as 84 production rules.

2. a constraint manager performing constraint-based reasoning for embodiment design purposes. This design knowledge sources creates a constraint network from an instantiated product model, and then simplifies and evaluates it. This design knowledge source also propagates any changes made by the user throughout the constraint network. In the process of embodiment design it acts as a main inference engine.

3. a simulated annealing program performing constraint-based search. This design knowledge source is being developed by modifying the program used in CADET so that: the control factors of the annealing algorithm can be adjusted by the user if necessary; and more complicated constraints can be dealt with.

4. a genetic algorithm program performing automatic search for feasible solutions in a potentially huge space of possible designs. This is used as an alternative to knowledge source 3 and it is a Lisp version of the genetic algorithms used in CADET.

5. a kinematic analysis program performing kinematic analysis of any design proposals returned by any of the above design knowledge sources. This design knowledge source is based on a matrix reduction method developed by Johnson in Cambridge University [4].

These design knowledge sources are integrated with the blackboard control system and a graphical user interface to complete a design session when an initial design requirement description in terms of input/output function is given by a designer.

4 IMPLEMENTATION

One of the major concerns of the implementation is to combine rule-based and object-oriented representations to represent and reason about product models of different complexity. The use of an AI-based tool ensures that quick prototyping can be achieved to provide feedback to the architecture and design knowledge source design. An initial prototype of the integrated functional modelling system has been implemented using a knowledge-based tool called GoldWorks III on a SUN SPARC station 5 running Solaries.

GoldWorks III is a Lisp-based tool for developing knowledge-based applications. It supports both rule-based and object-oriented programming. It supports knowledge representation via eight different objects, ie., *frame, instance, assertion, rule* (forward and backward chaining), *attempt* (backward chaining), *agenda item, relation,* and *sponsor*. It has an object-oriented graphic system for developing graphical user interfaces. GoldWorks III Sun version runs on top of Lucid Common Lisp which has interface to C or C++ applications.

Work is currently being focused on the consolidation of the above discussed design knowledge sources to achieve a high degree generality in mechanical engineering design. This initial prototype is to be further developed in order it to be connected with the knowledge base developed at the EDC's product modelling group. A knowledge base of mechanical components and reasoning methods have been developed by this group in C++. In order to combine the integrated functional modelling system and the knowledge base of the components. An interface between C++ and Lisp is being implemented.

5 CONCLUSIONS

Future design applications are likely to be supported by sophisticated design support system tools that employ the best available AI techniques. The integration of a blackboard inferencing control system, an ATMS-based design context management system, a design documentation system and a graphical explanation system forms the core of an intelligent design support system. The development of four domain independent inference engines enables it to be used as a tool to support constraint-based design applications.

Supporting the tasks in the early stage of the design process is always difficult and challenging. The integrated functional modelling system presented in this paper integrates a number of AI-based systems that supports knowledge representation and intelligent control, functional synthesis of design requirements, embodiment design as constraint satisfaction, and optimisation using simulated annealing or genetic algorithms. In the integrated functional modelling system, a more systematic and knowledge-based approach has been adopted in order to fully demonstrate the potentials of the current systems and to develop them further.

6 ACKNOWLEDGEMENTS

The work presented in this paper is currently being funded by the EPSRC and the author wishes to thank all EDC members, especially Dr. Nigel Ball, Dr. Amaresh Chakrabarti, and Dr. Tim Murdoch at the EDC for their support in the development of the presented system.

REFERENCES

[1] Ball, N. et al, 1992, "The Integrated Design Framework: Supporting the Design Process Using a Blackboard System", AI in Design, 92.

[2] Chakrabarti, A et al, 1994, "A Two-step Approach to Conceptual Design of Mechanical Device", AI in Design, 94.

[3] Hayes-Roth, B., 1985, "Blackboard Architecture for Control", Journal of Artificial Intelligence, 26:251-231, 1985.

[4] Johnson, A. L et al, 1993, "Modelling functionality in CAD: implications for product representation", in Proceedings of the 9th International Conference on Engineering Design, 1993.

[5] Smithers, T., Conkie, A., Doheny, J., Logan, B., Millington, K. and Tang, M., 1990, "Design as Intelligent Behaviour: An AI in Design Research Programme", International Journal of Artificial Intelligence in Engineering, 5.

[6] Smithers, T., Tang, M., Ross, P. and Tomes, N., 1993, "An Incremental Learning Approach for Indirect Drug Design", International Journal of Artificial Intelligence in Engineering, Special Issue on Design and Machine Learning, Vol. 8, 1993.

[7] Thornton, A., 1993, "Constraint Specification and Satisfaction in Embodiment Design", PhD Thesis, University of Cambridge, Department of Engineering, 1993.

[8] [de Kleer 1986] de Kleer, J., "An Assumption-based TMS", *Artificial Intelligence*, Vol. 28. 1986.

INTEGRATED PLATFORM FOR AI SUPPORT OF COMPLEX DESIGN (PART I): RAPID DEVELOPMENT OF SCHEMES FROM FIRST PRINCIPLES

R H Bracewell, R V Chaplin, P M Langdon, M Li, V K Oh, J E E Sharpe & X T Yan

1. INTRODUCTION

The engineering design process typically entails the construction of a description of an artefact that satisfies a formal (in most cases) functional specification, meets certain performance requirements and resource constraints; is realisable in a target technology; satisfies one or more criteria, for instance manufacturability, reliability, safety and testability; and, more recently, environmental friendliness. The common perception of engineering design is often taken as that of converting a need - expressed as an abstract concept in terms of general functionality - into a product fulfilling that need, and its process involving the mapping of a specified function onto a realisable physical structure - the designed product.

The engineering design process is, of necessity, a complex one that requires the design engineer to exercise initiative and inventiveness as well as deploy a wide range of skills and expertise in attaining a solution. In an interdisciplinary design environment, such as one involving a mechatronics[2] approach to design, the designer is also often required to function in a generalist mode with an eye on the possibilities of using a wide range of technologies. The increasing occurrence of interdisciplinary product development has not only removed many of the traditional constraints to design but has now given the designer a much wider freedom of choice as to the best solution to a particular design problem. However, given the time constraints frequently imposed on present day product designers as a result of the constant demand to reduce the time-to-market factor in product development, designers are often not able to develop and assess rapidly the various alternatives that may be made available through the use of interdisciplinary design.

There is, moreover, the problem of the "islands of knowledge" syndrome. Currently, much of engineering design is carried out by individuals or groups of individuals who have been trained in one engineering discipline with limited knowledge or experience of other fields of engineering. Consequently, these individuals, during design, often "lock" onto traditional and familiar technologies that may result in a less than optimal solution to their design problem. Logically, there are two obvious ways to enable interdisciplinary product design: (1) educate all designers with a generalists' knowledge of many technologies, and (2) co-locate technology specialists to encourage communication and co-operation in the design team.

With the advances made in computer-aided design and the infusion of AI outgrowths like functional reasoning, constraint processing and planning into traditional computer-based design tools, it is not difficult to see how a third alternative to the above approaches might

---

[2] Mechatronics may be considered as the integration, at all levels and throughout the design process, of mechanical engineering with electronics and computing technology to form both functional interaction and spatial integration in components, modules, products and systems. Common examples of products that are the result of a mechatronics approach to design are camcorders and auto-focus cameras.

arise - develop a computer system to support and guide the designer through the range of technological options available, aided with the appropriate tools to perform processing of the design data at the proper level for design abstraction, concept generation, technology selection and matching, model execution, optimisation, and visualisation. Unfortunately, there is a paucity of such systems available at present and the aim of the research project at the Lancaster University Engineering Design Centre is to research the design methodology required and embed this into such a system. In this we have achieved in developing a set of tools that provides highly integrated support for the rapid creation and evaluation of a wide range of outline schemes incorporating a range of technologies and to enable the comparison of technological alternatives to take place before large commitments are made and irrevocable decisions taken on the basis of biased information.

This paper - comprising two parts - describes an integrated suite of software systems called Schemebuilder and emphasises mainly the integrated set of software tools used in converting a qualitative description of a scheme into a quantitatively defined one whose performance can then be simulated and whose general layout drawn. This part of the paper, the first of two, provides an overview of Schemebuilder and shows how qualitative conceptual schemes and their alternatives can be created from first principles via the co-operative interaction of human and computer. In part two we will show how Schemebuilder is used to "firm up" or embody the conceptual schemes developed by the process described in this part of the paper and describe the computer-based mechanisms employed in achieving the embodiment process.

2. THE SCHEMEBUILDER PROJECT

2.1. Aims of Schemebuilder

The main remit for Schemebuilder is to assist the designer in the conceptual and embodiment stages of design, including problem analysis, for interdisciplinary systems, mechatronics being the main emphasis. The design of mechatronic systems, which combines mechanical, electrical, electronic and software systems, presents a challenging array of possible implementations, often based on proprietary equipment but with opportunities for elegant and original design solutions that require the deployment of advanced electronics and software techniques to provide the required levels of functionality. To take full advantage of these opportunities requires the ability to generate and evaluate quickly alternatives schemes from first principles to practical embodiment.

It is intended that Schemebuilder be a kind of "design workbench", where designers are guided through the range of technological options available, aided by the provision of tools for rapidly accessing relevant information. It, therefore, takes on a role as facilitator for the creation of a model of the system to be designed, and advisor on the appropriate means to achieve the design. More importantly it aims to facilitate the exploration of alternative conceptual *schemes* with an appropriate allocation of function between mechanical, electronic and software elements. Our definition of a scheme follows that of French [1] where a scheme is an "*outline of a solution to a design problem, carried to a point where the means of performing each major function has been fixed, as have the spatial and structural relationships of the principal components. A scheme should be sufficiently worked out in detail for it to be possible to supply approximate costs, weights, and overall dimensions, and the feasibility should have been assured as far as circumstances allow. A scheme should be relatively explicit about special features or components but need not go into much detail....*".

It must be emphasised that the philosophy underlying our work is a move from automated design using the expert systems approach [2] to one embracing a more co-operative relationship between man and machine as underlined by the work of Fischer [3] and described in [4]. We believe that a more appropriate approach to computer-based design tools is for the computer to provide decision support and allow the human designer to apply the judgement.

2.2. The Schemebuilder Environment

Schemebuilder provides an integrated assortment of design tools matched according to the specific needs of different design tasks. The provision of verification tools like those for simulation in the Schemebuilder environment (refer to figure 1) helps to tighten the link between design and analysis, and enables designers to predict the behaviour and performance of their digital "mock-ups" as early as possible. In short, Schemebuilder allows for the exploration of new ideas with minimal risks.

The Schemebuilder environment provides enabling tools to:
- assist in the production of specifications
- provide advice and browsing facilities on available technologies
- aid in the identification of 'technological holes'
- produce system models for dynamic simulation and parameter optimisation
- support embedded software design and development
- monitor system integrity, checking for continuity and matching
- produce preliminary layouts in the form of realistically sized 'solid sketches'
- design and model casings and support structures

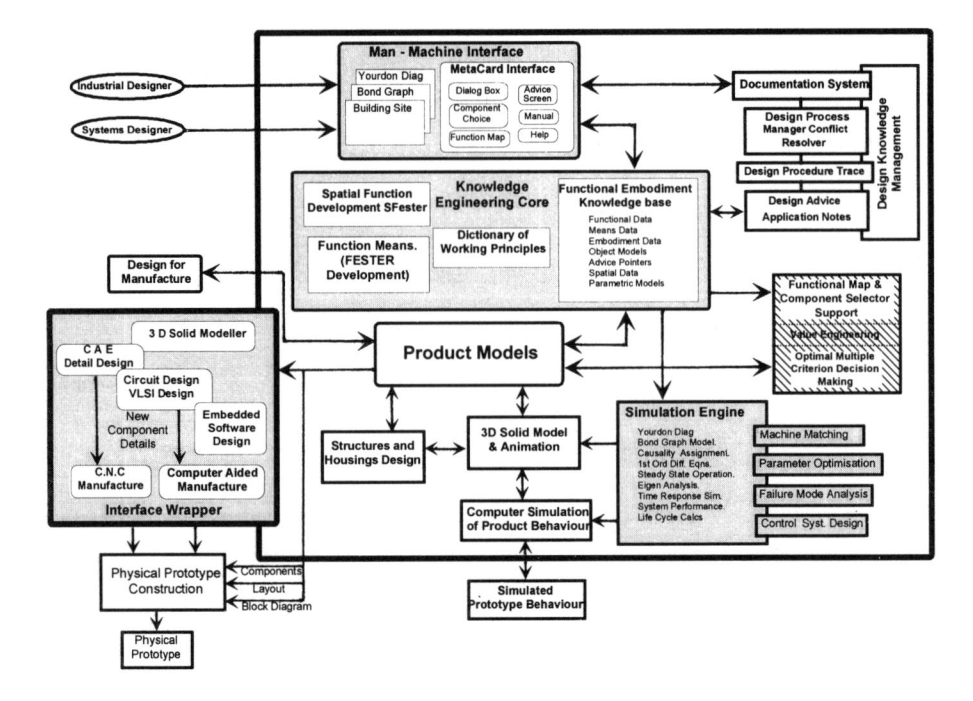

Figure 1. Overall architecture of the Schemebuilder environment.

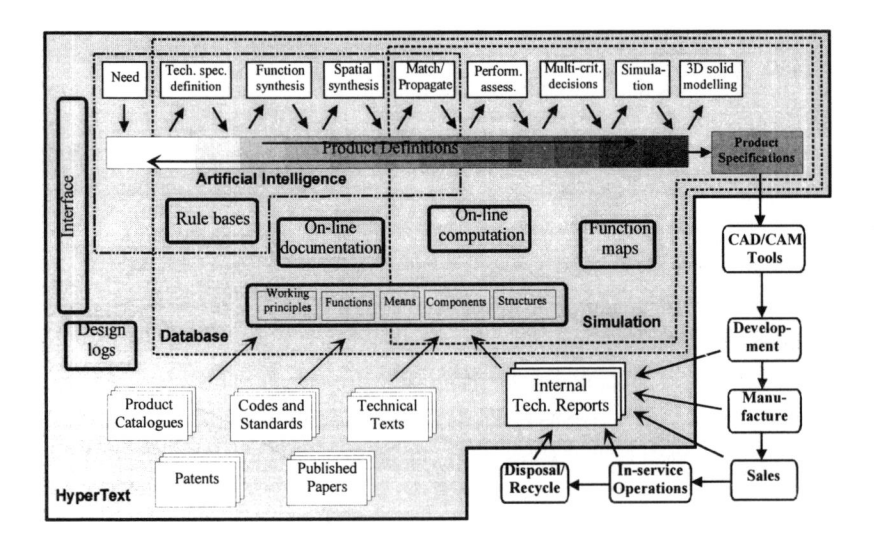

Figure 2. Schemebuilder conceptual design process.

2.3. Bond Graphs

As products become more complex and highly integrated, design teams will find it increasingly necessary to have a common language, independent of traditional engineering disciplines, in which to communicate. The use of *bond graphs* [7], [8] and [9], with its common structure and clear rules across engineering domains, for Schemebuilder's representation of the functional and behavioural aspects of energetic systems allows a very natural approach to design and allows the development of an integrated suite of object and rule-based computer aided conceptual design tools.

Bond graphs are a formal representational language for analysing physical systems developed for system dynamics. They have been used quite successfully in the representation of formal models for mechanical design, for example [10] and [11]. There has also been an increasing interest in the use of bond graphs for qualitative modelling in AI because of their simplicity and representational power [12], [13] and [14].

The modelling process is based upon the conservation of energy, with physical processes being linked in a labelled digraph through energy flows. Basic concepts like effort, flow, inertia, and capacitance are used in modelling, and their generality allow the method to be applied across domains like thermodynamics, rotational and translational mechanics, fluid dynamics and electronics. Thus, bond graphs cover essentially all classical macrophysical domains in an integrated fashion. Once a bond graph has been constructed for a physical system, traditionally one can generate a block diagram representation of the full differential equations that describe the system and then proceed with formal analysis. The bond graph can also support a useful set of qualitative inferences, including an analysis of the causal ordering that occurs within a system.

3. COMPUTER-AIDED METHODOLOGY FOR QUALITATIVE DEVELOPMENT OF SCHEMES FROM FIRST PRINCIPLES

3.1. Scheme Generation Conceptual Stage

The Schemebuilder design workbench described previously in [5, 15] has been enhanced so that schemes are developed using a variant of the Function Means Tree approach of Buur [16], guided by a structural set of design rules and working principles. The Function Means Tree approach, aided by the application of stored, rule-based conceptual design principles, allows for the progressive refinement and development of a design from required functions expressed abstractly to the eventual physical embodiment of those desired functions. Backed by an Assumption Truth Maintenance System [17], it is also possible for the tree structure to maintain the development of all alternative schemes, even though the designer may only progress a limited number of them at any one time. figure 2 illustrates conceptually the process of generating from a statement of need a set of alternative schemes, and the assessment of these schemes.

Although it is important to understand the mechanism whereby a design is developed from first principles, the development of a conceptual design may begin at a number of different starting points. Any satisfactory computer aided procedure must allow for this.

3.2. Scheme Generated from Experience

The designer may choose from past experience of a satisfactory similar design to use an established design as a starting point. Such a scheme may be laid out on the *Building Site* as a set of interconnected blocks representing existing types of components. The primary functions of those blocks together form an established working principle, with each component, represented by a block, being an embodiment of some working principle which then further defines another set of primary functions for that component. figure 3 shows such a block diagram for a metabolyte infuser, which will be our illustrated design exemplar in this paper.

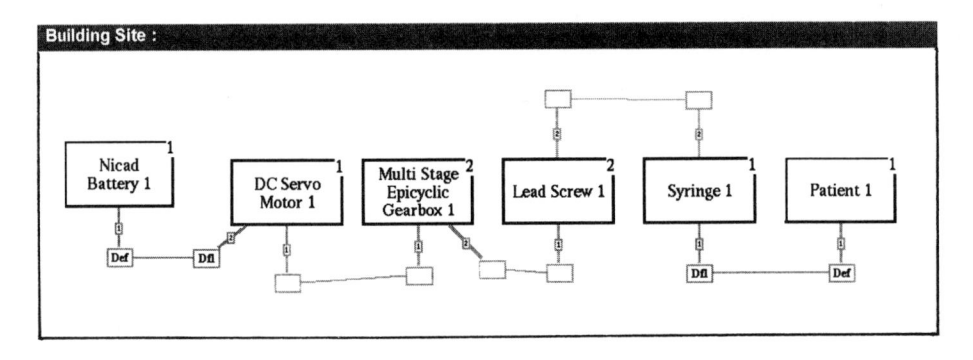

Figure 3. Block diagram of metabolyte infuser.

It may be of no consequence to the designer to know the nature of the primary function performed by the design. The designers concern is that the concept works and that it may be matched to the desired specification and designed in detail.

3.3. Scheme Generation from First Principles

Frequently, however, the designer wishes to develop a conceptual design from first principles, exploring the myriad of alternative schemes that are physically possible before choosing the best alternatives to pursue into detail. Such a process will begin with some 'expression of need' combining normal language and related numerical information. For example, the 'expression of need' for the metabolyte infuser could be defined as follows:

"The need is for a light weight portable infuser capable of providing a continuous constant flow of drug into a patient's artery for up to 48 hours. It must provide an indication of correct functioning. The flow rate is to be 2.5ml per day."

Employing techniques from AI, such as natural language processing and information retrieval in the implementation and processing of dictionaries, thesauruses and grammar, it is possible with the help of the designer, by means of prompts and iteration, to establish in Yourdon[3]-like terms, the context of the design, and the desired function. Furthermore, it is possible to use a thesaurus to search a dictionary of working principles to match those words appearing in the functional aspect of the desired need with the key words describing the application of one or more principles. We acknowledge that such an approach, although sufficient for our purposes at present, only maps surface features inherent in the requirements statement to the store of working principles and does not relate to deeper epistemological concerns of the structure of the working principles. For this we are considering the work and advances made in the area of case-based reasoning for augmenting the retrieval and storage of the working principles, which may be viewed as cases in the sense used in the case-based reasoning literature.

3.4. Functions, Means and Working Principles

3.4.1. Nomenclature Definition. A *function* can be classified as either a *data function* or an *energy function*. All functions which are used, must be pre-defined in the Function Embodiment Knowledge-base. Functions may have one or more defined attributes. Functions communicate with each other and with components via links.

The *FEST-ER* stands for **F**unctional **E**mbodiment **ST**ructure - **E**xtended **R**ecursively. It is a new information structure extending the capabilities of the Function-Means Trees of Andreasen [19]. The FEST-ER is developed by the designer's free choice of stored functional embodiments, which may be of either two forms: means or principles.

A *link* may carry energy and/or data, In a completed scheme, both ends of every link must be connected to a component within the scheme. The link may traverse up and down a number of levels within the FEST-ER, between its terminating components. Each link in the FEST-ER of a particular design project has a unique identifying number.

An *attribute* more precisely specifies the action of a function.

A *function embodiment* can consist of either a means or a principle.

---

[3] The Yourdon methodology is a popular structured technique commonly used in the analysis and design of systems, particularly software systems [18]

A *means* consists of at least one component and if necessary, one or more required subfunctions, which may have certain required attributes. Links joining components and required subfunctions within the means must be defined.

A *principle* consists of one or more required subfunctions, which may have certain required attributes. Links joining required subfunctions within the principle must be defined.

A *component* represents a family of actual, available physical items, for which quantitative information is stored in the Component Database, allowing sizing, matching layout and simulation to be performed later in the design process. Components possess ports which form the end-points of links, allowing input and output of energy and/or data from the component.

3.4.2. The Functional Embodiment Knowledge-base. The Functional Embodiment Knowledge-base is a tree structure of functions, in which energy flow types and data carrier types are progressively defined as you approach the levels of the tree. This allows principles to be stored at a high level of abstraction, with energy types undefined, then inherited into all appropriate energy regimes. A crucial aspect of the structure is the ability of single component types from the component database to appear in multiple means, embodying different functions at different points in the tree. This reflects the fact that components often are used in a number of alternative and sometimes unusual ways. A simple example of this is shown in figure 4, where it can be seen that the component "chassis" appears in means of embodying both functions CFJ.ROT and CFJ.TRAN, which are rotational and translational common flow junctions, respectively.

3.4.3. Working Principles and their Application. Consider the case of the *need* for a flow of chilled water at $5^{\circ}C$ for purpose of cooling power electronics. The flow is to be 1 kg per second within a closed circuit. The return water temperature is $15^{\circ}C$. Power is to be taken from a variable frequency voltage source.

Considering the word 'chilled', a search of the Dictionary of Working Principles identifies a group of Working Principles relating to refrigeration (refer to figure 5). These include: Air Standard Cycle; Absorption Cycle; Vapour Recompression; Evaporation; Seebeck Effect; Peltier Effect and the Joule-Thompson Cooler. Each of these alternatives of working principles has a required set of primary functions that may be satisfied by different means and these functions, together with their interconnections, are held in a related database. Thus for the chosen principle to be embodied, the functions are drawn on the highest level of the building site and placed at the head of the FEST-ER together with its context, in this case a closed circuit and the high temperature product stream.

For the refrigeration application the choice of the best working principle is not clear and therefore requires recourse to background information on the application of the different principles. This is provided through on-line hypertext links to appropriate handbooks or application notes and functional performance maps held in MetaCard, a hypermedia development environment.

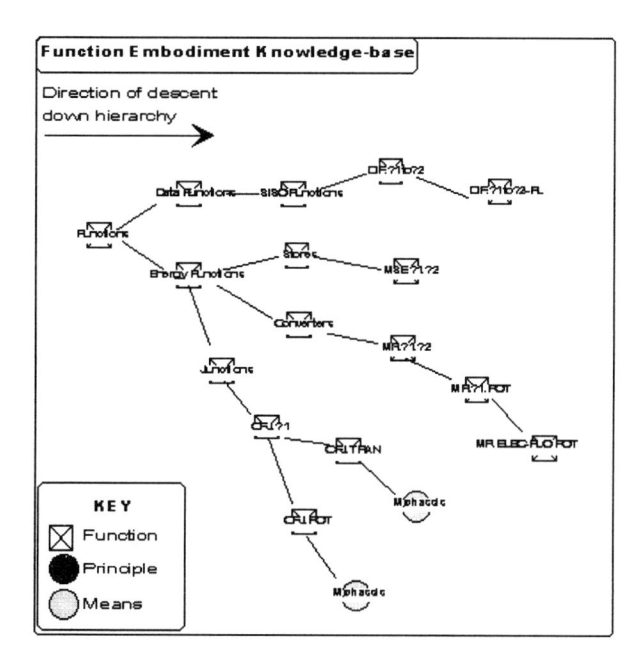

Figure 4. A example of the Functional Embodiment Knowledge-base.

Having chosen one or more probable applicable working principles it is necessary to determine the means of achieving the required functions. This is performed by searching the available means to achieve the function or alternatively decomposing the required function according to a set of rules, the Rules of Decomposition, and searching for available means of performing the generated sub-functions.

Each means may take a number of different physical forms depending on the working principle employed. Thus we find that if we choose the vapour recompression cycle embodiment of the required refrigeration means, the gas compression function may be achieved by different means and each of these means may itself have different embodiments depending on the working principle used, this may be a Reciprocating Piston, Rotary Vane, Screw Type or many others.

Suppose the designer chooses the Screw Type, this will require a number of functions. One of which will be to synchronise the rotation of the screw elements. There will be different means of achieving this, such as Gear Means, Toothed Belts, or Electronically Phase Locked Rotation. For each of these means there will be alternative embodiments. The gears may be helical or spur and will in turn lead to a further set of required functions which in turn lead to further required means and embodiments of working principles.

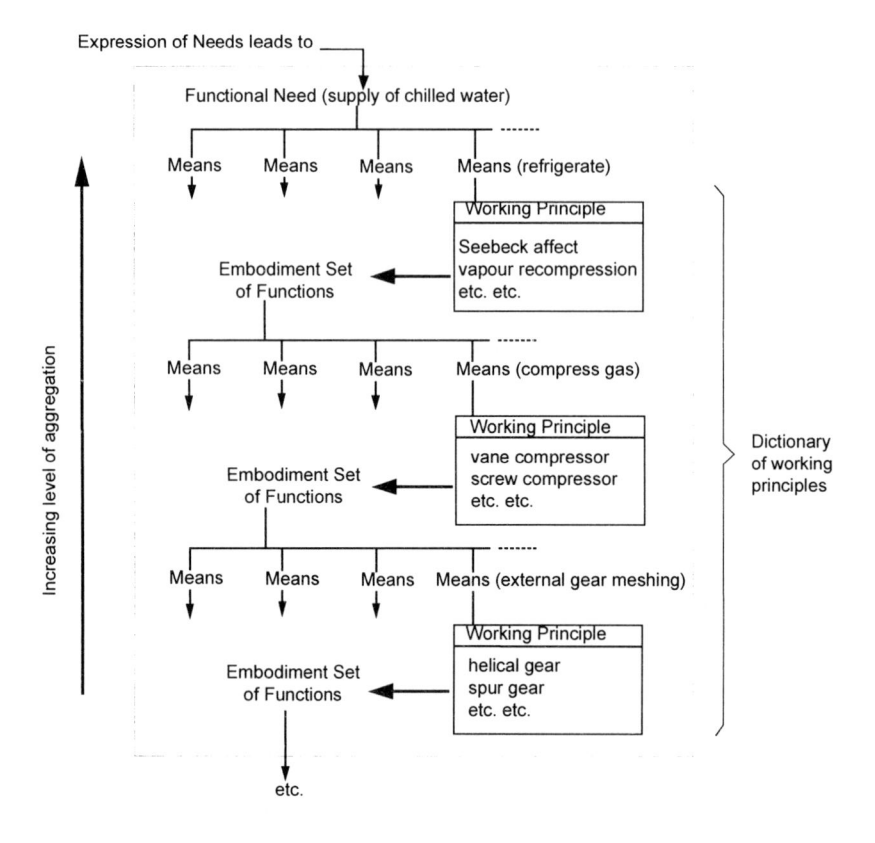

Figure 5. Working principles in qualitative development of a scheme.

In this example, there are some four levels of working principles, as indicated in figure 5. For a complex product there may be seven or eight levels of working principles and their embodiments, each requiring its own function means tree, and giving rise to a multiple level hierarchy of Means, Principles, Embodiments and Functions.

Spatial aspects of the functions, links to the environment and the need to provide physical structure may occur at any level in the hierarchy. Thus spatial principles of structures and kinematics may be invoked at any point. The starting part of the conceptual design decomposition is not restricted to any particular level of aggregation.

3.5. Bond Graph-based Principles

Bond graphs[8] provide a unique set of principles for the embodiment of energetic systems. These principles are based on the facts that only compatible energy ports may be connected to each other, that the causality and the dependency of the energy covariables must be correctly propagated and that there are a limited set of primary functions that may be used. These include the transformer (TF), the Gyrator (GY), the REsistor (R), the Compliant (C) and Inertial (I) energy stores, the Common Effort (0) and Common Flow (1) Junctions and Sources of Flow (SF) and Effort (SE). Each has a clear set of principles governing its function and how it may be decomposed whilst still performing the same overall primary function[20].

The following is a partial list of these Decomposition rules:
(i)   Search Function Means Database for Any Known Means;
(ii)   Insert Transformer to provide intermediate energy domain;
(iii)   Replace Source of Effort with a Compliant Store;
(iv)   Replace Source of Flow with an Inertial Store;
(v)   Replace Source of Effort with Source of Flow + Conflict Resolver[CRES];
(vi)   Replace Source of Flow with Source of Effort + CRES;
(vii)   Replace Inertial Store with Compliant Store + CRES;
(viii) Replace Compliant Store with Inertial Store + CRES;
(ix)   Choose CRES;

The designer is required to acknowledge the development of the FEST-ER by closing down branches that go nowhere or are currently of little interest. The decomposition may be taken to any level, although to go beyond 4 stages of decomposition is normally unnecessary within system design. The ability to split a function and introduce an intermediate energy domain allows the coupling of domains that are not normally possible. For example, it is not possible to transfer directly energy between electrical domain and the incompressible fluid domain, which will be illustrated in the metabolite infuser example where it is necessary to use either a translational or rotational stage as an intermediary.

The first decomposition rule is to see if the required function already exists as a primary functional means or as an embodied means based on some working principle, [existing design]. The FEST-ER is complete when the designer is satisfied that there are a suitable number of complete schemes of embodied means to choose from and to take into the next phase of the design process, the component choice, simulation and layout.

3.6. The FEST-ER

The FEST-ER differs from the well-known Function-Means Tree in that it supports two forms of non-hierarchical links that are very important for the efficient development and representation of competitive designs.

The first of these is the Embodiment Function Reference, which is used when a function, which had already been embodied at some point in the FEST-ER, occurs again in a location which is non-mergeable with the original occurrence. Instead of embodying the function all over again, an Embodiment Function Reference can be made, reusing the original work. There are many examples of this in figure 9, where reference functions are surrounded by heavy boxes and references to them by light ones.

The second form of non-hierarchical link occurs where a single means is, when appropriate, allowed to embody more than one function, which can often be the key to a more competitive design. This type of link appears in the Fin Actuator FEST-ER, where for example the microprocessor means and the absolute angular grey-scale encoder means both have a required subfunction SE.ELEC, an electrical effort source. A single nicad battery means is in this case able to embody both of these functions.

This technique is also the key to representing ASIC hardware code or embedded microprocessor software means, for implementing data processing functions. The Fin Actuator shows this as well, where a number of different data functions from widespread parts of the FEST-ER, are all mapped onto either a single microprocessor means or a single ASIC means. Once this mapping has been made, FEST-ER contains a complete Yourdon - style structured analysis of the information processing required of the microprocessor or ASIC.

If the individual data functions used, are stored in the Functional Embodiment Knowledge-base with their representation as High-level Petri Net fragments[21], then a complete net describing the desired functionality can be automatically assembled. The resulting net, if exported to the commercial software tool "SystemSpecs", will then allow automatic generation of simulation or implementation code, in C++, Occam or VHDL[22].

The FEST-ER information structure will equally support alternative embodiments of the same functionality in say analogue electronics or even a special mechanical linkage, thus allowing trade-offs of moving solutions either way across the "mechatronic interface" to be easily examined.

4. APPLICATION OF FUNCTION-MEANS DEVELOPMENT

4.1. Initial Function-Means Development

Having shown how the function means embodiment decomposition fits into a hierarchy of working principles and the features of the FEST-ER, attention will be focused on the development of a FEST-ER for the metabolite infuser example, which only requires one level of aggregation. This example may be decomposed into its alternative constituent components from a single functional requirement, a constant source of flow of an incompressible fluid, shown diagrammatically as SF.IFLU, within one level of the hierarchy (figure 6a). Figure 6b shows the further development of the FEST-ER from the single functional requirement of SF.IFLU, using the set of decomposition rules for energetic, bond graph systems.

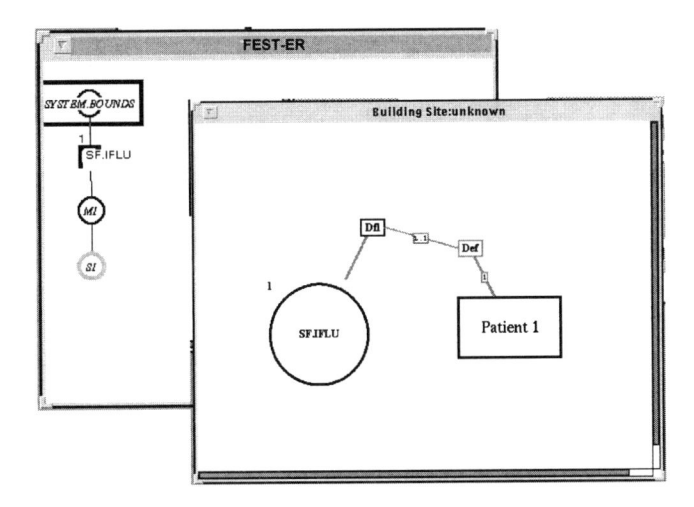

Figure 6a. Initial development of the FEST-ER.

The use of Bond Graph reasoning provides a unique set of rules for the decomposition of energetic systems. These rules are based on the facts that only compatible energy ports may be connected to each other, that the causality and the dependency of the energy covariables must be correctly propagated and that there are a limited set of primary functions that may be used.

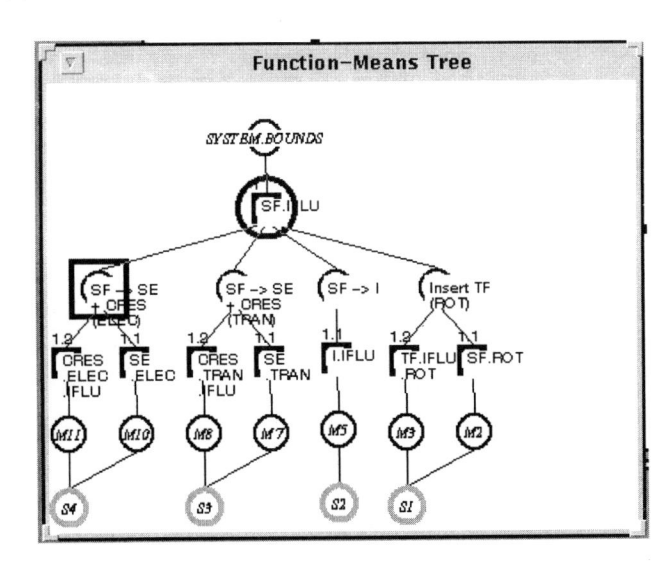

Figure 6b. Further development of the FEST-ER.

A Transformer, TF, may be replaced by two Transformers in series or any number of Transformers in series or two or any even number of Gyrators in Series. However, a Gyrator may only be replaced by an odd number of gyrators in series with any number of transformers.

This ability to split a function and introduce an intermediate energy domain allows the coupling of domains that are not normally possible. For example, it is not possible to transfer directly energy between electrical domain and the hydrostatic domain, thus we will see, in the example, that it is necessary to use either a linear or rotary movement as an intermediary. These intermediaries give rise to the development of the FEST-ER and the identification of the many alternative means of achieving a desired function.

In addition to the decomposition of the Transformer and Gyrator given above, in which there is no theoretical loss of energy, it is possible to maintain the same overall causality by using a dissipating element such as a Resistor to achieve the desired causality. This is known as a Causal conflict RESolver, CRES, which is used in the development of a FEST-ER shown in figure 6b.

4.2. The Complete FEST-ER for Metabolite Infuser Example

A good example of the use of Bond Graph-based principles is the Metabolite Infuser. The FEST-ER for the metabolyte infuser is shown in figure 7. This shows the total development of the alternative designs including the points at which the designer has chosen not to proceed. Any of these branches may be revisited at any time if required. The procedure begins with the definition from the expression of need of the desired function and its context. This is shown diagrammatically at the top of the FEST-ER structure.

In this case the required function is a Source of Flow in the incompressible fluid domain, SF.IFLU. The application of the first rule to search the Function Embodiment Knowledge-base returns 'no known means'. The insertion of a Transformer [TF] to give an alternative intermediate energy domain shows that, for the intermediate rotation, there are several alternative transformers from rotary motion to incompressible fluid flow TF.IFLU.ROT but that there are no matching sources of Rotation. Replacing the Source of Flow by an Inertial Energy Store yields 'no known means'. Replacing the Source of Flow by a Source of Effort or Compliant Store and a Conflict Resolver, which may be a Resistor or a Gyrator, yields a whole array of possible alternative energy domains and means. The alternative energy domains are electrical, rotation and translation, two stages of rotation, or two stages of translation.

Each transformation or gyration between the domains has several different means, which may give rise to some 20 alternative embodiments that have the potential to perform the desired primary function. Each represents a working principle and as such may be added to the dictionary of working principles. Each scheme represents the minimum functional structure and is at this stage purely qualitative.

4.3. Iterative Relationships of Function Means Development

The schemes presented from the Function Means decomposition represent assemblies of embodied means and therefore are automatically represented in Bond Graph form. They do not contain any knowledge about the spatial arrangement of the designs unless it has been specifically included in the definition of the desired function. As the chosen scheme is laid out in the 3D solid model it will be necessary to introduce additional functions, principally spatial transformers, to couple the differing functional axes (offset axes X1 and X2) of the primary functional components. This will introduce additional functional needs into the chosen scheme as shown in figure 7, in which the primary functional components and the initial context form the context for the development of the function means for the additional coupler.

FESTER

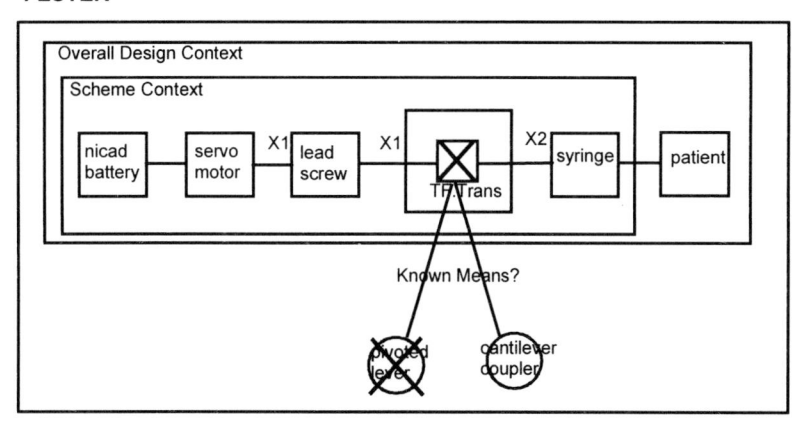

Figure 7. Function means interaction with functional axis.

The application of the decomposition rules leads to the choice of a cantilever slide or a pivoted lever. With the choice of the cantilever, the new modified scheme may be laid out in the building site.

As will be shown later the efficient matching of the DC motor to the load requires the introduction of a further rotary transformation as shown in figure 8. with common axes (X1) for input and output. A search reveals that the epicyclical and the multistage gearboxes are suitable means. The gear pair is not suitable because the input and output axes are offset. Thus there is a continued development of the FEST-ER and the alternative schemes as the detail matching and layout of the chosen schemes proceed.

The Bond Graph models generated for each scheme provide the basis for the simulation to be described later and, together with the defined functional axes, allow the forces and torques within the functioning scheme to be superimposed on the 3D solid sketch of the design.

FESTER

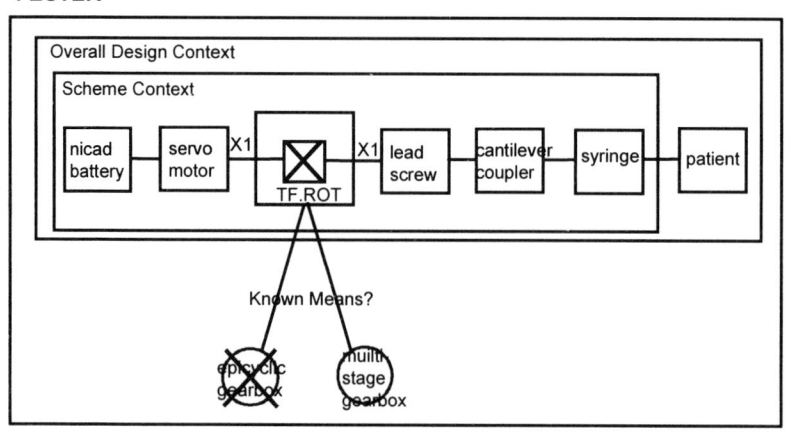

```
Overall Design Context
  Scheme Context
    [nicad battery]—[servo motor]—X1—[⊠ TF.ROT]—X1—[lead screw]—[cantilever coupler]—[syringe]—[patient]

    Known Means?
      (epicyclic gearbox) ✗    (multi stage gearbox)
```

Figure 8. Function means interaction with matching component.

5. CONCLUSIONS

Design is an integrated and integrating process during which the designer must call upon vast amounts of knowledge and experience in a systematic manner to develop the design of a product from first principles to a satisfactory prototype. It is a combination of rigorous intellectual discipline and creative thinking. In the development of the Schemebuilder environment, we have been very conscious of two important factors: (1) the need to allow the designer or designers if that be the case, to have total control over the design process at all times, allowing them to compare the alternatives in a clear and unambiguous way, and (2) to provide the underlying core structure of data and knowledge in such a way that the designer is not hindered in his actions and may freely develop new ideas and procedures whilst at the same time being equally free to revisit previous decisions whether in respect to the FEST-ER development or aspects of matching or simulation. The provision of a sophisticated audit trail of the designers action and the advice proffered or sought is important in this respect.

A further important aspect of the integrated set of AI tools is the manner in which the design process may be entered from a number of different points. This paper has largely followed French's model of conceptual design in describing how the different AI tools support the designer in the process. Practising designers often approach a problem not from first principles but using their experience. This may be an effective starting point especially if supported by good simulation and the ability to build and test rapid prototypes, which will be described in greater detail in part two.

The greatest importance of the work that has been described must lie in the manner in which it has been integrated to allow the complete design process to be undertaken in one homogenous computer environment. The example given throughout this paper has been of the design from 'first principles' of a portable metabolyte infuser for use with human subjects has demonstrated how the process may be used not only to ensure the creation of a number of workable alternative designs but how a design, once chosen, may be readily be constructed as a physical prototype from the list of chosen components, the block diagram of their interconnection, their physical layout and the spatial design of the special components and casing taken from the 3D solid model. This again is the subject of part 2 of the paper.

ACKNOWLEDGEMENTS

This paper represents a considerable effort over a period of years by a large number of individuals. The authors would like to thank Michael French, David Bradley, David Dawson and Martin Widden for their help and enthusiasm in this project, the Dept of Computer Science for the computational support and the Engineering Design Committee of the Science and Engineering Research Council for their financial support of the Design Centre.

REFERENCES

[1] French, M.J., *Conceptual Design for Engineers*, 2nd Ed, Design Council, London, 1985.

[2] Rychener, M.D., (Ed), *Expert Systems for Engineering Design*, Academic Press, Boston, 1988.

[3] Fischer, G., "Communication requirement for co-operative problem solving systems", *Journal of Information Systems*, Vol 15, No.1, 1990, pp 21-36.

[4] Oh, V., "Intelligent Design - Assistant Systems for Engineering Design", Technical Report EDC 1993/02, Lancaster University, Lancaster, 1993.

[5] Bracewell, R.H., Bradley, D.A., Chaplin, R.V., Langdon, P.M. and Sharpe, J.E.E., "Schemebuilder: A design aid for the conceptual stages of product design", *Proc 9th Int'l Conf on Engineering Design ICED'93*, The Hague, 1993, pp 1311-1318.

[6] Bradley, D.A., Bracewell, R.H. and Chaplin, R.V., "Engineering design and mechatronics - the Schemebuilder project", *Research in Engineering Design*, Vol 4, No 4, 1993, pp 241-248.

[7] Paynter, H.M., *Analysis and Design of Engineering Systems*, MIT Press, Cambridge, MA, 1961.

[8] Karnopp, D.C., Margolis, D.L. and Rosenberg, R.C., *System Dynamics: A Unified Approach*, 2nd Ed., Wiley, Chichester, 1990.

[9] Sharpe, J.E.E., "Bond graph synthesis of telechiries and robotics", *3rd CISM IFFTOMM Conf on Robotics and Manipulators*, Udine, Italy, 1978.

[10] Finger, S. and Rinderle, J.R., "A transformational approach to mechanical design using a bond graph grammar", *Design Theory and Methodology - DTM'89*, DE-Vol 17, W.H. Elmaraghy, (W.P. Seering and D.G. Ullman, Eds), 1989, pp 107-115.

[11] Ulrich, K.T. and Seering, W.P., "Synthesis of schematic descriptions in mechanical design", *Research in Engineering Design*, Vol 1, No 1, 1989, pp 3-18.

[12] Fishwick, P.A., "Qualitative methodology in simulation model engineering", *Simulation*, Vol 52, 1989, pp 95-101.

[13] Top, J. and Akkerman, H., "Computational and physical causality", *Proc 12th Int'l Conf on Artificial Intelligence*, Sydney, 1991, pp 1171-1176.

[14] Soderman, U. and Stromberg, J., "Combining qualitative and quantitative knowledge to generate models of physical systems", *Proc 12th Int'l Conf on Artificial Intelligence*, Sydney, 1991, pp 1158-1163.

[15] Bracewell, R.H. and Sharpe, J.E.E., "Application of bond graph methodology to concurrent conceptual design of interdisciplinary systems", *Conf IEEE Systems, Man and Cybernetics '93*, Le Touquet, Oct 1993.

[16] Burr, J., "A theoretical approach to Mechatronics design", *PhD Thesis*, Institute for Engineering Design, Technical University of Denmark, 1990.

[17] Filman, R.E., "Reasoning with worlds and truth maintenance in a knowledge-based programming environment", *Communications of the ACM*, Vol 31, No 4, 1988, pp 382-401.

[18] Yourdon, E., *Modern Structured Analysis*, Prentice-Hall, Englewood Cliffs, New Jersey, 1989

[19] Andreasen, M.M., "Syntesemetoder pa Systemgrundlag", *PhD Thesis*, Lund Technical University, Lund, Sweden, 1980.

[20] Sharpe, J.E.E., "Bond graph synthesis of telechirs and robots", *3rd CISM IFFTOMM Conf. on Robotics and Manipulators*, Udine, Italy, 1978.

[21] Jensen, K. and Rozenberg, G., (Eds), *High-level Petri Nets, Theory and Application*, Springer-Verlag, 1991.

[22] IvyTeam, *SystemSpecs 2.1 Reference Manual*, IvyTeam, Zug, Switzerland, 1993.

INTEGRATED PLATFORM FOR AI SUPPORT OF COMPLEX DESIGN (PART II): SUPPORTING THE EMBODIMENT PROCESS

R H Bracewell, R V Chaplin, P M Langdon, M Li, V K Oh, J E E Sharpe & X T Yan

1. INTRODUCTION

Engineering design is a process of generating a solution or a scheme to a design problem to satisfy a design requirement and its specification. Various processes and activities, such as concept generation, invention, realisation, evaluation, modification, refinement and detail specification and so forth are involved throughout the entire design. A generally well-accepted division of the early stages of engineering design can be expressed as conceptual design and embodiment design [1] (see figure 1). Conceptual design takes the statement of problem expressed in an abstract form and aims at generating broad working solutions to it in the form of schemes, whereas embodiment design concentrates on further specification of conceptual schemes in order to produce, on one hand, a set of detail specification drawings of components for a scheme and, on the other hand, to provide feedback and verification of the conceptual scheme. Although the design process is broadly divided into these two separate processes, overlapping of them is inevitable and encouraged in the context of concurrent engineering practice.

This paper describes the tools that are employed in translating a conceptual (or qualitative) scheme into a quantitatively defined embodiment scheme whose design parameters are mapped, performance simulated and general layout drawn with the help of the tools provided. The embodiment design process involves aspects of component function mapping and matching, simulation and 3-D modelling and layout design process, each of which is described in terms of its adopted methods, implementation and its use in the design process.

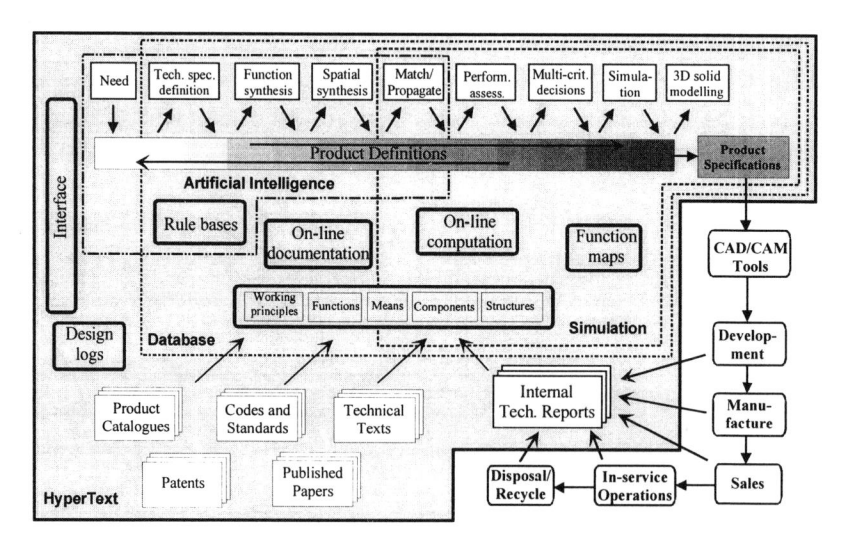

Figure 1. Schemebuilder's design process model.

2. THE EMBODIMENT DESIGN OF MECHATRONIC PRODUCT SCHEMES

The embodiment design is characterised by achieving the task of quantifying the important design parameters so that an optimum design scheme is obtained to fulfil the embodiment of a working principle from which a particular scheme is derived. It is very often necessary to evaluate several schemes or variants before a designer can make sensible judgement and comparison. At the embodiment stage, a comprehensive evaluation is made to the system function, spatial requirements, manufacturability, financial viability and overall design requirements. Embodiment design therefore involves various aspects of specification and evaluation and several computer aided supports are needed to enable a designer to carry out embodiment design effectively. In Schemebuilder [2,3], these requirements have been provided for by supporting tools that come largely under four major categories, these being component function mapping and matching, data serving, simulation and 3-D modelling and layout design tools, each of which is described in this paper.

3. FACILITIES FOR QUANTITATIVE DEFINITION OF SCHEMES EMBODIMENT STAGE

The quantitative definition of schemes essentially corresponds to the embodiment stage of design and the main facilities involved in enabling the designer to numerically quantify the schemes generated from the previous stage are discussed in the next few sub-sections. The typical processes involved at this stage are: (1) taking a product, component by component, at any stage in the overall Schemebuilder process and sizing them or selecting them so that their individual numerical data is defined sufficiently as to allow realistic simulation of that product; (2) production of layout drawings with leading dimensions given allowing estimates of cost and weight to be made. On the basis of these measures the designer is able to select the most suitable scheme or scheme-variant.

3.1. Dataserver

Each design primitive in a scheme is represented as a generic component object, which is an instance of a class of components, e.g. the class of dc-servomotors, that can physically realise the required function. Hence, if a generic dc-servomotor object, say "dc.servomotor.1" is the design primitive in a particular scheme, then "dc.servomotor.1" is an instantiation of the DC.SERVOMOTOR class and it represents the means by which to satisfy the required function "to convert electrical energy to rotary motion". Although "dc.servomotor.1" is in truth an instance of the DC.SERVOMOTOR class, it is not unique in the sense that it does not have specific values for its structural attributes, performance attributes, and so forth; "dc.servomotor.1" is not a specific instance. It is not a distinguishable object at this point in time and hence does not have a direct mapping between itself and the physically available object in the real world.

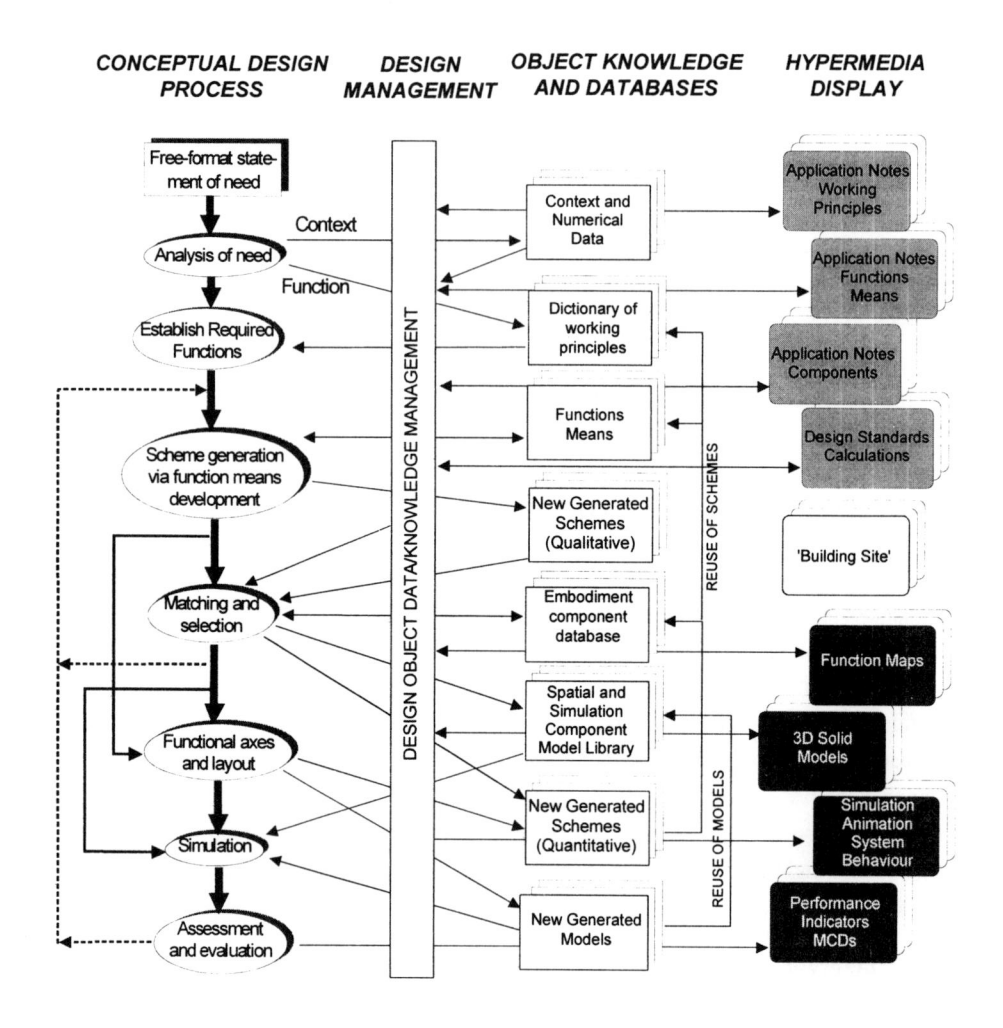

Figure 2. Schemebuilder's design process model.

However, each component class has an attached object called a *dataserver*, which acts as a data-providing agent to furnish the generic component object with real values in order to specify the object fully. Each dataserver contains a schema descriptor, parametric formulae, structural, functional, performance and cost data related to the component it is attached to. The structural, functional, performance and cost data are provided by a database, which is loaded at run-time and only when the dataserver is called upon to supply data. The database, structured as a one dimensional relational structure, stores component information taken from a manufacturer's catalogue and values for bond graph [4] constants.

The schema descriptor provides the mapping of labels to columns, e.g. "model.type" to column 1, "mechanical.friction" to column 2, "diameter" to column 10, and so on. It also provides further information like units that are associated with a particular attribute, e.g. "mNm" for mechanical friction. Mapping of attribute labels to parametric formulae is also provided by the schema descriptor.

When a particular model type is chosen, for example, from the Matching and Selection tool, all information, along with the pointer to that model, that is needed to specify fully the nature of the design component is transferred across to the generic component object. The generic component object is now said to be fully specified. The information can then be used to draw the physical representation of the component onto the Layout tool and provide more detailed information for simulation purposes. Some of the information transferred across to the component object is also taken up by the bond graph models.

3.2. Component Specification Matching

The facilities for matching implemented within Schemebuilder allow the user to proceed to a scheme which is sufficiently numerically defined to allow assessment by means of simulations and trial layouts. The facilities described below can be used in any sequence but are described in that of normal use.

3.2.1 Approximate Matching using the Dataserver. Tools associated with the Dataserver allow the user to select candidate components from the database of a component type such that they meet an inclusion set of performance-related requirements. The tool, moreover, allows the candidate components to be listed in order of least cost or weight, for example. In general the components are retrieved via SQL-like querying functions using any of the descriptive parameters in the database of that component.

3.2.2 Graphic Tool for Matching. Having selected sets of components as described above their respective suitabilities can now be examined. The graphic tool for matching shown in figure 3 displays the performance envelopes of the candidate components together with the working points derived from the Numerical Specifications in figure 2, which shows conceptually the process of generating alternative schemes, and the assessment of these schemes via the system supporting tools. These working points would represent a number of salient conditions e.g. continuous power, transient maximum torque at zero speed and the positions of these points on the respective envelopes of the candidate components would allow the user to judge their suitability and select one. The user would be guided to the relevant MetaCard advice.

The process described above would be repeated at each component-component boundary to propagate the specification requirements through the scheme. However, in order to do this the requirement has to be carried through each component. This is done using the bond graph constants stored in the component database: as a particular candidate component is added to (instantiated within) a scheme a demon object, specific to that component type, calculates the working points for the next component using the 'old' working points, the bond graph structure and data. (Thus taking account of internal losses in the component).

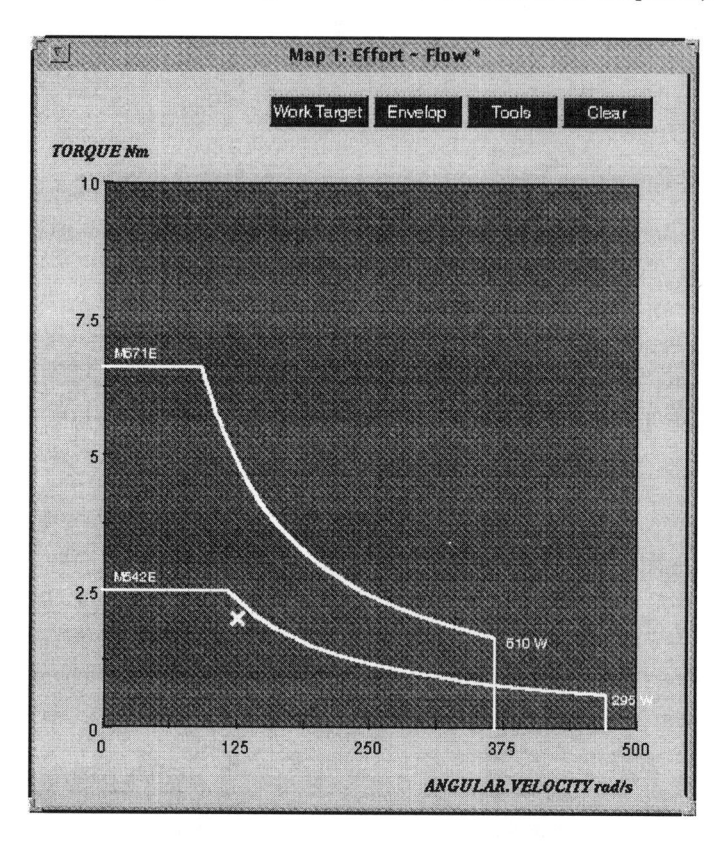

Figure 3. Graphical display of the matching tool.

3.2.3 Mismatches - Conflict Resolution.
In general, for a design to be economically viable, components have to be approximately matched in terms of power with individual components operating in their most efficient and economic regions of their respective operating envelopes. However, in the matching processes described above this may prove impossible unless a

transformer device is interposed, e.g. the gearbox between the leadscrew and the motor in the example case shown in figure 4.

The user will be advised to introduce these elements when such conflicts arise, either by consulting the Design Advice or by the system recognising and flagging gross mismatches in power, effort or flow capacity between neighbouring components.

3.3. Check calculations Facility

Particular components often have specific features that must be checked before they can be accepted for a scheme. In these cases the user would be offered advice to use such checks as part of the defining process. Specific check tools would include, in the case of lead screws for example, active formulae for the calculation of whirling speed; or in the case of an electric drive might demand the simulation of a motor in isolation, but under set conditions, and to assess its likely running temperature (using the simulation tool described below in Section 3.5).

FESTER

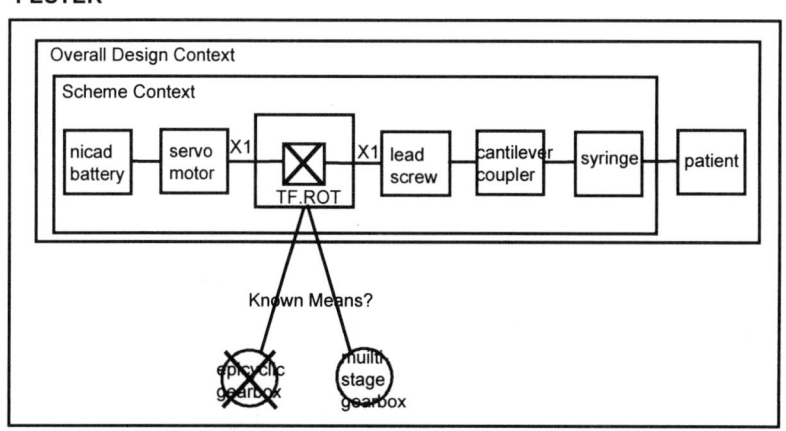

Figure 4. Function means interaction with matching component.

3.4. Inter-Component and Inter-Scheme Comparisons

To further assist the user in choosing between components and schemes a facility is provided which allows comparisons on the bases of cost, weight, working temperature range, maintenance interval etc. The tool presents a list of comparison variables for that component type, or scheme, and the user interactively selects items or combines them algebraically and displays the results as a series of bar graphs (figure 5). Thus power per unit cost and per unit weight could be compared if a design of least cost and weight were desired. The tool would use data from component databases when doing inter-component comparisons and from scheme files when doing inter-scheme comparisons.

3.5. Simulation Facilities

Bond graphs [4,5,6,7] describe system energy variables through their bond elements by denoting effort variable flows in one direction and flow variable flows in the opposite direction which are decided by bond causal structure. It is therefore very easy to obtain the dynamic response of any energy variable from a bond graph simulation method. However this approach does not support the modelling of information systems and a separate complimentary method to describe such systems is required. Modelling of information systems will not be discussed here as it is beyond the scope of this paper.

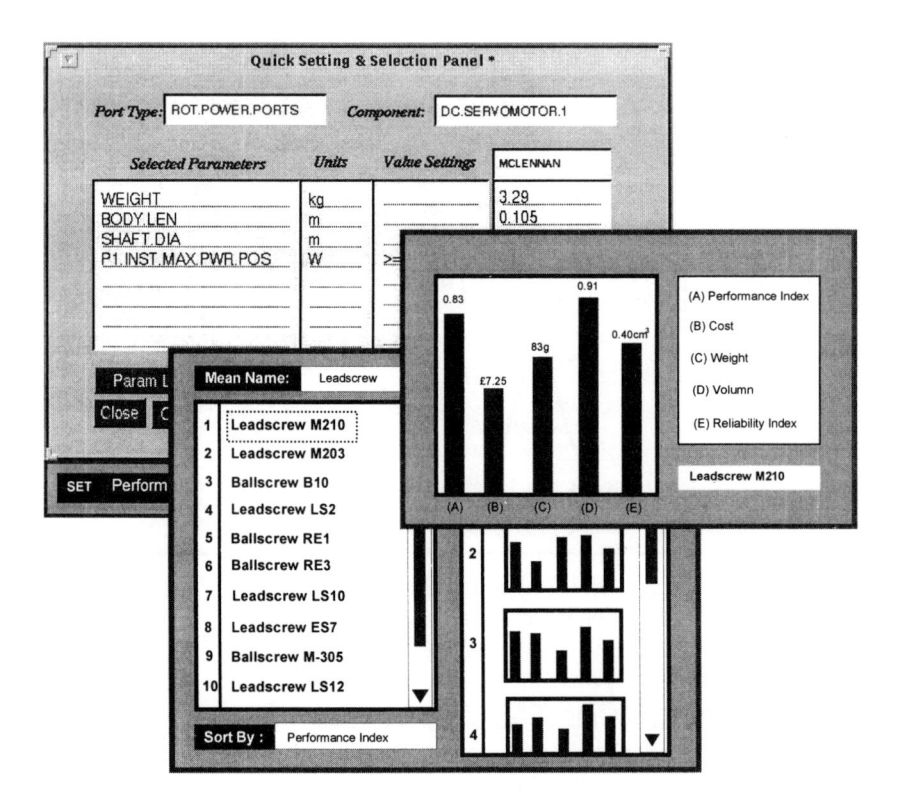

Figure 5. Component comparison tool.

The block diagram approach has long been successfully used to model dynamic systems and it has been proven to be reliable and easy to use. However due to its emphasis on the relationship between the output and input signals of a system, energy conservation is not considered in its system modelling and consequently important information is often not given in its model description.

It is however possible to combine the virtues of both approaches by converting a bond graph model into a block diagram model by separating effort and flow variables represented by one bond in a bond graph model into two separate flow lines. One of these two lines describes the variation and change from one type of effort to another by a governing bond graph element, and the other in turn depicts the relationship of flow variables before and after that governing bond graph element. In this way a model is constructed by employing basic block diagram elements to build a dual effort and flow variable block diagram model.

The facility described here uses "Simulink", a block diagram based simulation software developed by the Math Works Inc. It is used to derive modelling and simulation methods for mechatronic products by selecting and aggregating high level functional modular blocks. A detailed dual effort and flow variable relationship block diagram model is created for each component from basic block elements. This is then built into a high level-of-function block which is created for each component type by grouping them into a single block and masking it to give an interface of a number of inputs, output and a dialogue window for it. These blocks can be collected into a library and are used to create models to simulate the dynamic performance of mechatronic products.

Each component type can be described with different levels of model complexity. Correspondingly different objects have been created and collected in the library to represent them. Up to three levels of model complexity have been considered for each component and these models are stored and marked for different levels of system performance simulation. Figure 6 shows a bond graph model for a D.C. motor at its second level of complexity including its structure support. Figure 7 illustrates a converted dual effort and flow variable relationship block diagram model which can be thereafter grouped and masked into a high level D.C. motor object corresponding to its bond graph model at complexity level 2.

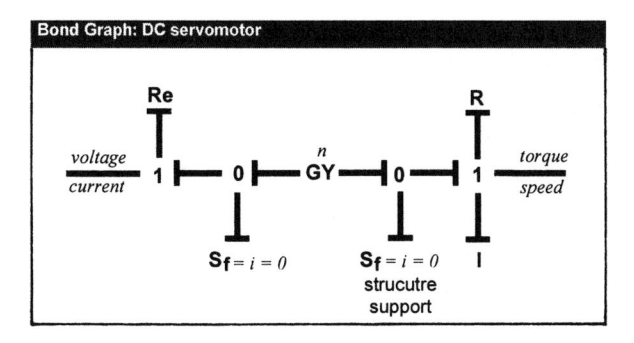

Figure 6. A Bond Graph model of a DC servomotor at its second level of complexity.

Importantly for Mechatronic and other interdisciplinary systems, information system simulation can also be performed by Simulink. Firstly a number of elementary function blocks are created, and similar ones collected into libraries, such as a logic operation block library. These operation blocks can then be used to create a control device. Various sensors can also be created to form a sensors library which allows a user to create a model incorporating sensor devices. Sensor and controller libraries can similarly be created. All these sub-libraries together form a central library for modelling and simulating mechatronic products in a unified manner.

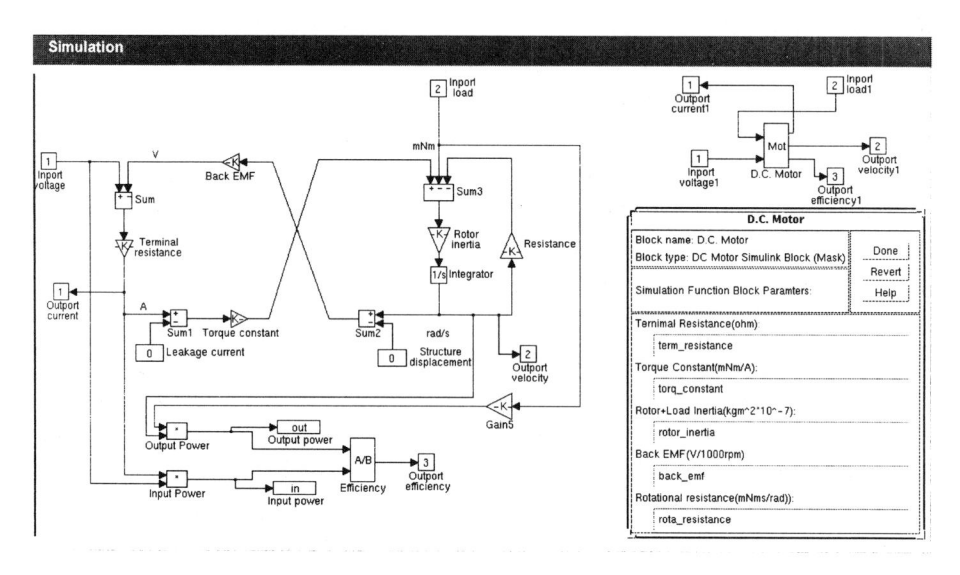

Figure 7. A converted DC servomotor dual effort and flow variable block diagram model and its graphical object with dialogue window.

3.5.1 A simulation model example. Consider again the metabolyte infuser example. This is composed of a controlled power driven injection system, which allows an authorised person to vary the flow rate. The drive system in turn consists of a battery, a D.C. motor, a gearbox, a lead screw with a special nut also acting as a coupler to push the syringe piston, a casing and a syringe. The control system is realised by using a number of integrated circuit chips with counters and logic gates. The motorised drug infuser system can be modelled as simple function blocks with their dialogue windows shown in figure 8, where each component is modelled in a simple function block which in turn consists of a detailed dual effort and flow variable block diagram model. Each of these function blocks normally has a pair of effort and flow input variables, a pair of output variables and a component efficiency variable. Bond Graph Causality is used to determine which is the input variable between a pair of energy variables. Two ways of specifying parameter data are implemented, namely auto-data retrieval through an intermediary data file (see figure 13) and user parameter specification through dialogue windows. The first method can be used to verify a chosen component scheme through simulating it with determined parameters at conceptual design stage, whereas the second approach can be used in the design of a new component by varying its determining parameters. (The second method can also be used to design a new component by modifying the component parameter values based on the evaluation of some existing components.)

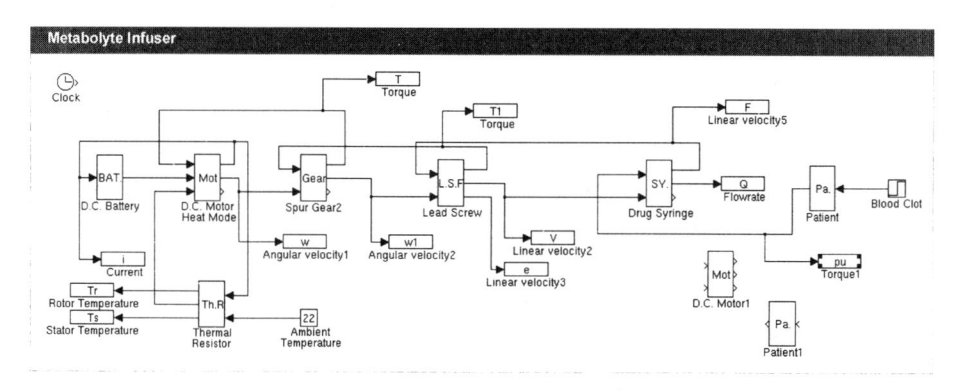

Figure 8. A high level functional block model for a metabolite infuser.

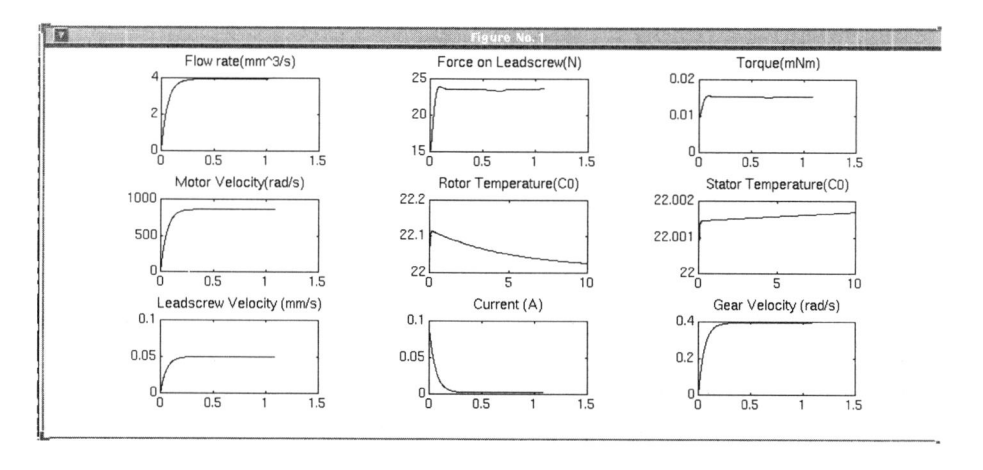

Figure 9. Simulation results of the response to a step input of voltage of a motorised drug infuser.

Figure 9 gives a graphic simulation output of motorised drug infuser system response to a step voltage input from the battery. Each of these simulation results can be easily obtained by linking an output variable block to the corresponding variable line to store simulation results. Thus multi-variable output simulations can be used to study the system's dynamic performance. Several simulation sessions can be conducted to evaluate the response of variant schemes within a product design.

3.6 Layout Facilities

The Layout tool supports the preliminary embodiment phase of design, producing quick 3-D 'solid sketches' of schemes. It provides a rough feel of how the components are to be assembled together giving a sense of their spatial layout. The intention is simply to 'reserve space' in a design for final components, so the drawings can be much simpler than those which eventually replace them. Features such as fixing holes, spigots and fillets as found in detailed drawings are omitted. We believe that at this stage of the design process omission of such details is justified as there is sufficient information available to give the designer a chance to assess the correctness of the layout and check for interferences between the components and other supporting structures.

The AutoCAD system is currently being used to accomplish the requirements of the Layout tool. The routines to enable the drawing of the 3-D solid representation of a component have been implemented using the AutoLISP macro language. The data necessary for providing values to the dimension variables describing the structural aspects of each component type is drawn from the set of values obtained from the dataserver and, in some cases, provided by the user.

Each component is represented geometrically by a combination of 3-D solid modelling primitives, which include cylinders, boxes, wedges, spheres, and cones. There are also procedures in which to produce solid models of more complex shapes via extrusion, sweeping, subtraction and union.

Figure 10 illustrates how the solid model of a typical dc-servomotor component is represented. The component is represented as a union of either two cylinders or a cylinder and a box, the selection of either one depending on the values transferred across from the dataserver. The actual positioning of each solid primitive with respect to each other is determined by constraint relationships, which are parametrically driven by the specifics of their structural properties. Hence, the positioning of the "motor body" is constrained by the length of the "motor shaft", which has a value supplied by the dataserver.

Figure 10 also illustrates other aspects of the solid modelling representation for a component object. The mating position, which is represented as a 3-D co-ordinate point, defines the point at which the next component in the scheme is to be assembled with the component in question. The functional axis defines the direction in which the next component is to be assembled for the valid functional interfacing between the two components. These two further aspects of representation augment the "vanilla" approach to solid modelling by allowing the system to assemble the components within a scheme automatically based on their functional interfacing as determined within the scheme representation. Based on the functional interfacing topology of the component models in a scheme, the Layout knowledge base generates a layout description of the assembly and sends it to AutoCAD to align and assemble the components automatically.

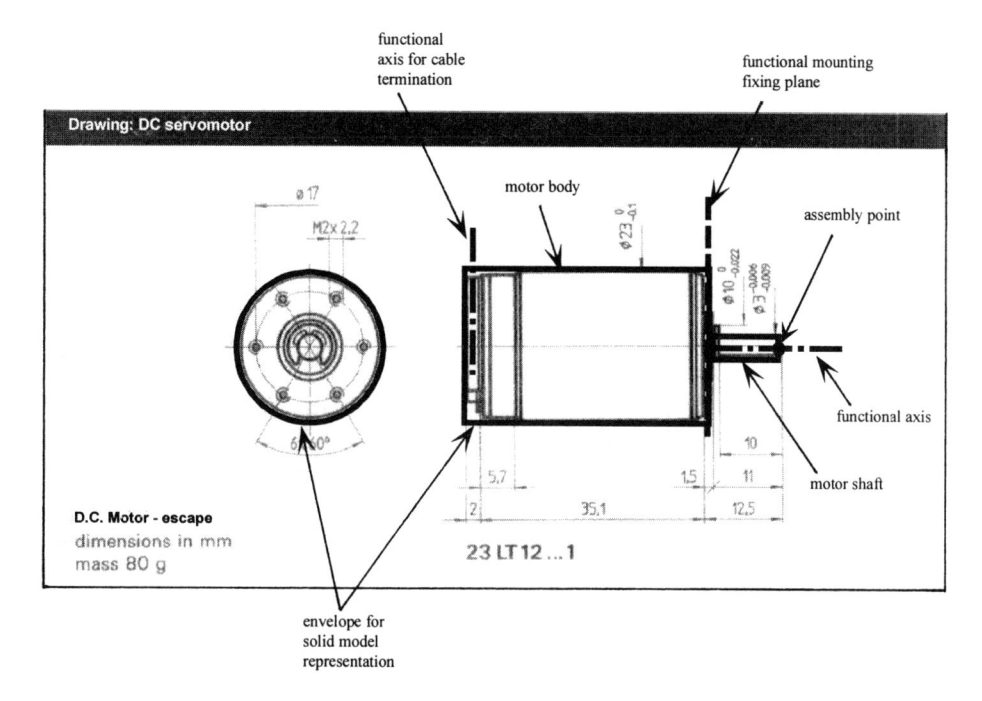

Figure 10. Solid model representation for a DC servomotor with the function axes shown.

The process of drawing a component in AutoCAD in response to a triggered action in KEE is enabled by an agent called the Draw Manager. The Draw Manager is responsible for storing object representations of the solid primitives and, if necessary, stores the list of instructions needed to create a more complex solid model representation using a series of solid modelling operations like union, subtract, chamfer and fillet. Information is sent across to AutoCAD which will then execute the relevant AutoLISP procedures to draw the representation of the component.

Figure 11 shows the solid modelling layout of one scheme from a set of alternative schemes for the design of the motorised drug infuser. The components have been aligned and assembled automatically after execution of a spatial reasoning process activated when an option on the system menu is picked.

4. INTEGRATION MECHANISMS

There are essentially three integration methods employed in the Schemebuilder environment:
1) use of a unifying modelling technique, namely bond graphs, to integrate heterogeneous concepts in different energetic domains;
2) use of computational agents called design facilitators to map between data and knowledge structures and models in one system onto another;
3) use of the underlying operating system mechanisms - UNIX pipes and intermediary text files - to integrate the various software systems.

As the bond graph modelling technique has already been described earlier on, it will not be discussed any further here. The next two sub-sections will describe items (2) and (3) above in greater detail.

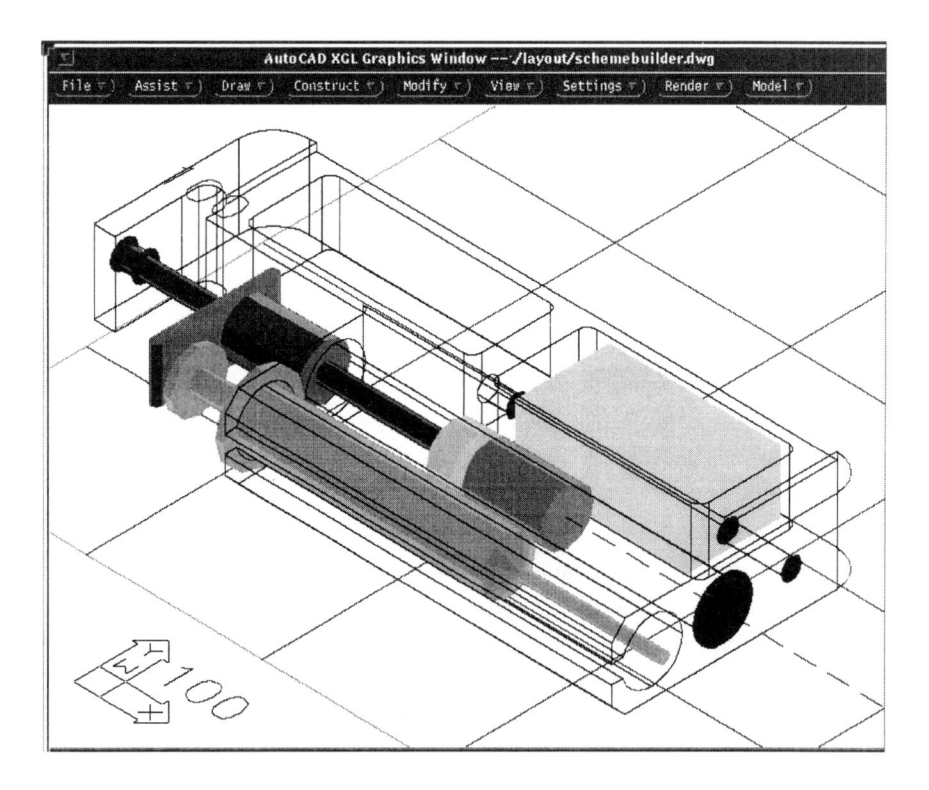

Figure 11. Solid model of a motorised drug infuser.

4.1. Design facilitators

Designers and engineers frequently carry out optimisation and simulation exercises, especially at the embodiment stage of design, to assess the viability of a chosen scheme. The tools that support such activities allow the designer or engineer to try out what-if scenarios and to change rapidly the values of design parameters in order to assess the effects of the changes. In interdisciplinary design, it is common to use a variety of heterogeneous sub-systems to model and analyse the different aspects of the design from different viewpoints. Design facilitators can help with the data translation between one tool and another, which if done manually may lead to considerable time wastage, and in the browsing and retrieval of information from electronic catalogues and libraries. The main job of design facilitators are to act as intelligent computational entities that work on behalf of the human designer to map between models and translate data between heterogeneous design tools in an integrated design environment.

The design facilitator in Schemebuilder is an agent that provides mechanisms for the mapping and conversion of different schema and model representations from one tool to another. Figure 12 illustrates one example of its use.

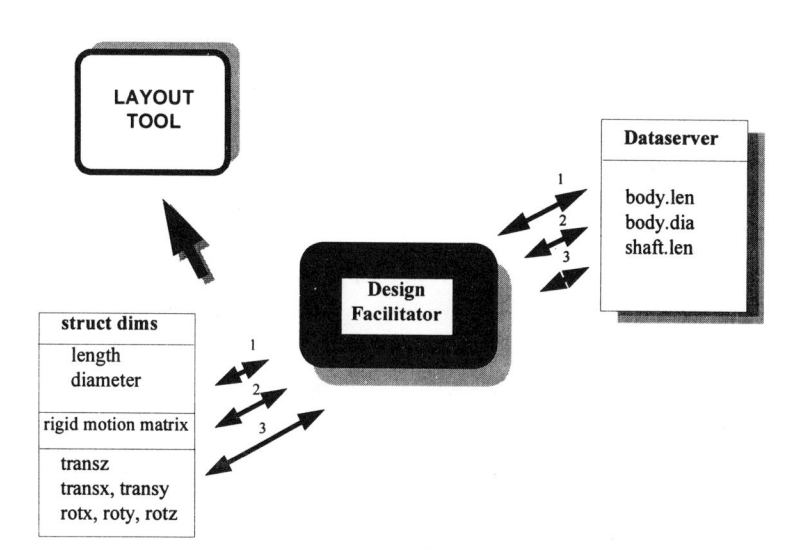

Figure 12. Inter-schema mapping via design facilitator.

The person responsible for constructing the schema description of the database of dc-servomotors for instance may describe the dimensions for the dc-servomotor as "body.len" (body length) and "body.dia" (body diameter). However, the knowledge base consisting of the classes of solid primitives may have the definitions different from that of the database. It is not practical and sometimes may even not be feasible to enforce congruency of the definitions in the database with that in the knowledge base. The design facilitator hence is an object wrapper that mediates between the two by providing a mapping mechanism to translate one schema definition to another so that one would correspond with the other when both systems are integrated.

Another responsibility of the design facilitator is to convert numerical values associated with conceptual notions used in one model to those in another model. In these cases the normal one to one correspondence between vocabularies does not hold. For example, the term "rotational resistance" and "mechanical rotational friction" of a dc-servomotor may at the outset seem to mean be semantically similar, the former being used in the Simulation tool and the latter in the dataserver, but in actual fact they are different - their relationship being formally defined by a mathematical one.

4.2. TOOL INTEGRATION

Figure 13 illustrates the integrated infrastructure of the Schemebuilder environment. As the figure shows, the KEE environment is the focal point for information movement and acts like a blackboard structure for the other tools.

The figure illustrates a typical flow of information when fully specifying a generic component object in a scheme. The Selection/Matching tool queries the dataserver for information about model types of a component and after a selection is performed the record number for the specific model type is transferred across to the generic component object, which will now send a message to its associated dataserver to download all necessary attribute values.

There is a slot called "dependants" in each generic component object which stores a list of all objects that have a dependency on the generic component object. Whenever the current record slot - represented as "current.rec" in figure 13 - is updated each dependent object is sent a message to update itself. This mechanism is essential to provide a co-ordinated update of models so that data consistency between models is maintained all the time.

5. CONCLUSIONS AND FUTURE WORK

Design is an integrated and integrating process during which the designer must call upon vast amounts of knowledge and experience in a systematic manner to develop the design of a product from first principles to a satisfactory prototype. Design is a combination of rigorous intellectual discipline and creative thinking. In the development of the integrated set of AI supported tools for use throughout the whole process of conceptual design and embodiment design described in both parts of the paper, we have been very conscious of two important factors.

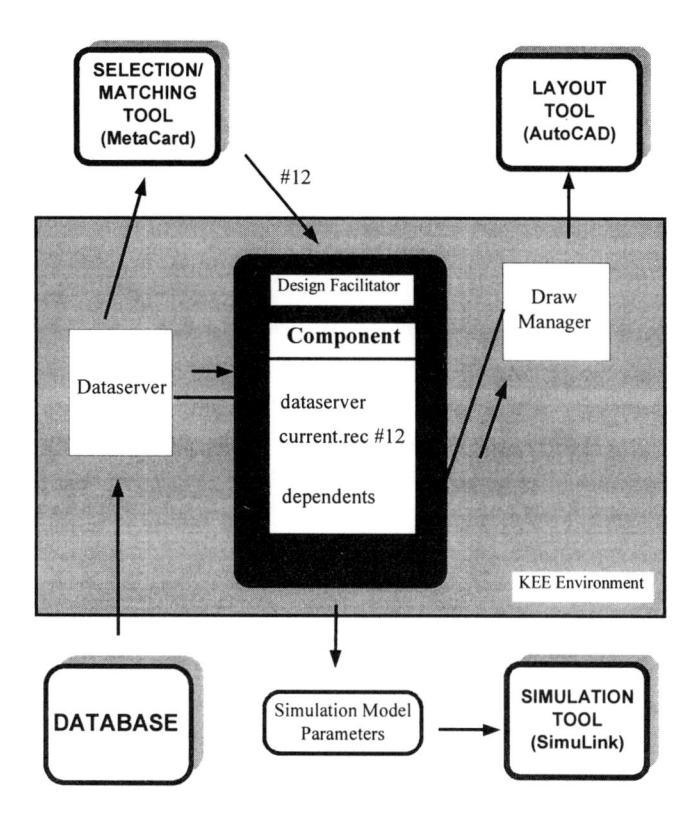

Figure 13. Schemebuilder's integrated infrastructure.

First, the need to allow the designer or designers if that be the case, to have total control over the design process at all times, allowing them to compare the alternatives in a clear and unambiguous way. Second, to provide the underlying core structure of data and knowledge in such a way that the designer is not hindered in his actions and may freely develop new ideas and procedures whilst at the same time being equally free to revisit previous decisions whether in respect to the Function-Means development or aspects of matching or simulation. The provision of a sophisticated Audit Trail of the designers action and the advice proffered or sought is important in this respect.

A further important aspect of the integrated set of AI tools is the manner in which the design process may be entered from a number of different points. This paper has largely followed French's model of conceptual design in describing how the different AI tools support the designer in the process. Practising designers often approach a problem not from first principles but using their experience. This may be an effective starting point especially if supported by good simulation and the ability to build and test rapid prototypes as has been demonstrated in the preceding sections.

The greatest importance of the work that has been described must lie in the manner in which it has been integrated to allow the complete design process to be undertaken in one homogenous computer environment. The example given throughout this paper has been of the design from 'first principles' of a portable metabolyte infuser for use with human subjects has demonstrated how the process may be used not only to ensure the creation of a number of workable alternative designs but how a design, once chosen, may be readily be constructed as a physical prototype from the list of chosen components, the block diagram of their interconnection, their physical layout and the spatial design of the special components and casing taken from the 3D solid model.

Arguably the single most compelling reason for an engineering enterprise to make the paradigm shift to concurrent engineering, difficult and sometimes painful as it may be, is to reduce as much as possible the time-to-market factor for a product: the earlier a product can be brought into the market the greater the market share would the company potentially capture. It is often observed that the concurrent engineering approach engenders a more protracted design phase than in the traditional serial engineering approach. This can be attributed to the fact that more design iterations occur at the design phase than traditionally would be: designing it right the first time is the prime concern. To reduce lead time further it is sensible that designers be provided with greater computer support at the conceptual design stage. Furthermore, an interdisciplinary approach to design should be encouraged so that the optimal allocation of functions to different engineering domains can be achieved more successfully. The Schemebuilder project addresses this problem by providing an integrated design workbench for the designer (novice or experience) to take an "expression of " need and developing it into schemes very quickly which can then be quantified and evaluated to assess the impact of the decisions made. This has been achieved by adopting a unifying representation via bond graphs for modelling of designs and supported by a set of integrated design tools made up of knowledge based systems, simulation packages, CAD packages and hypermedia systems.

The AI tools that have been demonstrated have concentrated on the energetic aspects of conceptual design and have built on the unique properties of Bond Graph Theory to provide robustness and rigour to interdisciplinary design. In our future studies we will be extending this thinking to information domains. Research into the nature of the information aspects of design is indicating that a similar approach to that for energetic systems will prove successful in providing the designer with much needed help to handle the information domain, in which ever form that information is transmitted.

In presenting this paper we have sought to demonstrate the potential for AI tools of many different forms to be integrated into a homogenous whole to provide support throughout the conceptual design process. This we believe we have been able to do. It is now for us to build on these foundations and extend to a level where the designer in industry feels comfortable and confident with this support.

ACKNOWLEDGEMENTS

This paper represents a considerable effort over a period of years by a large number of individuals. The authors would like to thank Michael French, David Bradley, David Dawson and Martin Widden for their help and enthusiasm in this project, the Dept of Computer Science for the computational support and the Engineering Design Committee of the Science and Engineering Research Council for their financial support of the Design Centre.

REFERENCES

[1] French, M.J., *Conceptual Design for Engineers*, 2nd Ed, Design Council, London, 1985.

[2] Bracewell, R.H., Bradley, D.A., Chaplin, R.V., Langdon, P.M. and Sharpe, J.E.E., "Schemebuilder: A design aid for the conceptual stages of product design", *Proc 9th Int'l Conf on Engineering Design ICED'93*, The Hague, 1993, pp 1311-1318.

[3] Bradley, D.A., Bracewell, R.H. and Chaplin, R.V., "Engineering design and mechatronics - The Schemebuilder project", *Research in Engineering Design*, Vol 4, No 4, 1993, pp 241-248.

[4] Karnopp, D.C., Margolis, D.L. and Rosenberg, R.C., *System Dynamics: A Unified Approach,* 2nd ed., Wiley, Chichester, 1990.

[5] Bracewell, R.H. and Sharpe, J.E.E., "Application of bond graph methodology to concurrent conceptual design of interdisciplinary systems", *Conf IEEE Systems, Man and Cybernetics '93*, Le Touquet, Oct 1993.

[6] Finger, S. and Rinderle, J.R., "A transformational approach to mechanical design using a bond graph grammar", *Design Theory and Methodology - DTM'89,* DE-Vol 17, W.H. Elmaraghy, W.P. Seering and D.G. Ullman, Eds, 1989, pp 107-115.

[7] Paynter, H.M., *Analysis and Design of Engineering Systems*, MIT Press, Cambridge, MA. 1961.

A COMPUTERIZED TOOL TO CREATE CONCEPT VARIANTS FROM FUNCTION STRUCTURES

D Brady & N P Juster

1 INTRODUCTION

1.1 Overview

Conventional computerized tools for assisting mechanical engineers in the design of new products tend to focus on the representation of complete geometry. However, the conceptual phase of the design process requires the manipulation of a mixture of incomplete geometry and informal information. Commercial computer aided engineering (CAE) systems are therefore of little use in the conceptual stage of design. It is recognised by companies mature in the use of CAE systems that future efficiency gains in new product introduction will depend in part upon the availability of tools that aid the conceptual design process. This paper makes a contribution in this area by proposing a computerized technique that allows designers to select and manipulate incomplete geometry. The geometries considered are those surfaces (features) of components that interface with the other components in the assembly, thus enabling the assembly to be held together.

The starting point for the work is the *Function Structure*. Many design methodologies [13], [17], [5] and [15] advocate the use of function structures at the conceptual stage, is an attempt to separate the required purpose of an object from any specific form. From the description of the problem in terms of its functional elements it is possible to create a set of outline concept variants. Because the function structure is independent of any physical form it enables many concept variants to be generated, all of which meet the functional objectives. Relationships between functions in a function structure can be manipulated, whilst maintaining the top level function. This increases, still further, the number of possible concept variants.

Figure 1 shows part of the design process as described by Pahl and Beitz [13]. The shaded area in the figure indicates the activities addressed by the work outlined in this paper, namely to:

- Search for solution principles to fulfil the sub-functions
- Combine solution principles to fulfil the overall function
- Select suitable combinations
- Firm up into concept variants

The technique and software, known as the conceptual design technique (CDT), to be described in this paper, addresses these activities by generating a set of concept variants based on a single function structure. The function structure can be generated by software developed at Leeds by Baxter [1]. The CDT is based on partial components stored

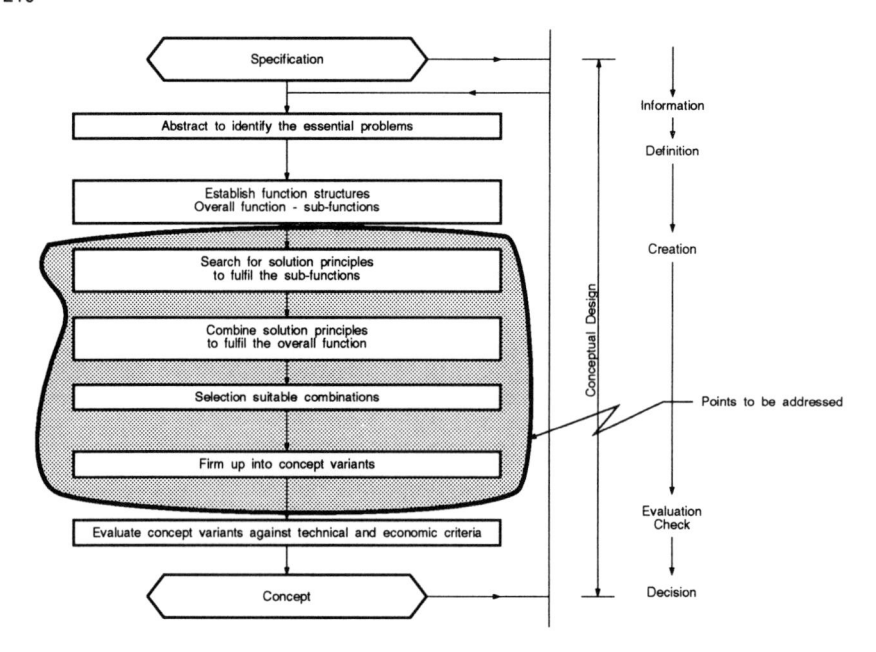

Figure 1: Pahl and Beitz model of conceptual design

in a library. The library, in its most basic form, holds information pertaining to the description, function and geometry of a each partial component. Allied to the library is an application program which searches the library of functional elements. This allows the designer to interrogate the library for a functional requirement and the CDT returns various functionally equivalent realizations.

What the tool does not do is comment on the suitability of any particular solution. The ultimate responsibly for the selection of good solutions remains with the design team. The major advantage of the software is that it presents the design team with an assortment of solutions from which to choose.

This paper is separated into four main sections this section continues with a description of the scope and background of the work. The second reports the development and definition of the CDT. Section 3 details the two case studies undertaken and the final section discusses some of the more interesting issues associated with the CDT.

1.2 Scope

Various researchers have attempted to classify assemblies based upon the way they act. Shah [16] describes *three informal types* of assembly: static, dynamic and kinematic. Whereas, research at Leeds has assumed that the functions within an assembly fall into two classes, based upon the way they 'act' within the assembly:

- **Assembly Functions** - are the functions which describe the requirements needed to hold the assembly together and these play no part in the dynamic behaviour of the finished object.

- **Product Functions** - are the functions which describe the requirements of the finished product when it is in-situ. Many researchers[14, 4] propose the use of energy flow, to model this type of function. Their use of bond graphs to model the product function seems a promising method but requires a complementary method to represent the non energy flow functions.

For example the nuts, bolts and washers which hold a gear box casing together, fulfil the function of attaching - an assembly function. In contrast, the casing fulfills the requirement (among others) of containing the lubricant which is a product function.

The research described in this paper concentrates upon assembly functions and their interaction with the function structure.

1.3 Motivation

At present the computerized representation of mechanical assemblies only occurs in a unified manner from the detailing stage downstream [6]. By creating a unified representation at the conceptual stage, all downstream activities may interrogate and propagate information. This will have the effect of:-

- capturing design intent within the design representation,

- improving the speed of iteration within the design cycle thus shortening product leadtimes [12],

- reaping large efficiency gains later in the product introduction process and

- geometric embodiment is deferred to its proper place in the design process.

1.4 Related Work

1.4.1 University of Leeds

Recent work at Leeds[1] has focussed on the development of product models and software applications that use these models to support the product introduction process. Within this framework researchers have developed data models that allow the representation and the subsequent manipulation of functional information, in particular:

- Baxter [1, 3, 2] presents a data model to represent the functions within assemblies, components and features. This work gives the application developer the opportunity to create, detail and reason with functions.

- Henson[9] has extended Baxter's functional data model to incorporate form and behaviour. The intention is to allow information to be shared between applications which manipulate function, form and behaviour.

The work detailed in this paper uses Baxter's functional data model as a starting point to create function structures. From the function structure the designer can use the CDT to interactively select functions. The CDT then returns abstract geometry which will fulfil the desired function. The designer can then select whether to develop the geometry.

1.4.2 Gui and Mäntylä

Gui and Mäntylä [8] report the development of the Δ modeller which aids *option design*. Option design is a method that facilitates the replacement of function carrying components by other components which fulfil the same function. Gui and Mäntylä's work will allow the change of components but does not suggest the changes. Moreover, the Δ modeller assumes that an assembly already exists and is therefore only applicable to re-design. The representation of the interaction between components within assemblies allows for change propagation i.e. when one component is replaced with another, any assembly inconsistencies could be pointed out to the user.

Gui and Mäntylä's work addresses many of the perceived problems of modelling mechanical assemblies but, the Δ modeller requires as its input a geometrical representation from which a function structure can be developed. For the system described in this paper, a function structure is the input from which abstract geometry is developed.

---

[1]The CAE group in the Department of Mechanical Engineering, University of Leeds, UK

1.4.3 Lin and Chang

Lin and Chang put forward a framework for automated assembly planning[11]. Of particular interest is the technique proposed for representing assembly relationships. Lin and Chang show that assembly relationships can be represented by a graph which relates the components in an assembly. Their work appears to be based on the relationships between complete components rather than the features which perform the assembly function. In the research presented in this paper the 'assembly features' are conceptually separated from the component thus allowing function to be *attached* to specific zones on the surface of the component.

2 THE CONCEPTUAL DESIGN TECHNIQUE

2.1 Description

The Conceptual Design Technique is based on three major elements:-

- the core application which interrogates the library of partial components,
- an object oriented library of partial components and
- an assembly representation technique.

2.2 The Application Architecture

The Conceptual Design Tool is controlled by the designer's wish to create a physical embodiment of a function in a function structure. The basic procedure for using the CDT is shown in Figure 2. The CDT elicits a function from the designer and then interrogates the partial component library in its function plane. Assuming the CDT finds a matching function in the library, it creates abstract geometry from the library. At this stage the designer can accept or reject the physical solution to the specific functional requirement. All positive solutions are stored to create concept variants. The designer then decides whether to search again for another physical solution (to that particular functional requirement) or to select another functional requirement from the function structure. The concept variants can then be selectively assessed and those that merit (further) embodiment can be transferred to a conventional 3D solid modeller. The geometry is represented in the CDT using STEP data types[10].

2.3 The Library of Partial Components

Information about common mechanical engineering components is stored in a library. The library classifies *partial components*, where a partial component is an area on the

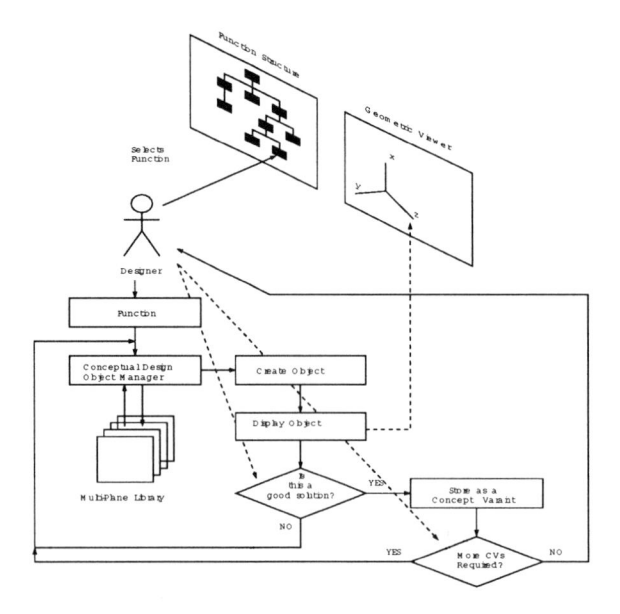

Figure 2: Architecture for the CDT

surface of any arbitrary component which may be joined to another partial component. This allows for the manipulation of elements which are not whole components and therefore may not have fully defined or complete geometry. The library has two interesting characteristics:

- Multiple planes of information
- Inheritance

These characteristics are described in more detail below.

2.3.1 Multiple planes of information

In its basic form the library consists of *three planes* (or related data structures) of information. Once each plane had been described the planes can be combined and the structures rationalized. In Figure 3 the planes are shown underneath an object manager which interrogates the planes. In brief, a plane is an hierarchical data structure where information is inherited by each child from its parent, grandparent, etc. The three base planes are:

- **The Description plane** - represents the partial component in terms of phraseology that engineers are familiar with. For example threads, keyways, flanges etc.

The inspiration for this plane came from the work of Gindy [7] in which he discusses a taxonomy of form features.

- **The Function plane** - allows each partial component to be described in terms of its function within the assembly. At present the functions shown in Table 1 are used. In the table the input/output columns detail the features which generate the function.

Function	Input	Output	Action
Attach	performer to attach	attach performer	attaches one entity to another
Fix	performer to fix	fix performer	reduces the number of degrees of freedom between two entities
Locate	performer to locate	locate performer	locate one entity relative to another
Seal	performer to seal	seal performer	seal the interface between two entities
Support	performer to support	support performer	support one entity by another

Table 1: The functions which have been examined

- **The Geometry plane** - describes the partial component in terms of abstract geometry.

The base planes may be augmented with *specialist* planes. At present, investigation continues into using an assembly rule plane. The assembly rule plane will inform the designer whether two partial components can be assembled.

The use of individual planes allows different applications access to only the information they require. While there are only three planes, allowing all the information to circulate to an application is not a major problem. But as the number of planes is increased more and more irrelevant information has to be passed to an application which is clearly inefficient.

2.3.2 Inheritance

The hierarchies in each plane have the same underlying structure. This allows the conceptual design object manager (CDOM), shown in Figure 3 to create a conceptual design object (CDO) by:-

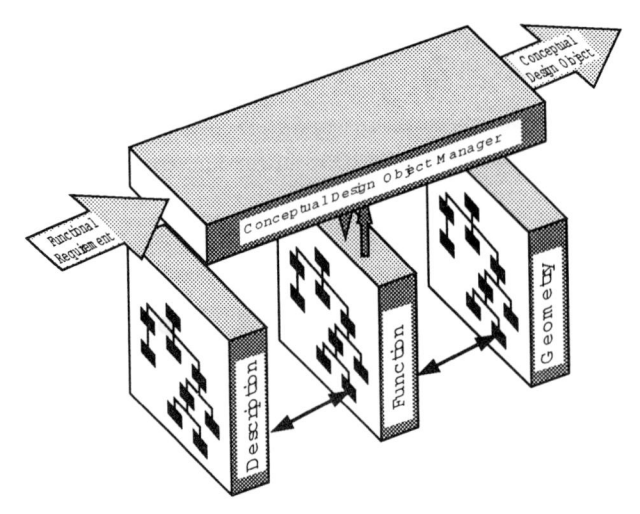

Figure 3: The conceptual design object manager interrogating the multi-plane library

1. Searching the function plane for the desired function[2],

2. Creating the function inheritance by stepping back through the tree to its root,

3. Moving incrementally through the planes obtaining the search node information and the information it inherits in each plane. The route through each of the structures is identical.

4. Gathering all the information into a single CDO.

The mapping from the function plane to any of the other planes is ensured to be one to one by having more than one instance of each function where necessary.

Figure 4 shows a small section of the partial component library detailing the description and function planes of information. For simplicity, the geometry plane has been excluded and the dashed lines represent branches which continue in the full version of the library. For example, using this simplified library a CDO could be created thus:-

1. Search for required function - assume a request is made for a **Seal** function to be realised. A search of the function tree returns the seal[3] function at the centre as highlighted in Figure 4 with a thick black box.

---

[2] Assuming that the desired function is present in the library the CDOM will undertake the remaining steps

[3] The CDT would on subsequent searches find the remaining seal functions in the function tree

2. Create function inheritance - **Seal** → **Attach** → **Function Root**. This is shown in Figure 5.

3. Move to the description plane and create the description inheritance - **Tapered** → **Thread** → **Description Root**

4. Gather all this information to create the complete object.

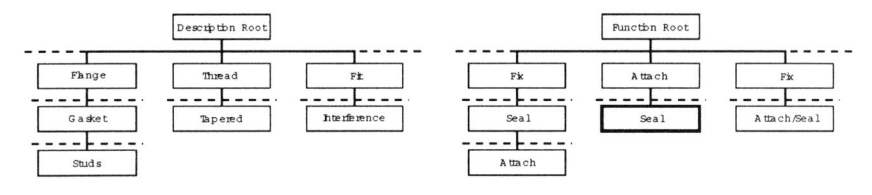

Figure 4: Simplified Library highlighting the **Seal** function

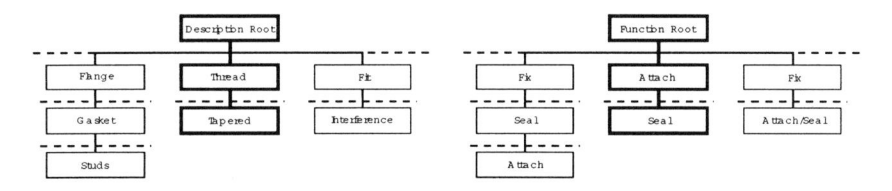

Figure 5: Simplified Library showing the inheritance pattern

2.4 Creating Assemblies from Conceptual Design Objects

A single CDO is of little use unless:

1. it can be collected together with other CDOs to form abstract components and

2. the relationship between two CDOs which are located on different components can be depicted thus allowing representation of assemblies.

The first requirement, to collect CDOs together to form components, is possible by identifying which CDOs from those stored is part of any particular component. The components can then be given labels so that component manipulation is possible.

The second requirement may be addressed by using a graph to represent the interconnections. This is similar to the technique proposed by Lin and Chang[11]. In Figure

6 an assembly relationship is shown for a centrifugal pump. The nodes of the graph are CDOs (labeled using the description plane information) and the curves represent the assembly interconnections.

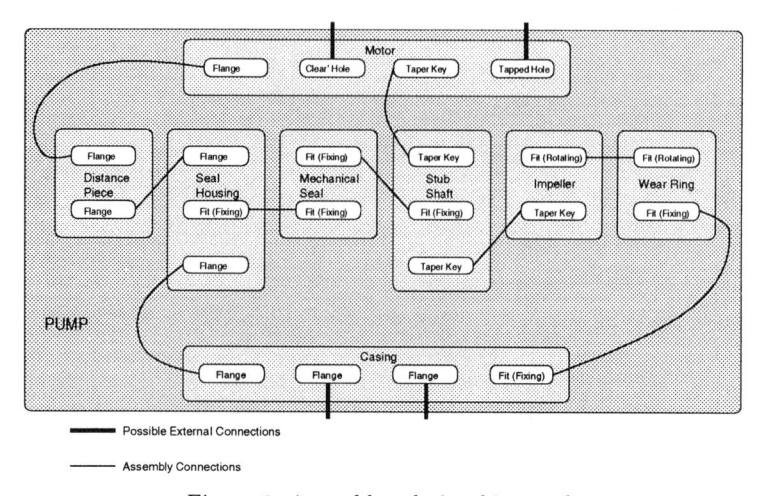

Figure 6: Assembly relationship graph

3 CASE STUDY

To illustrate the methodology outlined above two case studies are given below. The first is a top down design task where the designer is required to produce an article capable of transferring a fluid from an input pipe to an output pipe.

The second case study examines the problem of redesign. A functional description of an existing reduction gear box is used to generate new concept variants.

3.1 Case Study 1 : Top Down Design

In this case study it is shown that from a description of the problem space a function structure can be created and finally a set of concept variants can be generated automatically under the control of the designer. The case study shows that, given a function structure but no preconceived ideas about the geometric representation of the solution, various assembly forms can be created.

3.1.1 Description

The design task is to create an object[4] or collection of objects that will constrain all the fluid to flow from pipe A to pipe B as shown in Figure 7.

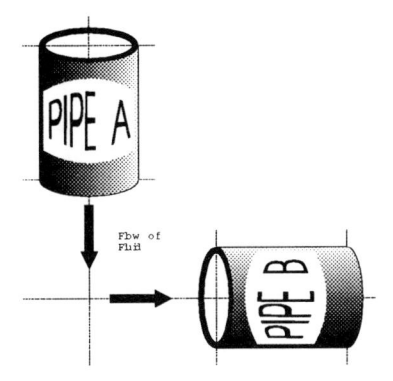

Figure 7: The problem space for the first case study

This is a somewhat simple design task, but is useful to show how the CDT works and the results that are possible. The procedure employed in this case study is:-

1. the designer creates the function structure for the design task

2. the CDT searches the partial component library for solutions for particular functional requirements

3. the designer creates the concept variants from the suggested solutions.

3.1.2 The Function Structure

The construction of the function structure is undertaken manually by the designer and requires careful thought. The overriding function of the object is to **Transfer Fluid** from A to B and is thus the primary or root function. At the next level of detail one considers:

- the **Transfer** of **Fluid** through the object. This is a product function and is therefore ignored at this stage,

- the **Attach** of pipe A to the object,

- the **Attach** of the object to pipe B.

---

[4]A physical object

At the next, and in this case the last, level of detail the **Seal** function is considered. The seal function was identified by the designer from the problem description viz. ...an object that will constrain *all* the fluid to flow...

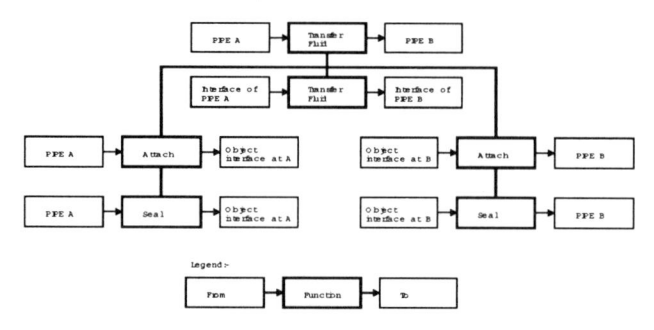

Figure 8: The function structure for the first case study

3.1.3 The Generation of Concept Variants

At this stage the designer selects a function from the function structure to generate form. For example from the function structure shown in Figure 8 consider the function **Seal**. If one selected the Seal function the Attach function is inherited into the selection criterion and therefore a search is made to find a seal and attach branch in the library. If the simplified library shown in Figure 4 is *asked* to create some CDOs based on the **Seal** branch various CDOs would be created as shown in Table 2.

Description	Inherited Description	Function	Inherited Function	Geometry	No. off	Form	Inherited Geometry	No.	Form
Studs		Attach		Cylinder	n	+			
	Gasket		Seal				Cylinder	1	+
	Flange		Fix				Cylinder	1	+
Tapered		Seal							
	Thread		Attach				Cylinder	2	+
Interference		Attach							
			Seal		Cylinder	1	?		
	Fit		Fix				Line	1	

Table 2: The results of a library interrogation based on a **Seal** function

An interesting point to notice is that the Flange partial component description has a function of Fix. This at first sight may appear to contradict the selection criterion (seal and attach) but a Fix function reduces the degrees of freedom of the component in a specific direction. Obviously, if at a later stage the component is completely fixed by an attach function a previous Fix function has little importance.

Table 2 shows that from the seal branch of the function structure various CDOs can be created. The table shows inherited information, the geometry of the 'assembly feature' created and whether it is a positive or negative feature.

From these CDOs a series of geometrical illustrations can be created as shown in Figure 9. In the Figure the grey filled areas represent geometry which is created manually by the designer and the solid lines show the CDO geometry. Note that the CDT makes no comment on the quality of the concept variants.

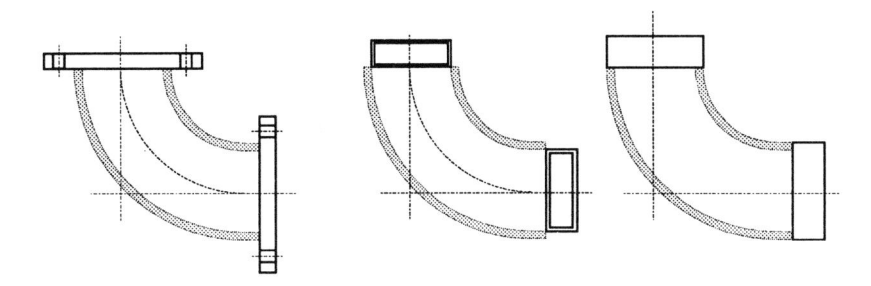

Figure 9: Some solutions to the functional requirements of Figure 8

3.2 Case Study 2 : Re-design

To show that the CDT can operate on assemblies which already exist a case study of re-design was undertaken. For this case study a Fenner[5] gear box was used using the functional and geometrical description detailed by Baxter [1]. The procedure employed for the re-design process was to:

1. use the CDT to create a concept variant which is similar to the 'real' assembly and is based on the functional description. This assumes that the original assembly was not created using the CDT

2. ask the CDT to find different solutions to various functions in the function structure

3. make a new concept variant with the new solutions.

This process is similar to that reported by Gui and Mäntylä [8]. However Gui and Mäntylä do not examine the underlying functions in the assembly.

---

[5]Fenner Power Transmission, Hull, UK. kindly made available drawings and other information of their D20 speed reducer used in this case study

3.2.1 Description

Figure 10 shows the input shaft arrangement for the Fenner gear box. In order to simplify the gearbox for presentation the oil seal and ball bearing are not described in the function structure shown in Figure 11. This function structure is an edited version of the structure presented in [1].

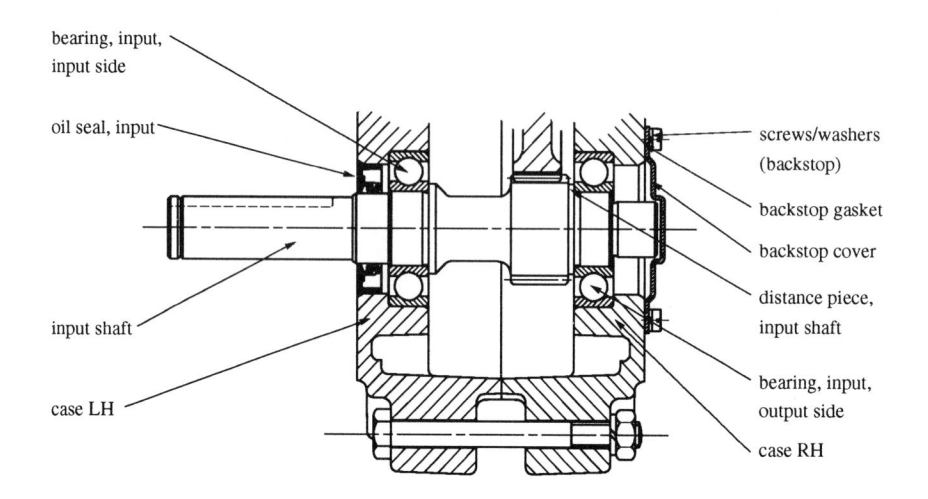

Figure 10: A segment of a section through the Fenner gear box

3.2.2 The Generated Concept Variants

In Figure 12 and Figure 13 the concept variant is shown on the left and an exploded version to the right. The exploded view clarifies the CDO interfaces, shown as thick black lines. The dashed lines represent the boundaries of the components. These boundaries are generated by placing a hull 'over' the CDOs of any particular component. The second concept variant, shown in Figure 13 was created by:-

1. Searching the partial component library based on the function **Fix** between the left casing and the shaft. The CDT returned a circlip, which fulfills the same function.

2. Searching the partial component library based on the function **Attach** between the left and right casings. The CDT returned glue as a solution to the attach function.

Note, once again, that the CDT does not comment on the quality of the solution in the context of any particular design task. The designer still has to make the final decision as to whether any suggested solution is practical and preferably.

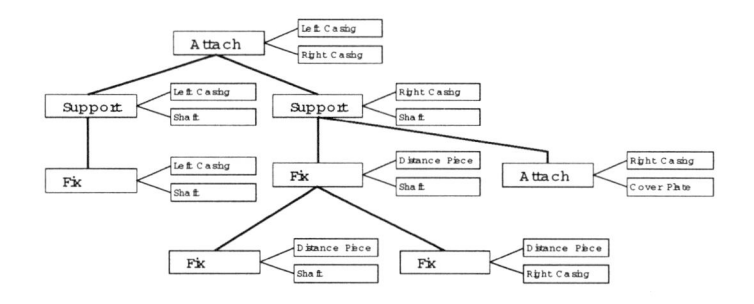

Figure 11: The function structure for the second case study

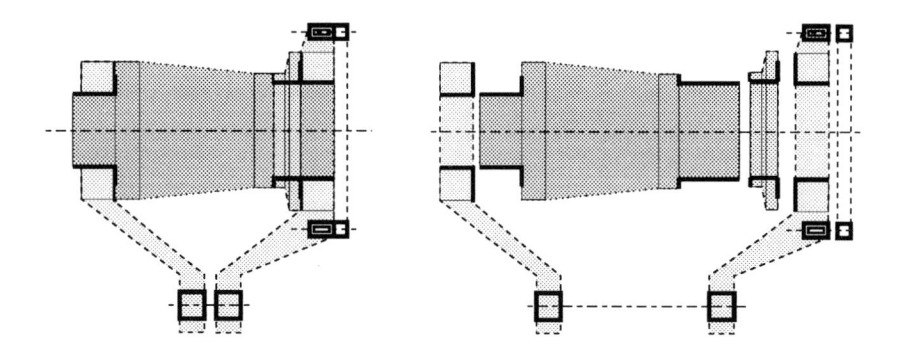

Figure 12: A concept variants based on the function structure shown in Figure 11. (The left hand side shows the product fully assembled and the right shows an exploded view)

4 DISCUSSION

At the current point in development the CDT appears to offer a technique which allows designers a starting point for the creation of computerized models of mechanical engineering artifacts. Moreover, the CDT facilitates a link to the next stage of design both theoretically and practically. The next stage is further embodiment, which is possible by transferring the abstract geometry of concept variants to a solid modeller via a STEP interface. The solid modeller can then create and represent the more detailed geometry that is required to manufacture the product. However the following points are raised by the work:-

- **Extensibility** - splits into two categories:

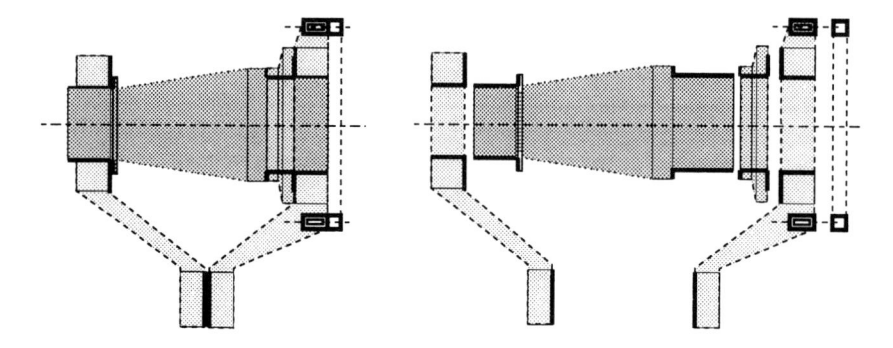

Figure 13: Another concept variants based on the function structure shown in Figure 11. (The left hand side shows the product fully assembled and the right shows an exploded view)

1. Vertical - This is the extension of the library by adding new partial components. The library is relatively simple to expand in this direction. For example the procedure for adding a new partial component is :-

   (a) Describe the new partial component in terms of the three base planes,

   (b) Add information about the partial component for any specialist planes,

   (c) Identify a location for the new partial component in the library hierarchy.

   This is possible because all the information in the library is stored in an external data base which is read at run time.

2. Horizontal - This is the extension of the library by adding new specialist planes. This is more complicated than adding new partial components because one has to describe the reasoning method so that the CDT can ensure correct operation. At present this can only be accomplished by re-compiling the software.

- **Back Propagation** - Using the *Conceptual* Design Technique it is, only possible to move from the function plane to the other planes of the library. Present investigations are concerned with whether given geometry or description information, it is possible to create a function structure retrospectively. This could be a possible technique for analysing design intent given geometry.

- **Explosion of Solutions** - As the size of the library grows the number of generated solution principles increases dramatically. This is not a major problem while the library is developing, but if at some stage the library is to cover more than just a narrow domain of engineering components, this issue will need to be addressed.

- **Energy Methods** - This work will create concept variants from the assembly functions and therefore does not address the product functions of the assembly. These product functions are inherently dynamic and can be modelled and analysed using Bond graphs. Current work at Leeds is examining the possibility of using Bond graphs to model the dynamic behaviour of assemblies which eventually could be a route to a full description of partial components. This would permit the creation of complete concept variants given a function structure.

Acknowledgements

The work presented here would not have been possible without the provision of a scholarship from the Keyworth Institute and is gratefully acknowledged.

References

[1] **J.E.Baxter** *A Functional Data Model For Assemblies* PhD Thesis, University Of Leeds, Department Of Mechanical Engineering May 1994.

[2] **J.E.Baxter, N.P.Juster and A. de Pennington** *A Functional Data Model for Assemblies Used to Verify Product Design Specifications* Proceedings of the Institute of Mechanical Engineers Vol 208 1994.

[3] **James E Baxter, Neal P Juster and Alan de Pennington** *Verification of Product Design Specifications Using a Functional Data Model* Lancaster International Workshop on Engineering Design. 1994.

[4] **S. Finger and J.R. Rinderle** *A Transformational Approach to Mechanical Design Using Bond Graph Grammar* Designs Theory and Methodology - DTM 1989, DE-Vol. 17. Book No. H00507 pp 107-116 ASME 1989

[5] **Micheal J. French** *Conceptual Design for Engineers* The Design Council 1985.

[6] **Allan Gardam** *The impact of design tools on the engineering design process in practice* Computer Aided Conceptual Design 1994.

[7] **N.N.Z. Gindy** *A Hierarchical Structure for Form Features* International Journal of Production Research Vol. 27 No. 12 pp 2089-2103 1989.

[8] **Jin-Kang Gui and Martti Mäntylä** *Functional Understanding of Assembly Modelling* CAD Vol 26 Number 6 June 1994.

[9] **Brian B. Henson, Neal P. Juster and Alan de Pennington** *Towards an Integrated Representation of Function, Behaviour and Form* Lancaster International Workshop on Engineering Design. 1994.

[10] **Industrial Automation Systems** *Product Data Representation and Exchange - Part 42* ISO 10303-42 1993.

[11] **Alan C. Lin and T. C. Chang** *A Framework For Automated Mechanical Assembly Planning* Journal of Mechanical Working Technology 20 pp 237-248 1989.

[12] **Vincent Oh, Patrick Langdon and John Sharpe** *Schemebuilder: An integrated computer environment for product design* Computer Aided Conceptual Design 1994.

[13] **G.Pahl And W.Beitz** *Engineering Design - A Systematic Approach* Springer-Verlag 1988.

[14] **D.R. Prabhu and D.L. Taylor** *Synthesis of Systems from Specifications Containing Orientations and Positions Associated with Flow Variables* Automated Design 1989.

[15] **S.Pugh.** *Total Design.* Addison-Wesley. 1990.

[16] **Jami J. Shah and Mary T. Rogers** *Assembly Modeling as an Extension of Feature-Based Design* Research in Engineering Design 1993 5:pp218-237.

[17] **N.P.Suh** *The Principles of Design* Oxford University Press 1990.

INTEGRATED INNOVATIVE COMPUTER SYSTEMS FOR DECISION SUPPORT IN BRIDGE DESIGN

C J Moore & J C Miles

1 INTRODUCTION

Many concepts and ideas have resulted from Artificial Intelligence research. How best to use these ideas to assist designers has grown into a major area of research. Initial efforts to achieve these aims were inflexible and fragile but more recent work is starting to overcome these problems; for example, the generic spatial reasoning system of Coyne and Subrahamin [2]. The authors have concentrated on the development of innovative systems for conceptual design which are of more immediate benefit, ([17],[11],[5],[8]). The research has progressed from the development of relatively large expert systems which covered the entire domain, to more flexible systems which aim to cover smaller, more focused sub-domains. Although the aims of these latter systems are more pragmatic and therefore easier to achieve, they could be criticised for being too limited in their coverage. To overcome this, the systems currently being developed are highly interactive, both with other complementary systems and with the user. The merits of this approach form part of the discussion in this paper.

The provision of sophisticated decision support software for designers is becoming accepted by many as a desirable goal. A number of large scale projects whose aim is to create comprehensive design environments which incorporate CAD, various KBS, databases and analysis capabilities are either in progress or have been attempted. These systems offer the advantage of compatibility. However, to date, the success rate of these projects has been disappointing. Generally throughout the software industry, it is recognised that large systems are difficult to develop, demanding a disproportionately high number of man hours compared with the development of smaller systems [10].

The alternative to large complex projects is to develop separate, readily compatible systems which can be successfully linked. This leads immediately to the concept of linking technologies such as product models [18] which facilitate the transfer of information between different systems. However, the development of product models is still in its infancy and furthermore their development is a relatively involved process. These difficulties have hampered the acceptance of product models by the construction industry. Hence, funding is difficult to obtain and it is likely that progress will be slow. In the absence of such sophisticated methods of linking, it is pragmatic to develop methods which exploit the available technology and can therefore be more immediately employed. One such method is described here.

This paper describes four complementary sub-domain design systems; their utility and the findings of the evaluation of these systems. The paper also discusses the mode of interaction chosen by the developers, both between the systems described and with recognised, commercial software. The application domain is the conceptual design of bridges. The subsequent discussion describes the underlying philosophy of the work, the linkages between the systems themselves and also between the system and the user.

2 SUPPORTING WORK

Work at Cardiff on the development of innovative computer support systems has been underway for nine years. During this time, the approach to the research has changed dramatically, primarily due to the input and suggestions of the industrial collaborators [8]. Initially, research at Cardiff aimed to develop 'standard' knowledge based systems for conceptual design. That is, the systems developed were of a prescriptive, question and answer format, utilising rigid rule based knowledge bases which provided definite solutions based on limited user input. These systems were well received by the evaluators as they offered a new approach to programming and engineering software. However, on closer examination, understandably, the software was limited and some aspects of the systems frustrated the evaluators.

This preliminary development and evaluation work showed that the end products being produced by accepted approaches to innovative system development were not being adopted by industry, as they tended to take over the design process, resulting in an approach to design which was over prescriptive and inflexible. Our evaluation studies showed that the engineers involved with the system development largely resented being ' replaced' by an 'intelligent' system and felt that a system of this type would be damaging to their work environment and skill development.

The evaluation also showed that the systems developed to cover the bridge design process would be more useful if the approaches adopted by system developers were more graphical and flexible and allowed easy addition of new information.

These sentiments have since been expressed by a number of researchers who feel that there is a danger of over automating the design process and reducing the quality of the design produced [8].

These findings led the Cardiff group to begin developing innovative computer systems which focus on specific parts of the design domain and which assist designers in areas which they find difficult, as opposed to attempting to substitute them in areas which they already perform well [8]. Initial work in this area showed that such systems were more readily accepted by industry and hence this approach has since been adopted by the research group.

The evaluation primarily highlighted a number of specific areas which an engineering designer could be supported effectively. The applications described in this paper are aimed at four of these areas.

3 BRIEF DESCRIPTIONS OF THE SYSTEMS

Four systems associated with different aspects of bridge design which are currently being developed are described. All four systems are being developed in Microsoft Visual C++ and operate under a Windows environment. The Case Based Reasoning system is also currently using Remind but it is anticipated that the final system will also be written completely in C++.

Windows was chosen as it is rapidly becoming increasingly accepted in engineering design offices.

C++ was chosen instead of a conventional AI language due to the greater power and flexibility it offers, albeit at the expense of increased programming effort for certain parts of the work. In particular, the version of C++ used offers considerable control over the user interface which past work [15] has shown to be important. In the original system of Moore [11], because of the limitations of the software used, it was not possible to incorporate sophisticated graphics. Instead the user interface was predominantly text based. Tests of this software showed that designers prefer to reason about designs in a more pictorial way, presumably because such information can be more readily handled in short term memory [8]. Thus, wherever possible, the systems provide information in a graphical format.

The domain of the work is small to medium span road bridges, as this is where the experience of the research group lies [11],[6]. In addition, as these are currently the most common form of bridge built in Britain there are large volumes of accessible data.

As with all of the authors' work, these systems are all being developed by collaborating closely with practising designers who are used to evaluate the work. This helps to ensure relevance as well as providing and ideas and impetus for further research.

3.1 System One: A Non Prescriptive Conceptual Design System

The original conceptual bridge design system [11] underwent extensive industrial evaluation. This revealed that the users did not feel in control of the design process because the original system was too prescriptive. As described earlier, like many other KBS developed in the mid eighties, the system controlled the decision making process, asking the user only to input rigid and limited information. This form of reasoning may be appropriate for some diagnostic domains but when used for an innovative and demanding design domain, the approach resulted in a system which was inflexible, particularly when the user wished to incorporate non-standard, case specific information [8].

Despite the criticisms levied by the evaluators, the original system was, in principle, well received as it provided correct answers and useful advice. This was because, despite the style of user interaction being flawed, the knowledge base was sound. Based on an analysis of the reactions to this initial system, it was decided to replace some of the heuristics incorporated in the knowledge base with more sophisticated forms of reasoning (as described in Systems Two and Three below) and for the initial conceptual design a far more flexible, user driven, knowledge-based decision support system has been developed. This system solely undertakes conceptual design and does not venture into preliminary costing or member sizing as was the intention of the original system. These areas are now catered for by separate systems, as described below.

The new system has been developed from the initial knowledge base, although this has been rewritten in an object oriented format. In addition the entire structure of the knowledge base has been altered so that the "rules" are clustered in small groups with no more than 10 rules per group. Each group is associated with a daemon which only fires when the user violates certain constraints. Constraint violation can occur for a number of reasons, for example when the user chooses an uneconomic structural form or when there is a locational clash say between a water main and a foundation. Thus rather than controlling the design process, the knowledge base observes and interacts only when necessary via a message on the screen. On receiving a message, the user is left to decide what action (if any) to take. The initiative is left with the user because given the impossibility providing of knowledge bases to cope with every situation, one has to allow human judgement and common sense to be included in the design procedure. This obviously permits the user to make mistakes but is a vast improvement on current procedures where no checks are made. This style of user interaction we call non-prescriptive because the system does not prescribe an answer; it only suggests alternatives [7]. By leaving the user in control of the design, the benefits of computers are maximised (i.e. memory, computational power and reliability) without stifling the capabilities of human beings (creativity, flexibility and innovation) [8].

A further facility allows the user to access and amend the knowledge base through a purpose built knowledge manager. This is possible because the knowledge base has been fragmented into short and separate constraint trees, which is in turn possible because of the non-prescriptive nature of the interface. This facilitates access by expert bridge designers who are not familiar with the system, hence allowing design consultancies to modify the knowledge base to suit their own practices. Further work is in progress on allowing users to add to (rather than to amend) the knowledge base and progress to date is encouraging. Knowledge can now be added to the knowledge base and the effects of adding this knowledge is made apparent to the user. This acts as a checking procedure and helps to prevent unforeseen knock on effects of adding new information to a well structured knowledge base.

At present the system is undergoing its first design office trials. Initial findings show that the designers see a distinct role for the system, providing a quality assured design procedure. They particularly like the knowledge manager and those people who had used the previous system [11] are appreciative of the improved style of interaction.

3.2 System Two: A Preliminary Costing System for Bridges

Following the evaluation of the first bridge design system, it became apparent that engineering consultants found the costing of alternative bridge designs a major problem. Currently, bridge designers use very simple heuristics (typically a cost per m^2 of deck).

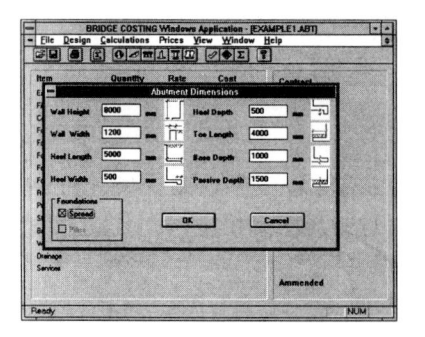

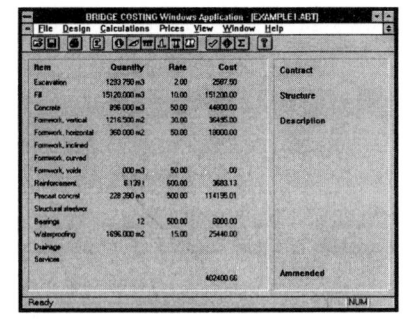

Figure 1

Obviously such a method is very crude. When a more detailed search of the design space is required then the usual procedure is to design a limited number of bridge options in some detail, and take off quantities in order to cost them. This can take several man weeks, the amount of work involved effectively prohibiting a proper search of the design space. To overcome these problems, a design costing system has been developed. This provides a cost estimate which is far more accurate than that reached by using current heuristics by sizing the components of the bridge to a level of accuracy which is close to that achieved by using a full analysis. The system is then used to take off appropriate quantities to obtain a preliminary cost.

By combining improved estimating techniques, approximate contingency factors and heuristic replacement [8] with the power and speed of computers, an effective system which can rapidly and accurately cost a bridge has been produced. The system enables bridge costings to be produced in few minutes, which in turn provides the designer with a tool which can rapidly cost alternatives and assess the impact of small changes, thus enabling the design to be 'fine tuned'. This means that the designer is able to search the design space for an optimum solution: something which has not previously been possible.

This system has also been undergoing intensive evaluation and testing using case studies. To date the results have been remarkable. All of the reviewers have stated that they would find this system extremely useful if it were to be made available in a completed and fully robust form. Further development suggestions have been made which are currently being taken on board, such as provision of life costs, the weight of the bridge and the total reinforcement weight. In addition to this, work is currently underway on risk assessment: that is, identifying those components of the bridge which are most difficult to cost accurately, helping the designer to appreciate the impact of certain changes to the design on the security of the cost. Work is also underway to investigate the commercial development of this system.

Costing is a problem which is apparent in most areas of engineering and many of the reviewers have recognised the generic qualities of a system of this type and have identified alternative fields of engineering where a system of this type could be useful.

The principle of heuristic substitution and the developed system are described in detail elsewhere [7] [14]. Two example screens from the system are shown in Figure 1.

3.3 System Three: An Advisory System for Bridge Aesthetics

Another decision variable in bridge design is aesthetics. This is an area of engineering design which is highly subjective and therefore difficult to investigate. However, the need to elicit information and provide assistance to designers in this area was identified during the evaluation of Moore [11]. The knowledge base for this project is being developed with the help of a number of expert bridge designers and architects. The opinion of the general public is also being included, via sophisticated questionnaires. Some work in this area has already been undertaken, notably by Crouch [3] and the work of the authors has extended this.

The system does not aim to provide a definitive set of rules which must be adhered to for all road bridges. Instead, it provides advice and assistance with the benefits of visualisation. As with the costing system, this system enables the designer rapidly to evaluate options. It is important that the use of such a system should not lead to the standardisation of road bridge design, therefore it only provides suggestions for improving the aesthetics of a bridge and, as with the other systems in the suite, control of the design stays ultimately with the designer.

At present, the system allows the user to enter criteria about a design: the proportions, its position, its span and so on. A 2d version of the bridge is then represented on the screen. The system will then 'review' the design using an in built knowledge base, containing a large amount of information on bridge aesthetics. The system will only produce comments which are relevant to the design and there is provision for various designs to be compared. In addition to this, the system allows the user to make small changes to the bridge to review the impact of these change on the aesthetics of the bridge. All of these features allow the user to explore different designs and ideas to enable them to make a guided design choice.

Initial evaluation has again showed that the system is liked by the reviewers and that they see a worthwhile role for the system in creating bridges which are aesthetically pleasing.

3.4 System Four: A Case Based Reasoning System for Bridge Design

Case Based Reasoning (CBR) is an important new technology which allows the more effective use of databases; initiating problem solving techniques which are based on the utilisation of past, recognised solutions [16]. Engineering design is a complex task. However, previous research has shown that much of the design conducted on a day to day basis consists of modifying past designs; as this approach is economic. In the construction industry, there are large collections of design data which are traditionally stored on paper or more recently in computer aided design packages. When considering a new scheme a designer will typically want to locate previous designs which can be used as the basis for a new design. Currently, this process is done manually. Many design offices recognise the need for more efficient search and retrieval techniques and CBR presents a possible solution to these problems.

The project does not aim to compile detailed design information on all types of bridges. Nor does it aim to develop a design standard which removes the creative side of design. Inevitably, bridge designs are complex and to aim to store all information about them would be unrealistic. Therefore, a sub section of bridge designs are considered (small to medium span road bridges as mentioned above). In addition, the CBR system aims to capture conceptual design decision information as well as specific design and maintenance criteria.

The project is still in its early stages, but industrial collaborators have already given a number of suggestions and bridge designs, which have provided the basis for the formulation of a prototype system which incorporates a preliminary case breakdown [13]. Work is also underway on the creation of a cleansing mechanism for the growing case base and ways of easily incorporating feedback from the construction of the bridge which can be used to assist in future designs.

4 THE INTERACTION MODEL

The four systems outlined above are being developed as independent systems. However consideration has been given to their interaction to form a complementary set of design systems. The mode interaction of the systems is shown in Fig.2. Also shown is the proposed interaction with AUTOCAD. Further work has taken place to investigate linkages to packages such as MOSS which could be used to input directly topographical data. The initial findings are that information can easily be passed using file protocols such as DXF but obviously this involves a substantial loss of information. It is also recognised that Fig.2 does not show any links to analysis software. However previous work [17] has shown the feasibility of such links and no problems are anticipated in providing them.

It can be seen that the user can enter or exit the system at any point. For example, the user can enter the system at the conceptual design stage, create a design, cost it, obtain advice on its aesthetics and leave the system with the completed conceptual design, with the option of storing the new design in the CBR system. Alternatively, the user could retrieve a design from the CBR system and check its cost, receive advice on its aesthetics or both. These are only two examples of many possible modes of consultations. The proposed connections

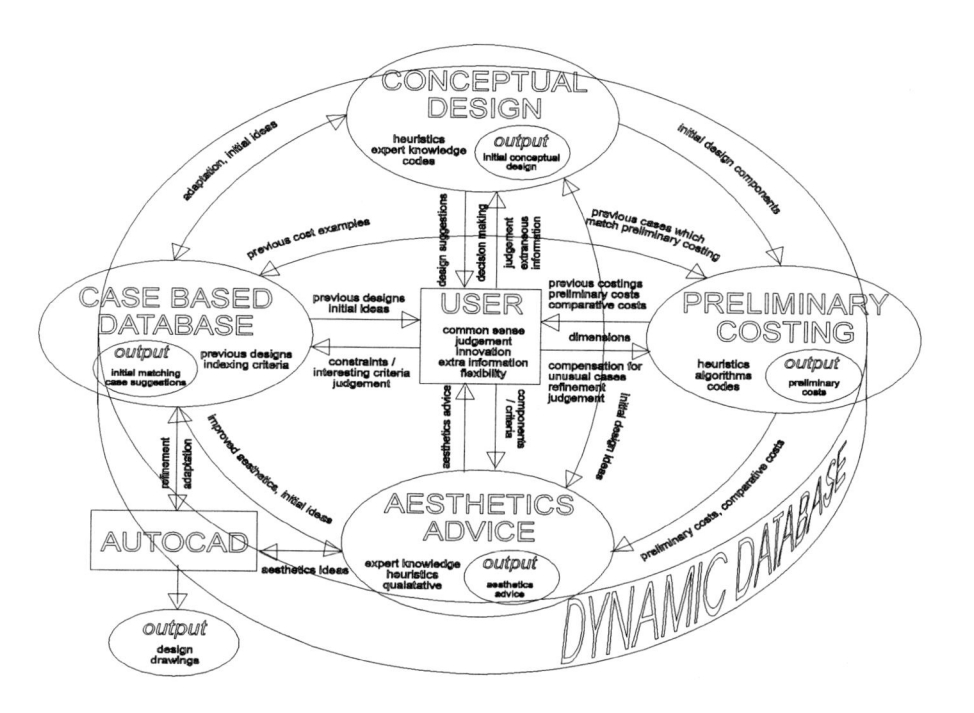

Figure 2

(shown by the arrows) have been deliberately limited, (that is, they are one directional in places), so that the prototype architecture is realistic in its aims.

This suite of systems will be supported by a dynamic 'database' of the information accumulated during a consultation (Fig.2). This facilitates the transfer of information between the interacting systems as well as providing an easy to access record of previous user input, enabling them to 'tweak' the input data to see the affect of changes in the design criteria. From this it can be seen that the systems will operate on an equivalent basis, with the user maintaining control over the entire consultation. The architecture adopted is reminiscent of a blackboard principle [1]. It also exhibits some properties of agents [4]. However, it differs markedly in that there is no central controller or system 'manager' as is the common case with these alternative architectures. Instead, the user takes control of the consultation and acts as the suite manager. The aim of the arrangement is to provide decision support for the user, with the user controlling the interaction and the design process. This maintains the sense of user control, which previous research has shown to be important [8].

The mode of interaction adopted allows the user to supplement the areas in which the computer systems are currently deficient, such as common sense, judgement, innovation and flexibility: all qualities which computers currently find hard to emulate and which are recognised as being important in design. As discussed above, the philosophy of the research at Cardiff is to support designers in areas which they find difficult, leaving them to cover tasks which come naturally and which are difficult to incorporate in computer programs. Using the user/designer as a system component enforces this philosophy and ensures that the systems operate successfully whilst maintaining their support role. This user centred approach is felt to be vital for the future success of design systems. As mentioned above, the authors believe that there is the temptation in AI to "over-automate".

There is a potential disadvantage to the above approach because the user exhibits such human failings as inconsistency but the supporting systems have been designed to help to mitigate such problems and it is believed that the gains far outweigh the losses.

5 TRANSFER OF INFORMATION BETWEEN THE SYSTEMS

The manner in which information is transferred between the various systems in Fig. 2 requires careful thought to ensure flexibility whilst avoiding excessive complexity. Given the inadequate development of suitable advanced coupling technologies such as product models, it was necessary to devise a schema which satisfies current needs and allows for future expansion. The importance of the inter-system linkages and the user interaction is such that a great deal of effort has been expended in planning and forethought. It is not possible to present all this work here but an outline of the approach can be given.

The information which needs to be transferred between the systems has been studied using a variety of techniques to show the types of information that are involved and the possible modes of interaction between systems. Firstly, the input and output of each system was listed. The outcome of the study of the input is summarised in Table 1. Numbers 1,2,3,4 represent the systems as described in the previous sections. Only those data used by more than one system are shown. The numbers in brackets indicate that the input is optional as opposed to essential (again enhancing the flexibility of the systems). The numbers showing an asterisk (*) indicate that the user can choose whether to input his/her own criteria or to let the system make the choice.

Many of the items in the input column are single facts or datum but others are more complex. For example deck type can describe the material(s), the form of construction and the shape of the cross section.

CATEGORY	INPUT	SYSTEMS	TYPE OF INPUT
Topological	Location	1,(2),3,(4)	descriptive
	Bridging Width	1,3,(4)	descriptive
	Clearance	1,3,(4)	numerical
	Skew	1,2,3,(4)	angle
	Inclination	(2),3,4	angle
	Horiz. Curvature	1,2,3,4	radius
	Vert. Curvature	(2),3	radius
Super Structure	Deck Type	(1), 2,(4)	descriptive
	Largest Span	(1),2,3,(4)	numerical
	Deck Depth	1,2,3	numerical
	Width	(1),2,3,4	1,4 - descriptive 2,3 - numerical
	Constr. Technique	1,2,(3),4	descriptive
	Deck X Section	2,(3)	descriptive/numerical
	Soffit Type	1,(3)	descriptive
	Superelevation	(2),(3),4	angle
Sub Structure	End Support Type	(1*),2,3,4	1,3,4 - descriptive 2- descriptive/ numerical
	Wing Wall Type	(1*),2,(3),4	1 - angle 2 - descriptive/ angle 3,4 -descriptive/angle
	Int. Support	1,2,3,4	1,4 - descriptive 2,3-descriptive/numerical
Earthworks	Embank / cutting	1,(2),3,4	1 - descriptive 2,3,4-descriptive / numerical
Foundations	Ground Conditions	1,2	1 - descriptive 2 - numerical
	Type	(1*),2,4	descriptive
Miscellaneous	Bearings	1,2,4	descriptive
	Services	(1),(2),4	descriptive

Table 1: Summary of Input Study

Table 1 is also interesting as it shows the overall data requirements for the domain. The data which are common between the systems are, as one would expect, topographical and basic dimensions. There are some variations in format between the different systems but these are fairly minor and should be easy to cope with. In addition to table 1, a series of Venn diagrams showing the overlap of information between various system combinations have been created. Space limitations preclude their inclusion but they have proved to be useful for planning modes of interaction and also they provide a new insight into the design domain structure. To further clarify the workings of the linked systems, a number of tree diagrams (or basic semantic networks) have been created to represent the hierarchy of the domains studied and their inter-relationship. An example of these is shown in Fig. 3.

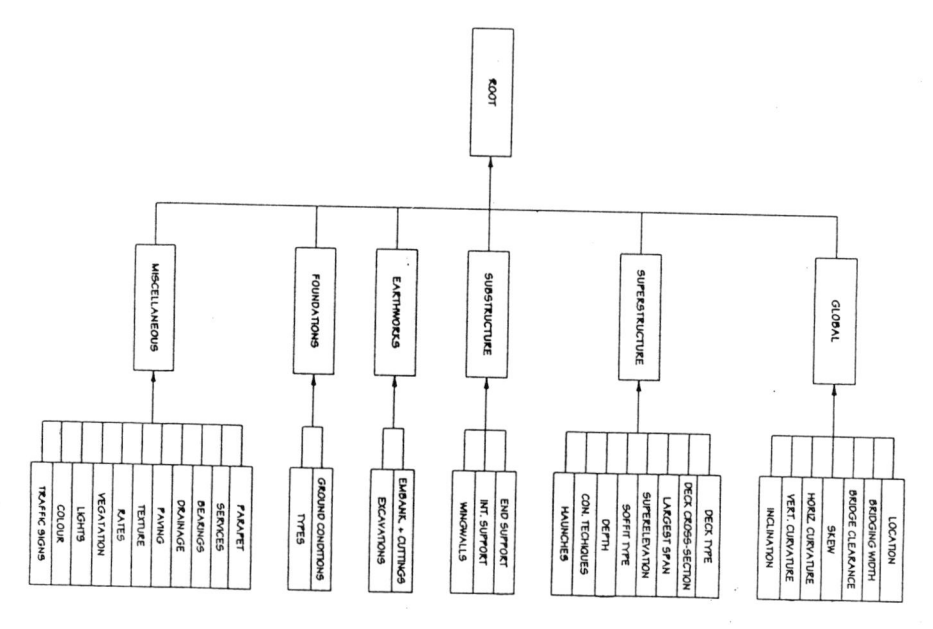

Figure 3

6 FUTURE WORK

In the immediate future, work will concentrate on developing the individual systems and with an increasing emphasis on the dynamic database. The industrial evaluation of the systems is underway and it is anticipated that this will produce some changes and new ideas which will be included in the development of all the systems. As well as evaluating the systems, opinions will be sought on the interaction between the systems. Depending on these findings, and in conjunction with the development of the component systems, the interaction architecture shown in Fig. 2 will be further developed.

As has already been mentioned, further additions to the above four systems are planned and work has already started on knowledge acquisition for a foundation design system. Further systems are also planned, particularly in the area of costing but as yet funding for these is not available. Further work on linkages to external software is also planned but as the research element in such work is minimal, it is not planned to go beyond feasibility studies.

7 CONCLUSIONS

A suite of systems for the design of bridges has been described. The systems all deal with sub-domains of the overall bridge design problem.

The systems are designed to allow them to integrate to form a complementary set of design systems which will cover the major parts of the preliminary design process. The user of these systems will be able to use any number and combination of these systems, hence offering maximum user control and flexibility and ensuring that control of the design remains with the user. This level of flexibility will be facilitated by adopting the user as the vital 'fifth component' in the design set, and by supporting the user and the systems with a dynamic database, which stores and transfers information as and when necessary. This avoids the tendency to 'over automate' the design process: a problem which has been apparent in recent years with the advent of more advanced computer technology. It is this proposed automatic approach to design which our experience has shown tends to alienate practising designers as they have no confidence in the reliability of this approach and hence they are encouraged to reject the innovative design systems which are suggested.

The systems described here are innovative and individually offer considerable benefits. However, they are also able to interact, not only with each other but also potentially with other commercially available software such as AUTOCAD and MOSS. It is this style of interaction and the advantages offered by this, which the authors believe to be completely novel and to offer enormous benefits

A study of the data requirements of the various systems has been outlined. This is part of ongoing work into the formulation of a dynamic database which will be used to control and facilitate the interaction between the current systems and future systems. Linkages to other common design software have been investigated and it is not anticipated that these will present any significant problems although with current technology the types of data which can be exchanged are somewhat limited. However it is felt that despite these limitations, such linking has much to offer and to wait until more sophisticated methods are available would waste the benefits of current technology.

REFERENCES

1. CORBY, O., Blackboard architectures in computer aided design. Civ Eng Systems. Vol. 4. pp 14-19/1986.
2. COYNE, R.F. AND SUBRAHAMIN, Computer supported creative design: a pragmatic approach, In Gero, J.S. & Maher, M.L. (eds) , Modelling creativity and knowledge-based creative design, Lawrence Erlbaum, USA, 1993.
3. CROUCH, A.G.D., Bridge aesthetics - a sociological approach, Civ. Eng Trans, Inst. Engs. Aust., CE16(2), 138-142/1973.
4. HUANG, G.K. AND BRANDON, J.A., Agents: Object Oriented Prolog for Co-operating Knowledge Based Systems. Knowledge Based Systems 5, (2), 125-136/1992.
5. HOOPER, J.N., A Knowledge Based System for Catchment Planning. Ph.D. Thesis. University of Wales, 350pp, 1994.
6. MILES, J.C. AND MOORE, C.J., An expert system for the conceptual design of bridges. in Topping, B.H.V. (ed), AI Techniques and Applications for Civil and Structural Engineers, CIVIL_COMP Press, Edinburgh, UK, 171-176, 1989.
7. MILES, J.C. AND MOORE, C.J., Conceptual Design: Pushing Back the Boundaries with Knowledge Based Systems. In: Topping, B.H.V. (Ed.): AI and Structural Engineering. Civil-Comp Press/1991.
8. MILES, J.C. AND MOORE, C.J., Practical Knowledge Based Systems in Conceptual Design. Springer Verlag,250pp, 1994.
9. MILES, J.C., MOORE, C.J. AND EVANS, S.R., Deriving Rules for Bridge Aesthetics. Civil-Comp '93. Edinburgh. August 1993.
10. MONTGOMERY, A., Combining SE Discipline With AI Creativity, in Software Engineering and AI, IEE (London), Digest 1992/087, 1/1-1/5/1992.
11. MOORE, C.J., An Expert System for the Conceptual Design of Bridges. Ph.D. thesis. University of Wales. February 1991. 350pp/1991.
12. MOORE, C.J., EVANS, S.N. AND MILES, J.C., Establishing a Knowledge Base for Bridge Aesthetics. To be Published in Structural Engineering Review/1994.
13. MOORE, C.J., LEHANE, M.S. AND PRICE, C.J., Case Based Reasoning for Decision Support in Engineering Design. IEE Colloquium on Case Based Reasoning. London. March/1994
14. MOORE, C.J., MILES, J.C. AND PRICE, G.E., Computational Decision Support and Heuristic Substitution in Preliminary Bridge Costing, In Smith, I.A.C.,(ed), IABSE Colloquium, Bergamo, 1995.
15. PHILBEY, B.T, MILES, C. AND MILES, J.C., The development of an interface for an expert system used for conceptual bridge design. In:Topping, B.H.V(Ed.): AI and Struc Eng, Civ-Comp Press, 87-96/1991.
16. SLADE, Case Based Reasoning: A Research Paradigm. AI Magazine 1991. pp 42-55.
17. SOH, C.K. (1990): An Approach to Automate the Design of Offshore Structures. Ph.D. Thesis. University of Wales. 330pp/1991.
18. WATSON, A. AND BOYLE, A.,Product Models And Application Protocols, in Topping, B.H.V. (ed), Information Technology for Civil and Structural Engineers,121-129/1993.

'MODESSA', A COMPUTER BASED CONCEPTUAL DESIGN SUPPORT SYSTEM

T Kersten

1. BACKGROUND

Unilever is one of the world's leading companies in the fields of food products (eg. ice-cream, soups, margarines), personal products (eg. shampoos, skin care products) and detergents (eg. wash powder, wash softeners). When developing new innovative products for these areas, the means of manufacture are often not available on the market. Therefore, Unilever designs and builts prototype production equipment in a manufacturing research department called the "Technology Application Unit" (TAU).

A typical example of a TAU design is the production line for the "Boomy" water ice-cream product. This line produces "Boomy" using injection moulding technology similar to the technology that is used for manufacturing plastic products. Its attractive shape made "Boomy" a sales success throughout Europe, while the novel production technology protected Unilever's competitive advantage. Examples of other production equipment designed by the TAU are automated flexible filling and case-packing production lines.

These automated flexible production lines can no longer be designed only using traditional mechanical engineering solutions such as cams and bar-mechanisms. They require extensive use of electrical, control and software engineering, involving components such as servo-drive systems, robots, sensors and micro-controllers. As a result we see a shift from purely mechanical design by mechanical designers towards so called "mechatronical design" by multi-disciplinary design teams.

2. INTRODUCTION

Research [6] has shown that up to 85% of the life-cycle costs of a product can be committed at the end of the conceptual design phase, while only about 5% of the actual life-cycle costs have been spent. Therefore, it is very important that in the conceptual design phase, the solution space of a design problem is thoroughly explored so the chance of overlooking cheaper or better design alternatives is reduced. The extra time that might be spent looking for better design alternatives will eventually result in considerably lower life-cycle costs.

The next important step in the design process is to decide which of the many design alternatives is the best. Traditionally, the mechanical designer would evaluate all the design alternatives on their mechanical merits and select the best alternative based on mechanical engineering calculations and/or his experience. Lately many new mechatronic components have become available resulting in many new and sometimes better mechatronic design alternatives. Unfortunately it has also become more difficult to select the best design alternative. Not only the mechanical aspects have to be considered, but also the electrical and control aspects have to be taken into account. This requires the participation from mechanical, electrical and control engineers. Each of these specialists will use his own calculation methods, design tools and experience to evaluate the design alternatives. All these opinions need to be 'integrated' into the evaluation process to achieve an objective selection of the best design alternative.

A recent survey [1] among 200 UK engineering designers showed that they typically spend 23% of their time on paperwork. Most of the recorded design information however is written in personal diaries, memos, reports and logbooks. These records are not structured and therefore not suitable for retrieval and reuse in the future by others. So although a designer spends a large amount of time on producing design documentation the knowledge in this documentation can hardly be reused by others. As a result it is likely that "the wheel is reinvented".

A good documentation of the reasoning behind the design alternative selection has become even more important, because from 1 January 1995 all production equipment sold within the EC must have the so called "CE certification". To obtain the CE certification a machine supplier must prove that the design of the equipment is such that safety-hazards are minimized. The supplier is obligated to document all safety related design data and to store this data for at least 10 years.

Therefore, the challenge is to provide the design team with support in:

1. Finding all the potential design alternatives.

2. Rapidly and objectively selecting the best design alternative(s), considering mechanical, electrical and control aspects.

3. Documenting design alternatives and design decisions in a structured manner.

3. THE MORPHOLOGICAL DESIGN METHOD

The TAU developed its own design method to structure the design process called "the morphological design method". This design method has been derived from the systematic design methods proposed by Pahl & Beitz [8], Van den Kroonenberg [5] and the German Engineering Society (VDI 2222) [13]. Key element in the morphological design method is the "common language". This common language is used to discuss and document all design alternatives and design decisions. It is simple and can be understood by all members in the design team regardless of their discipline. The vocabulary of this common language consists of so "morphological overviews", "info sheets" and "weighting tables".

Morphological overviews were first proposed by Zwicky [14]. A morphological overview is a matrix that shows on the vertical axis the design functions and sub-functions that have to be solved. The design alternatives that could possibly solve these design functions are visualized by means of small pictures (icons) on the horizontal axis. Figure 7 shows an example of an morphological overview.

Both the design functions and the design alternatives in the morphological overview have attached to them "info sheets" that contain detailed information. The function info sheet contains the design function description and the requirements a design alternative should fulfil to solve the design function satisfactorily. Figure 3 shows an example of a function info sheet. An alternative info sheet contains a description of the design alternative and information on its areas of use, options, critical aspects, advantages and disadvantages. An example of an alternative info sheet is shown in figure 6.

The most suitable design alternative for a design function is selected using weighting tables. In the weighting table each design alternative is given a score per requirement. This score is based on the level to which a requirement is met by the alternative multiplied with the "weighting factor" of that requirement. The weighting factor shows the relative importance of a requirement compared to the other requirements. Important requirements (demands) have a high weighting factor whereas less important requirements (wishes) have a low weighting factor. Several methods can be used to determine the weighting factors. Often the simple "binary dominance matrix" method [12] is used. Finally all the scores are added and the design alternative with the highest total score is selected as the best design alternative for the design function. Figure 9 shows an example of a weighting table.

The morphological design method has been used in the TAU for some years now and we have found [9] that:

a) The "common language" (morphological overviews, info sheets and weighting tables) offers all the members of the multi-disciplinary design team a good understanding of the total design problem and each others opinions.

b) The selection by means of the weighting tables 'forces' the design team to find objective arguments to select between design alternatives and makes the selection more transparent.

c) The method produces better structured design project documentation that allows a reader to understand the design decisions made.

However we also experienced that the paper-based implementation of the morphological design method suffers from some drawbacks:

a) The morphological overviews are time consuming to create and they often have to be changed because of new ideas and insights into the design problem. In practice this has lead to the use of "post-it morphological overviews": A3-sized posters with the design alternatives drawn on "3M post-it memos". Only the final version of the morphological overview is created using a standard graphics editor like "WordPerfect".

b) Although the design information is now stored in a structured way the retrieval and reuse of this paper-based information in future design projects is not efficient. Many reports have to be searched and pictures and texts have to be redrawn and rewritten.

4. COMPUTER BASED DESIGN SUPPORT SYSTEMS

A decision was made to look for a computer based design support system that would eliminate the drawbacks of the paper-based morphological design method by offering an user-friendly way with to generate, store and retrieve morphological overviews, info sheets and weighting tables.

Two applications named "Morphos" by Schlicksupp [11] and "Wysikon" by Fisher et al. [2] were identified that incorporate the idea of supporting the design process by computer using morphological overviews. Unfortunately both systems allow only limited textual description of design alternatives and do not further document the design alternative with a picture or detailed information. Also the possibilities to store and retrieve design data are rather limited.

More advanced computer based design support systems were identified such as "Schemebuilder" developed at Lancaster University EDC by Oh et al. [7] and a system developed at the University of Vermont by Hundal et al. [4]. These systems contain databases of design functions and alternatives and therefore can automatically retrieve applicable design alternatives. Both systems describe a design problem using physical functions descriptors such as conversion, change magnitude, store and supply. These physical functions operate on energy, material or signal flows such as force, heat, pressure, stress, displacement etc.

The advantage of these systems is their ability to directly retrieve physical descripions of design alternatives from the database. These physical alternative descriptions can then be used to calculate the dynamic behaviour of the design alternatives which is very useful for the evaluation.

Their disadvantage however is the very detailed way in which design functions and alternatives have to be described. Unfortunately the necessary detailed information is often not available at the start of the conceptual design. For example a design problem could be described as: "Design a production line for the production of a new ice-cream that costs less than 1000.000 and operates at higher than 95% efficiency". When dealing with this kind of abstract design problems many design decisions have to be made before arriving at a point where design support systems based on physical function descriptors can be used.

5. MODESSA

Not being completely satisfied by the functionality of existing computer based design support systems to develop our own system which was given the name MODESSA: MOrphological DESign Support Aid. A simplified flow chart of Modessa is shown in figure 1. It displays how Modessa supports each of the design activities in the "basic problem solving cycle" [10]. The design team goes through this cycle at least once for each design problem (function) that needs to be solved.

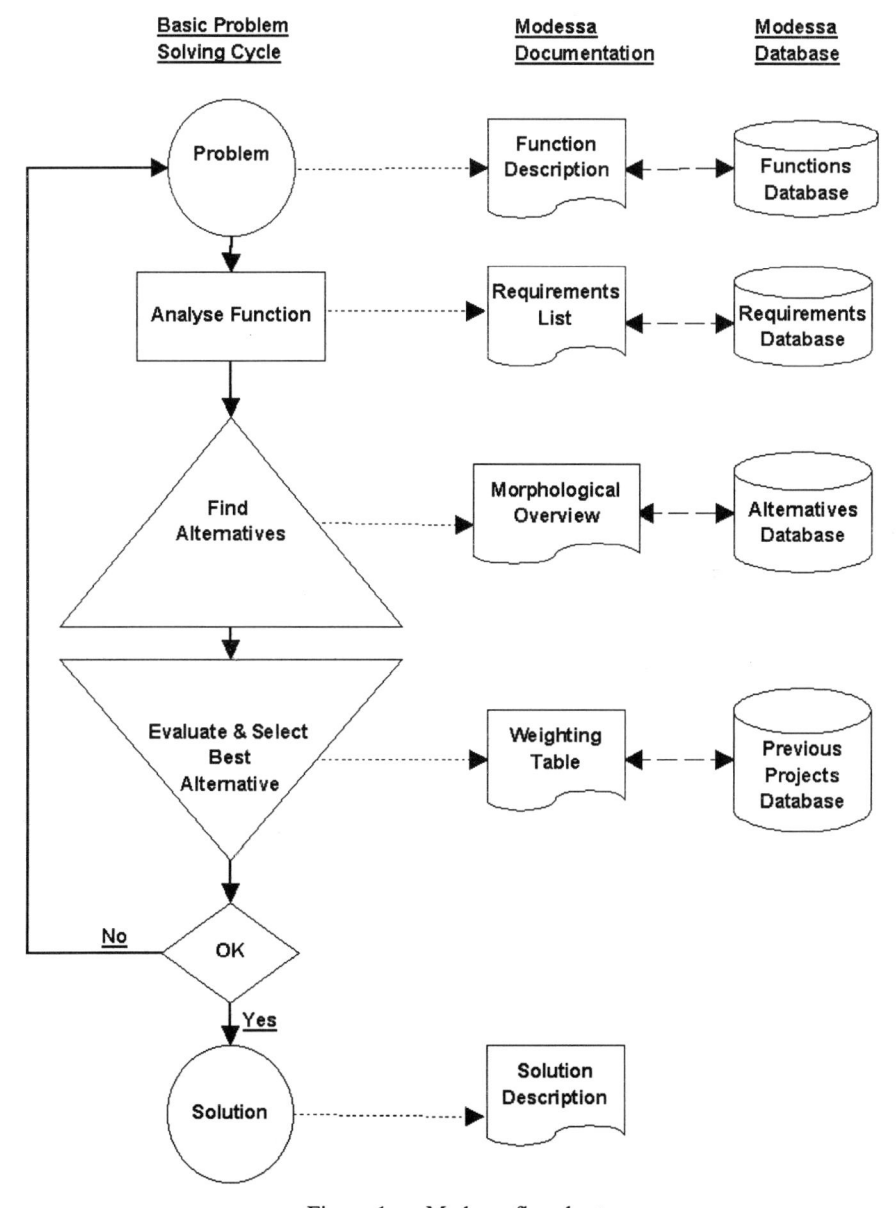

Figure 1: Modessa flowchart

As an example of MODESSA we will consider a typical TAU design problem: "Design a flexible case packer that can fill a case with tubs of various sizes and shapes; Before putting the tubs into the case the tubs first have to be arranged into a layer".

First the design problem description is translated into a one or more design functions or sub-functions. A function in Modessa is described using two words: the "action" that should be done and the "material" that is handled. A function can either be created or retrieved from the functions database. In this example we will assume that a similar case packer has been designed in a previous project and we can select a function titled "fill case" from the design function database using the "select functions screen" shown in figure 2. In the "select functions screen" the designer can click on the "info about function" button to view the "function info sheet" of an interesting function. The function info sheet contains a description of the function, a specification of the material handled, and an empty requirements list. Figure 3 shows the "fill case" function info sheet.

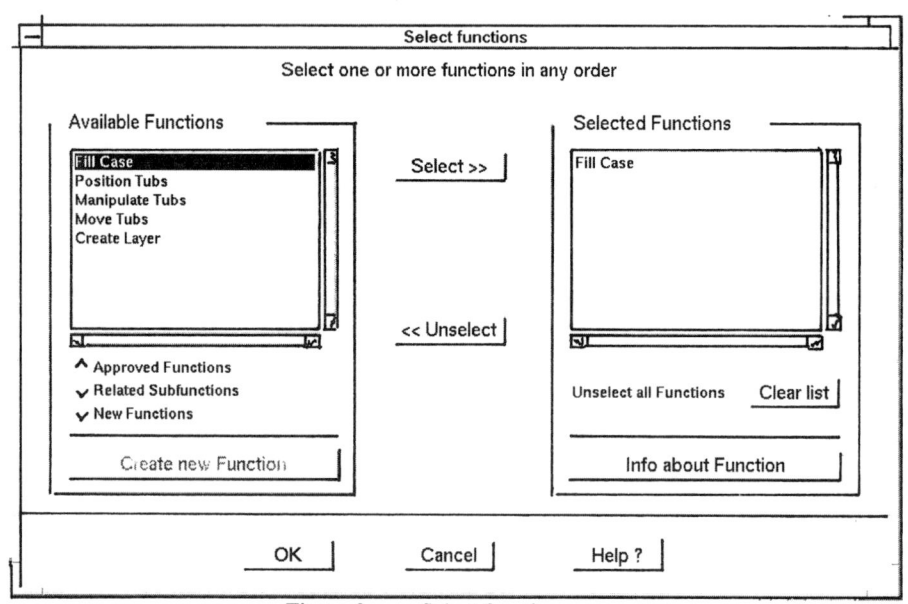

Figure 2: Select functions screen

Requirements can be created by the designer or retrieved from the requirements database. From the problem definition it follows it must be possible to fill the case with tubs of various sizes. This implies that layer pattern must be flexible because the case size is fixed. Therefore we have a "pack flexibility" and a "pattern flexibility" requirement. These two requirements are included in the requirements list of the "fill case" function as shown in figure 4.

Figure 3: Function info sheet of the "Fill Case" function

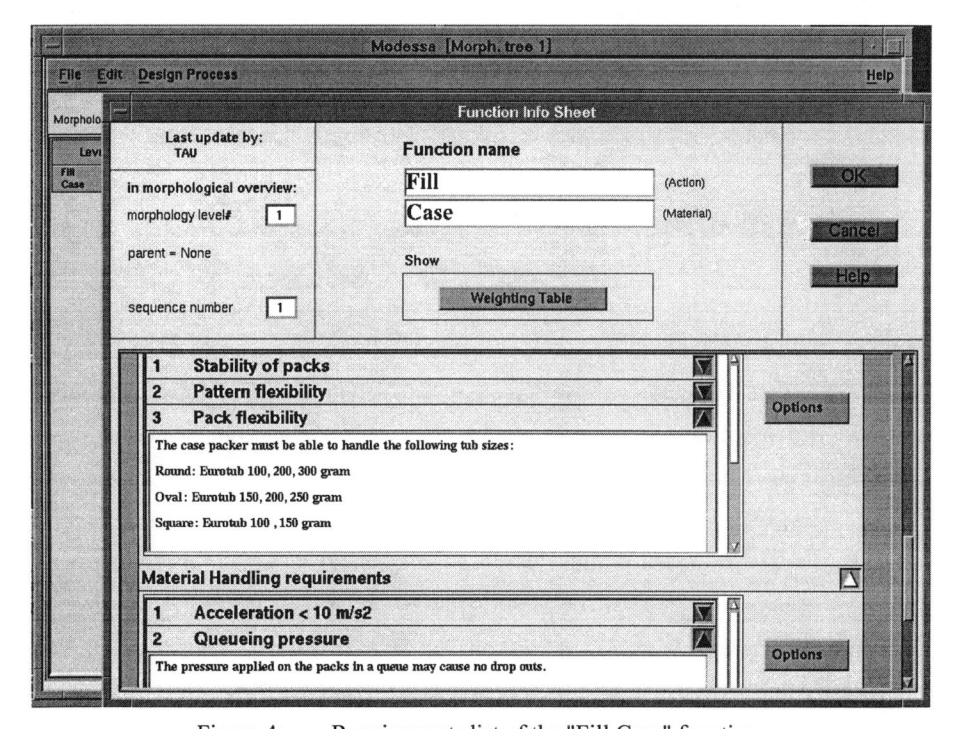

Figure 4: Requirements list of the "Fill Case" function

The next step is to search for design alternatives that could possibly solve the "fill case" function given the requirements. These design alternatives can be retrieved from the design alternatives database or they can be created by the designer using the "select alternatives screen" (see figure 5). In the "select alternatives screen" the designer can click on the "info about alternative" button to view the "alternative info sheet" of an interesting alternative. The "side loading" alternative info sheet is shown in figure 6. The new alternatives that are created are temporarily put in the "new alternative" category (see left side figure 5). Before the "new alternatives" become "approved alternatives" they will be checked by the database manager to avoid duplicate alternatives or mistakes. The selected alternatives are automatically inserted into the morphological overview. Figure 7 shows the morphological overview of the "fill case" function.

After all design alternatives for the "fill case" function have been identified the best alternative(s) has to be selected. In Modessa this can be done in 2 ways: 1) A rough selection by means of the so called "M/E score". This scoring method allows the designer to give each alternative two scores, one score for mechanical performance and one score for electrical/control performance (see figure 8). 2) A thorough selection by means of the previously discussed weighting table. The weighting table functionality is still under development but a preliminary screen layout is shown in figure 9. When the evaluation is completed the order of the alternatives in the morphological overview will be changed such that from left to right the total score of the alternatives decreases. This is shown in figure 10.

As we can see in figure 10 the "top loading" alternative was selected as the best alternative for the "fill case" function. Having made this decision, we go more into detail and focus on the design of a flexible case packer based on the "top loading" alternative. Doing so we find that the next important design problem we have to solve is the "position tubs" sub-function. Going through the "basic problem solving cycle" of figure 1 again we create the morphological overview for the "position tubs" sub-function shown in figure 11. This "level 2 morphological overview" is displayed as a tab-sheet of the "top loading" alternative with a different background colour. By clicking on the maximize button the tab-sheet is expanded to full screen as in figure 12.

The "layer two step" alternative was selected as the best alternative to solve the "position tubs" function. Figure 13 shows how the "layer two steps" alternative is further detailed with three sub-functions and several alternatives in the "level 3 morphological

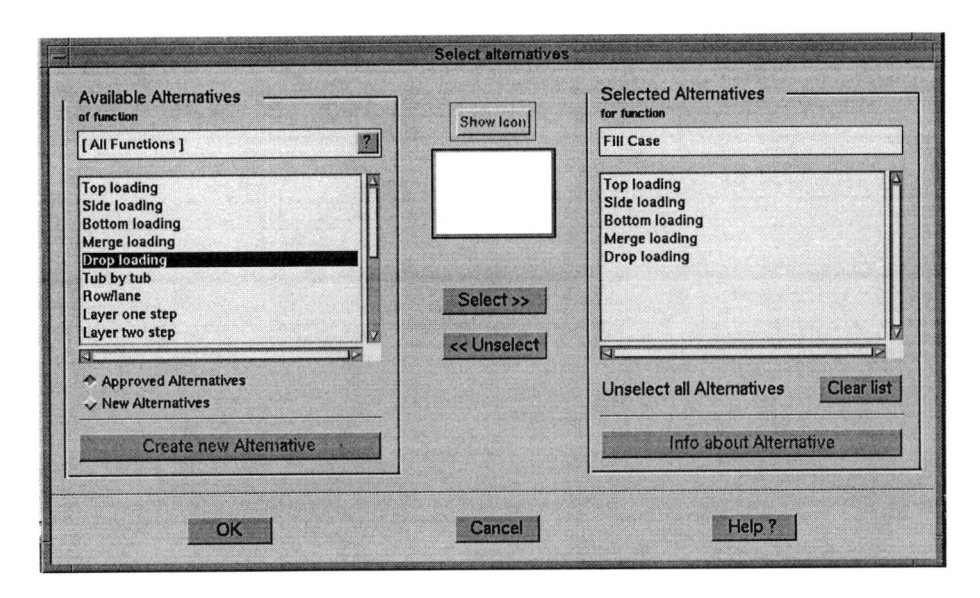

Figure 5: Select alternatives screen

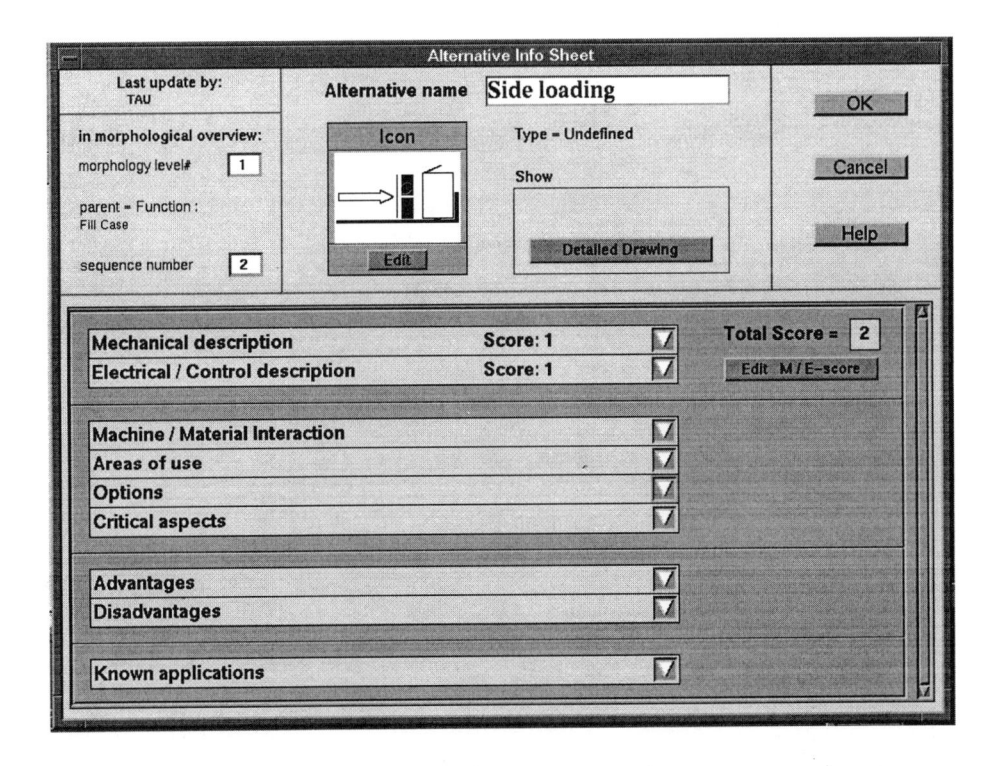

Figure 6: Alternative info sheet of the "Side Loading" alternative

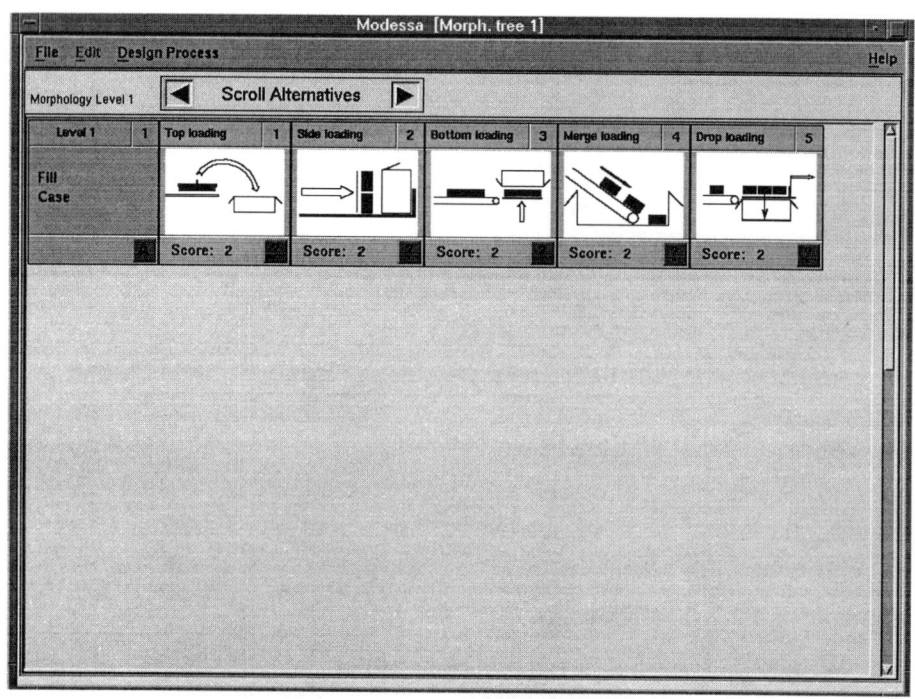

Figure 7: Morphological overview of the "Fill Case" function

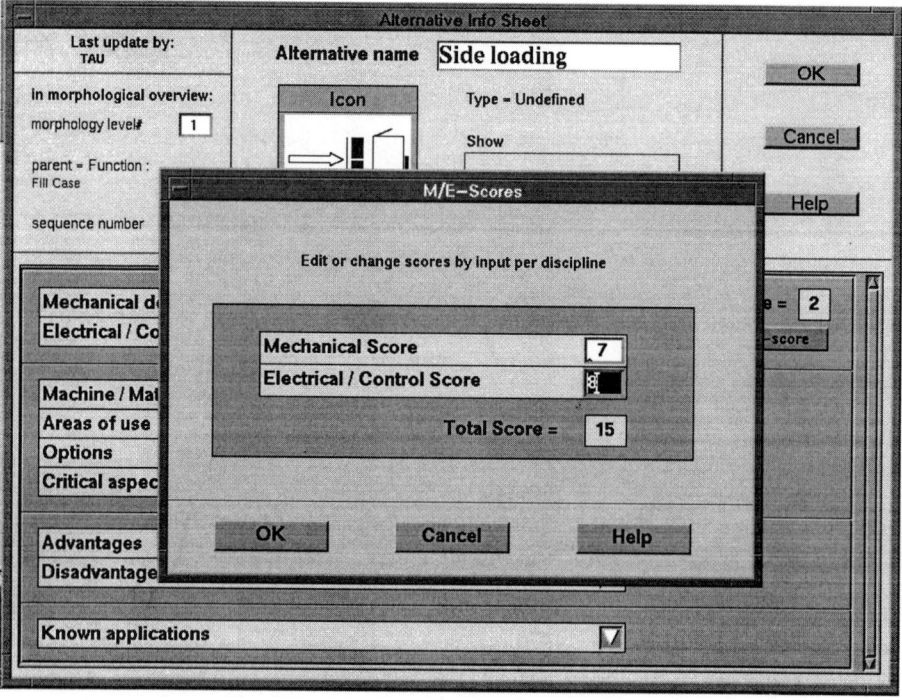

Figure 8: M/E Scores screen

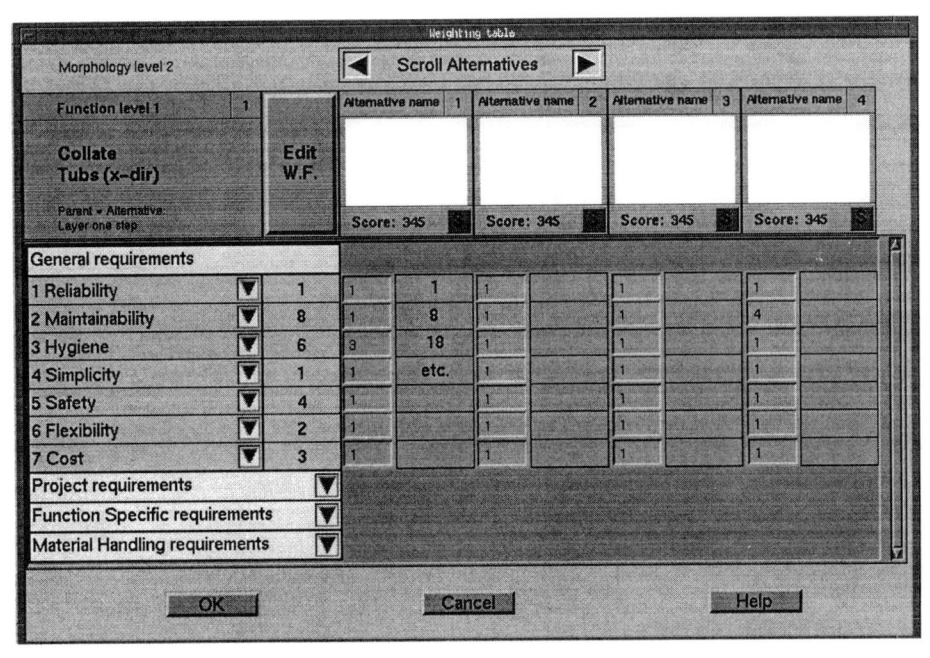

Figure 9: Weighting table screen layout (not yet implemented)

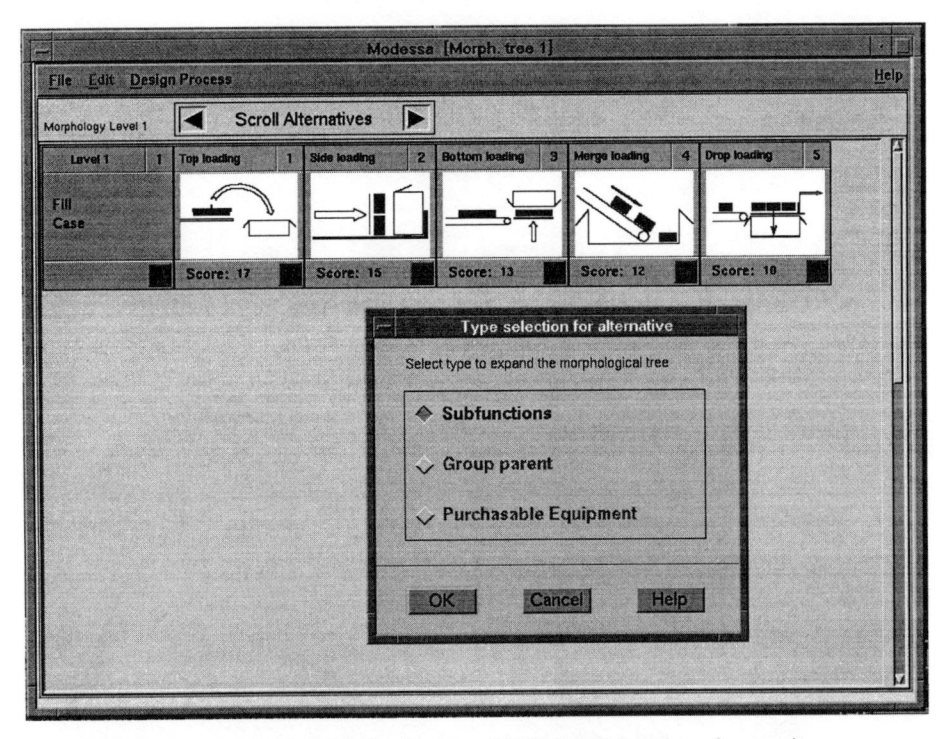

Figure 10: Morphological overview of "Fill Case" function after scoring

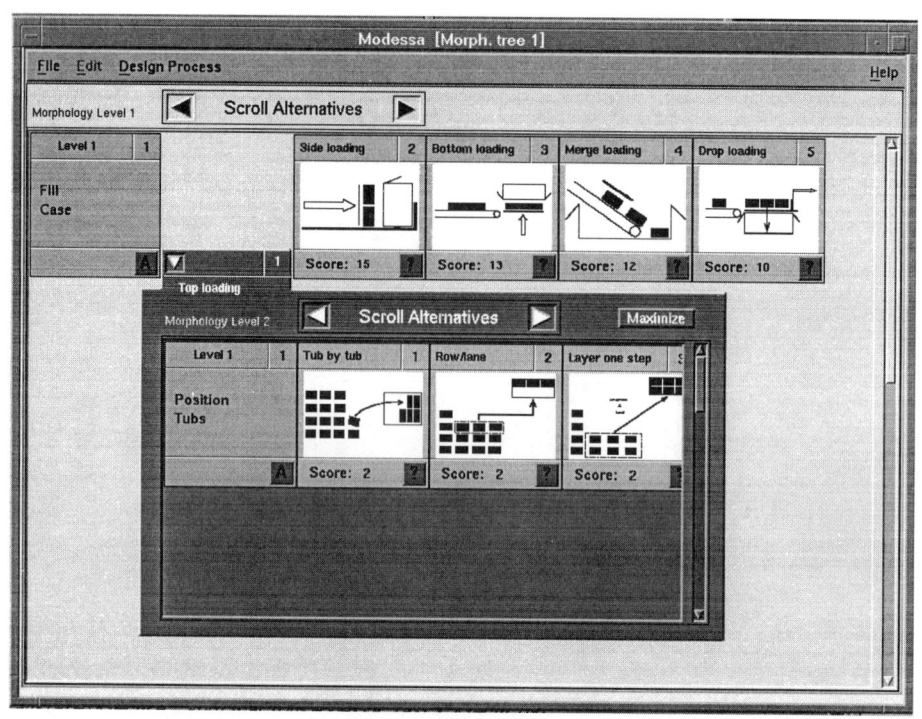

Figure 11: "Top Loading" alternative expanded with "Position Tubs" sub-function

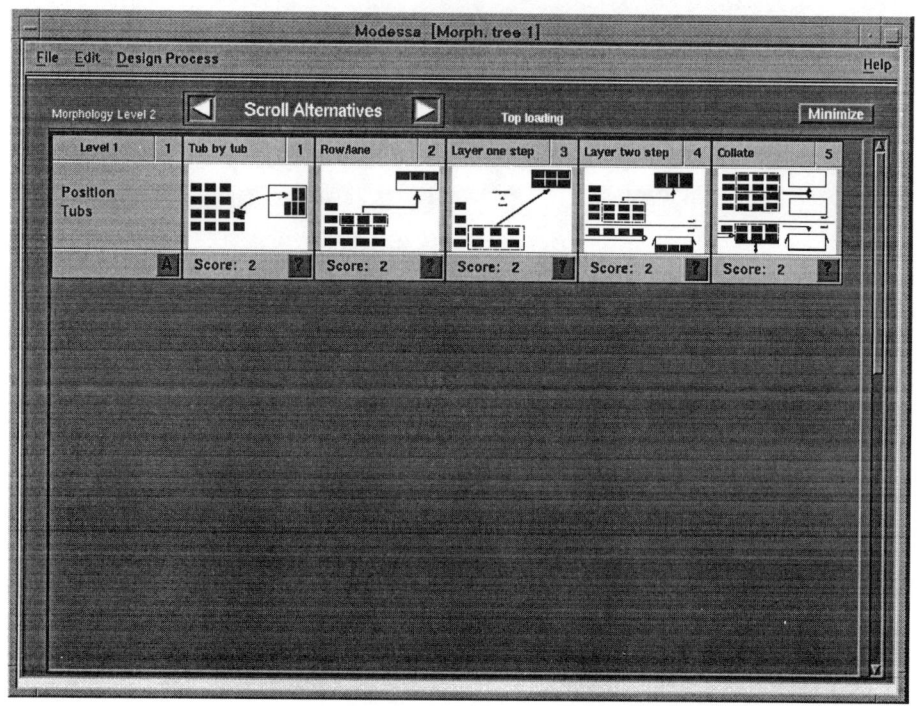

Figure 12: Maximized morphological overview of the "Position Tubs" sub-function

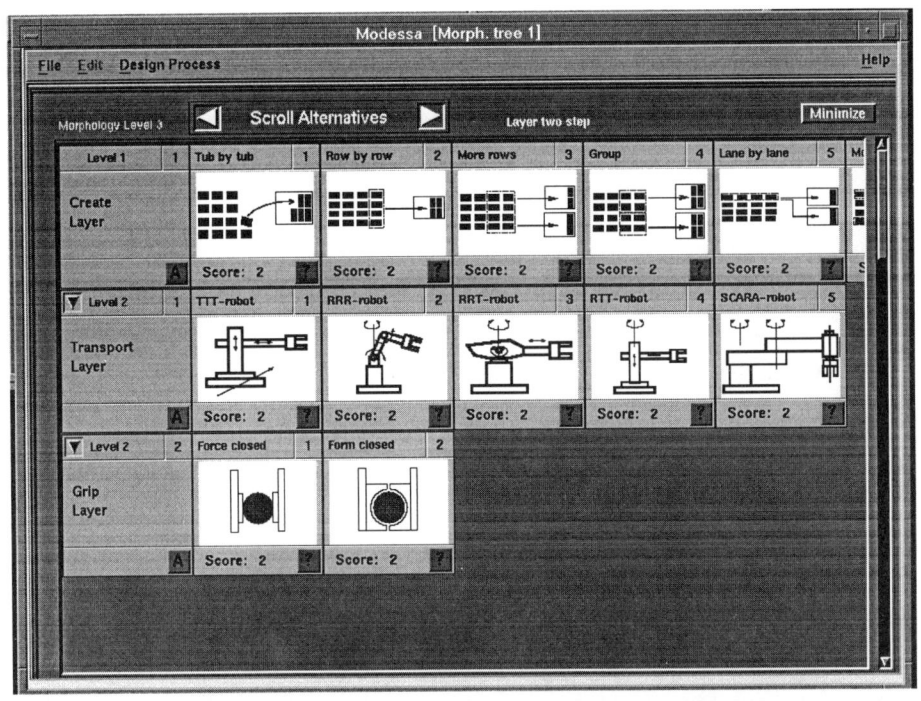

Figure 13: Morphological overview of the "Layer Two Step" alternative

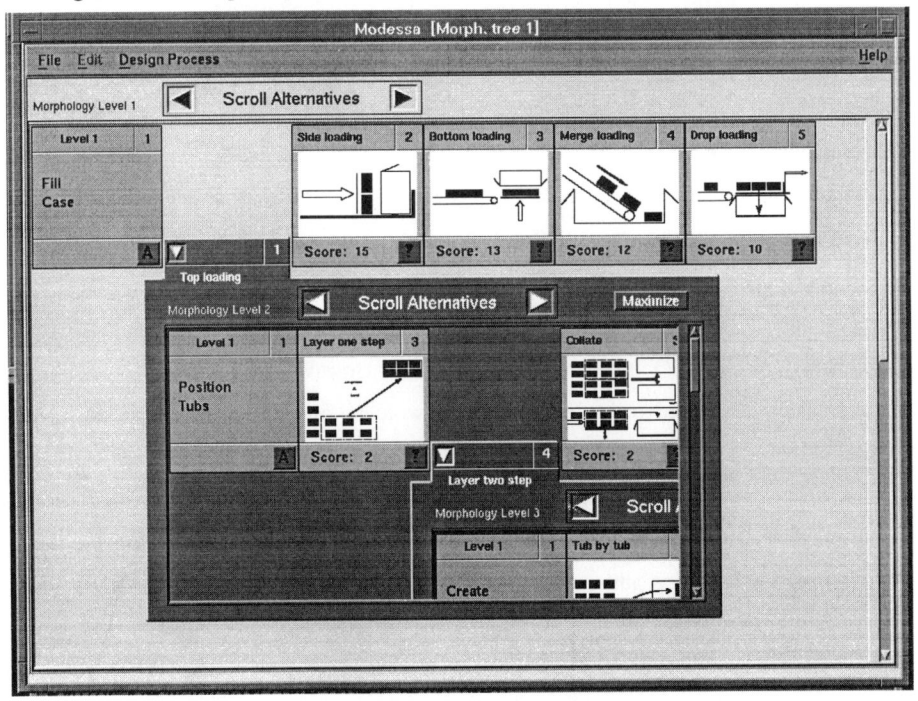

Figure 14: "Flexible Case Packer" morphological overview

overview". In this fashion we can continue to detail the design problem upto a level where a suitable design alternative has been identified for each relevant design function. Figure 14 shows the completed morphological overview of the "flexible case packer" design project.

6. MODESSA SOFTWARE & HARDWARE STRUCTURE

6.1 Software Structure

The Modessa software has been developed according to the so called "three-tier model" consisting of a user-interface, application and database tier.

- The user-interface tier

The GUI is being developed according to the OSF/MOTIF standard using Hitachi's ObjectIQ GUI builder [3]. The graphical user interface enables the user to control Modessa by means of pull-down menus, pop-up menus and buttons. The aim is to enable users to use the system with minimum training requirements. Therefore the GUI is designed so that no commands have to be remembered and all actions can be started with the mouse.

- The application tier

The functionality of each of the GUI Objects (button, menus etc.) is implemented in the application tier using the ObjectIQ application development tool for knowledge based system and expert systems. The ObjectIQ language is a 4th generation language that compiles into C. It allows the use of both objects and rules. At this moment the rules functionality of ObjectIQ is not used in Modessa. However, a future AI-extension of Modessa using rule-based reasoning is foreseen with the objective to automatically select suitable design alternatives for a function based on the requirements list.

The drawing functionality to visualize a design alternative is provided by the user-friendly object oriented drawing editor "IslandDraw" launched from within Modessa. The alternative drawing is automatically converted to an icon in the so called TIFF pixel format so it can be imported in ObjectIQ. A file reference to the original "IslandDraw" drawing file is kept in the database so it can always be edited when needed.

- The database tier

An Oracle relational database application is developed to store, search and retrieve design data and to ensure the security and the integrity of the data stored. Within the relational database application we can distinguish four sub-databases: 1) The "Functions Database" containing the function info sheets, 2) The "Requirements Database" containing requirements indexed in generic, problem specific and material specific groups, 3) The "Alternatives Database" containing alternative info sheets and 4) The "Previous Projects Database" containing the documentation of previous design projects (project description, requirements list, morphological overviews and weighting tables).

6.2 Hardware Structure

The Modessa hardware structure is based on a client-server architecture using a UNIX server and both Unix workstations and Windows PC's as clients. The Modessa and the Oracle database applications run on a Hewlett Packard 715/80 server using the HP-VUE X-windows based graphical user environment. The mechanical engineers in the TAU all have access to a personal HP 712/60 CAD workstation. They run Modessa on their own workstation but access the Oracle database on the server.

The other design team members however have personal computers that run under Microsoft's Windows 3.11. They use Modessa by means of a remote login on the server using the HCL EXCEED 4.1 X-windows emulator. Unfortunately this emulation results in a lower graphical performance and increases the server workload.

At start-up Modessa asks the user for the screen size that is used. The size of the morphological overviews is adjusted so both PC users having 17 inch monitors and HP or PC users having 20 inch monitors have a full screen display.

7. DISCUSSION AND CONCLUDING REMARKS

A computer based design support system called Modessa for use in the conceptual design phase has been described. It is intended for use by multi-disciplinary design teams in designing new production equipment. However Modessa is general purpose in nature and can therefore can be used for a wide variety of design applications. On the one hand Modessa can be used to support the design of a new factory where the problem is to find for each product (function) the best production equipment among those offered by many suppliers (the alternatives). On the other hand Modessa can also be used for the design of an end-effector as we saw in the example.

The knowledge base in Modessa increases continuously as the design team adds new design functions (problems) and applicable alternatives (solutions) to the database. This design application specific knowledge base allows Modessa to work more efficiently in a specific area such as the design of packaging machinery.

However, Modessa will not replace a gifted designer or model creativity. Rather it can support a good designer or even average designer to improve the quality of his design and increase the speed of the design process by:
- Suggesting known design alternatives for design functions leaving more time for the creation of novel design alternatives.
- Offering a "common language" to objectively evaluate the design alternatives.
- Simplifying the creation and reuse of design documentation.

REFERENCES

[1] Court A.W., Culley S.J., "A survey of information access and storage amongst engineering designers", Univ. of Bath, UK, ISBN 1857900049, 1993.

[2] Fisher U., Fisher D., "Wysikon- Ein wissensbasiertes system zur integration methodischer hilfsmittel im konstruktionsprozeß", CAT '91, Stuttgart, Germany, 1991.

[3] Hitachi, "ObjectIQ User Guide", Manual number 3050-7-353-20(E), Hitachi Ltd., Japan, 1993.

[4] Hundal M., Langholtz L., "Computer-Aided Conceptual Design: An application of X-windows, with C", Proceedings of the ASME Design Technical Conferences DTM '92, DE-Vol. 42, New York, USA, ISBN 0-7918-0936-6, 1992, pp 1-9.

[5] Kroonenberg H van den, "Methodisch ontwerpen", Mech. Eng. Dept. Univ. of Twente, The Netherlands, WB.83/OC-12461, 1983.

[6] Nicholls K., "Getting engineering changes under control", Journal of Engineering Design, Vol 1, No 1, 1990, pp 5-15.

[7] Oh V., Langdon P., Sharpe J., "Schemebuilder: An integrated computer environment for product design", 1994 Lancaster Int. Workshop CACD '94, Eng. Design Centre, Lancaster University, UK, ISBN 0-901800-37-6, 1994, pp 339-362.

[8] Pahl G., Beitz W., "Engineering Design a systematic approach", The Design Council, London, UK, ISBN 85072239X, 1988.

[9] Prins J.A.A., Olthoff M.M., "Morphology as a documentation framework", ICED '93 Vol 1 WDK 22, Heurista, Swiss, ISBN 3856930272, 1993, pp 158-164.

[10] Roozenburg N., Eekels J., "Produktontwerpen, structuur en methoden", Uitgeverij Lemma, Utrecht, The Netherlands, ISBN 90-5189-067-2, 1991, p. 79.

[11] Schlicksupp H., Fahle R., "Morphos - Methoden systematischer problemlösung", Vogel Buchverlag, Würzburg, Germany, 1988.

[12] Starkey C.V., "Engineering design decisions", Edward Arnold, ISBN 340543787, 1992, p. 110.

[13] VDI Richtlinie 2222, "Sheet 1. Konstruktionsmethodik, konzipieren technischer produkte", VDI-Verlag, Düsseldorf, Germany, 1973.

[14] Zwicky F.,"Entdecken, erfinden, forschen im morphologischen weltbild", Droemer-Knaur, Munich, Germany, 1976/1971.

INTERACTIVE KNOWLEDGE SUPPORT TO CONCEPTUAL DESIGN

L Candy, E A Edmonds & D J Patrick

1 INTRODUCTION

In the work described in this paper, we are addressing the question of how to support the designer with appropriate knowledge during conceptual design. An important aim is to provide that knowledge support to the designer as the design activity proceeds. The knowledge is used to enable the exploration of possible design solutions by the provision of generative and evaluative capabilities in the system. These requirements have been demonstrated in an architecture and a support system that applies to the task of vehicle design.

We have developed an approach to design support that uses a new kind of environment called an interactive Knowledge Support System. Support for interactive knowledge-based design requires fluent interaction between designer and knowledge in a way that does not impede the conceptual design process. From a computer support perspective, this implies a number of needs such as, multiple and parallel viewing of information and flexible interactive facilities to handle the information. The key activities include:
* knowledge acquisition from the designer *during* the design process
* knowledge maintenance that is a *continuous* process of refinement and updating
* graphical interaction techniques that draw upon *domain specific terms*.

The LUTCHI Vehicle Packager Knowledge Support System (VPKSS) is a demonstrator system for aiding designers at the conceptual stage of the vehicle design process. It supports the creation of new designs by way of a solution generation and evaluation process that relies upon co-operation between the designer and the knowledge system.

The paper begins with a discussion of support for the conceptual phase of the design process and knowledge-based approaches to design. We then propose a set of requirements for knowledge support to conceptual design that have been identified from previous work. This is followed by a scenario account of the use of a knowledge support system for vehicle concept design. The graphical features of the system and its architecture are then described. Finally, future research developments are proposed, in particular, the extension of the system to a multi-user platform to support team working.

2 CONCEPTUAL DESIGN AND KNOWLEDGE-BASED APPROACHES

2.1 Conceptual Design
There have been many attempts to define conceptual design and the variation in perspectives often depends upon very fundamental issues, such as whether it is based upon a scientific rationalist tradition or it invokes non-rationalist views [1]. In practice, approaches to the whole design process itself are undergoing revision rapidly. This may be partly a result of changes in

the customer requirements or market imperatives and partly a result of new technological support tools and methods. In that context, conceptual design may include the requirements specification as well as the early design solutions. For example, Bowman and Cooper characterise the task of the conceptual designer as, "to understand the customer's need, analyse it and produce a model of the possible solutions and present these for the customer's choice/acceptance". They also distinguish it from detail design thus, "The object, therefore, of concept design is to describe the total product, that of detail design to describe how that product might be made." p12 [2].

In Engineering Design, conceptual design has been characterised as that part of the design process which succeeds the stage when the essential problems are identified and the requirements defined [3, 4]. The designing activity involves the formulation of solutions or concepts that meet those requirements before going on to embodiment and detail design. Here, there is consensus amongst the so-called "prescriptive" design models [5]. However, the exact nature of the negotiation that occurs between the various stages from the initial problem definition and concept design stages through to detail design and the different strategies adopted by designers remains a research issue.

The domain-orientation of design process models in current use has not been the explicit subject of research although it is plain from the models themselves that their character is influenced by such factors. Thus, for example, differences between the engineering and industrial design are evident in respect of the importance placed on visual design activities, which are more prominent in the latter case [6]. The advent of concurrent engineering has longer term implications for the integration of all aspects of design from the visual (and spatial) design processes to numeric aspects. In respect of vehicle design, the vehicle package provides a total product model to which a knowledge-based engineering approach can be applied.

2.2 Knowledge-Based Design

The approach to conceptual design reported in this paper is one set within a framework of Knowledge-Based Engineering (KBE). Claims that KBE systems offer significant improvements in the time taken to engineer complex assemblies have been made and, although the gains have not been documented in any detail, the approach is gaining ground rapidly [7]. In KBE, an important aim is to capture knowledge about the design, analysis and manufacture of a product so as to represent the engineering intent behind the design in a systematic product model [8].

A primary goal of KBE is to maximise the synergy between the design concept and the manufactured product. In the lengthy process from conceptual design through specification to manufacturing, it is often the case that the original design is not carried through because it turns out not to be feasible at the implementation stage. The changes or compromises made in manufacture, however, may not reflect the designer's original intentions, because their full implications are not (and probably cannot be) transmitted through the process.

Knowledge-based techniques are being applied in order to support concurrent design engineering with the aim of improving overall cost effectiveness in design [7]. These techniques provide an opportunity to integrate all the elements of the design process in one model. The use of rules makes it possible to incorporate many diverse design considerations. e.g. costs, manufacturing capability, legal requirements etc. When all the significant considerations have been expressed in rule form, the designs may be checked and, where not feasible or legal, may be amended at an early stage before additional commitment is made.

In respect of vehicle design, it could be argued that there is no advantage in providing computer support to the concept design styling stage because the time-scales are very short and no significant lead time can be gained. In the design development stage of vehicle design, there are, by contrast, obvious opportunities for adopting a knowledge-based approach [9]. This stage lasts longer and overlaps with computerised manufacturing procedures. However, if the goal is to tighten the loop between the concept design and detail design stages, there is a need to facilitate a total design approach.

The case for paying attention to optimising the quality and accuracy of the concept design rests upon the heavy resource commitments made at this stage that have significant cost implications for changes downstream. In vehicle design, the basis of a total or concurrent design approach is already in place in the form of the vehicle package where the attention to body styling and structure, (including the passenger compartment) takes place in parallel with engineering considerations. New developments are underway in a number of design applications [7].

One such approach to knowledge-based support for concurrent engineering design is that used in the ICAD system [10]. At Lotus Engineering, a Vehicle Packaging System (VPS) based upon ICAD [11], has been developed which supports the conceptual design of the whole vehicle. In that version of this system, the constraints are agreed first and then the designer is provided with a solution. Such an approach requires the designer to identify and input all parameters and variables beforehand. This is demanding, especially for the less experienced designer. However, the advantage is that the inter-dependencies are well-defined and for that reason, the designs are more reliable. Most significantly, the feedback from constraints to solution is fast enough to enable many iterations to take place.

One of the key aspects of the work described in this paper is the nature of such interactions between the designer and the support system. It is necessary, we believe, to consider these interactions within the context of the design activity that the individual designer is engaged in. It is important therefore, to be clear about the role of the individual designer in the design process as distinct from the organisational procedures within which that activity takes place.

3 KNOWLEDGE SUPPORT AND THE DESIGN PROCESS

We have argued in a previous paper [12] that the variation in design process models can be understood best in terms of the different levels of concern of the models: in other words, whether the intention is to reflect or prescribe the organisational process or the individual designer's process. Often the two are not distinguished where the models are highly abstract. We have represented the individual designer's process as a spiral, in which the emphasis shifts as the design ideas develop and the total range of activities range from initial idea generation to the production of physical models (see Figure 1 below). In certain respects, the spiral can be seen as a synthesis of other models of the design process. However, for the purposes of the discussion here, the main focus is upon the individual designer process.

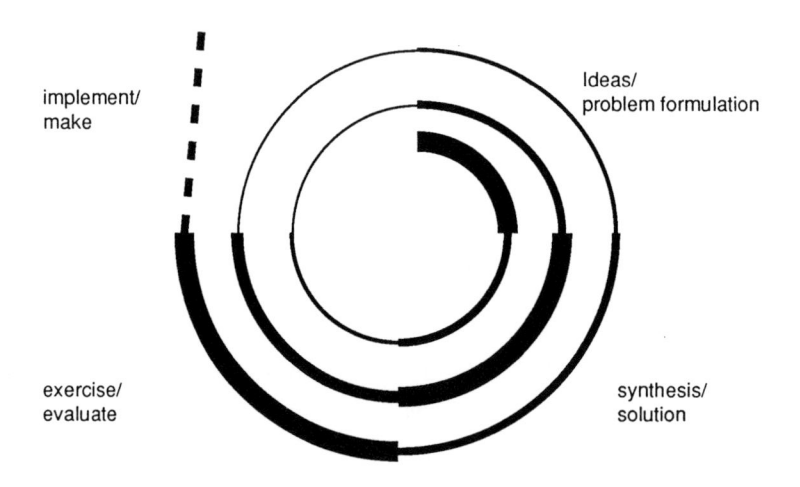

Figure 1 - The spiral design process model

3.1 Requirements for Knowledge Support

If designers are to be provided with support that will enable them to generate a set of possible designs and to evaluate them before committing to a particular solution, they require knowledge that is provided early on in the process and at the appropriate point during the design activity rather than at the end. For example, the Alias surface modelling system requires that the geometric constraints be checked and modified afterwards rather than during the actual constraint specification activity [13].

In the early stages of design, the designer's initial work is not easily characterised by formal procedures. Hence, computer systems that provide support to formally defined processes are unlikely to be helpful to the conceptual and exploratory stages. Flexible and usable support implies taking account of the cognitive attributes of the designer during the process. A number of such issues in relation to the implied requirements have been identified in previous work [14]. From that and related work, core research requirements for knowledge-based support to conceptual design have been identified as follows:-.

i) Access to Knowledge: Acquisition and Evaluation

A key issue is interactive knowledge acquisition and validation. Interfaces that enable designers to express their knowledge to the system are needed for that purpose. Visual and graphical interaction techniques based upon appropriate domain models enable direct interaction between designers and the knowledge represented in the system. Being able to interact with the knowledge enables the designer to apply it in design and to use it for evaluation.

ii) Development of Knowledge : Problem Formulation

Methods are necessary to support the designer in identifying and formulating the design problems to be solved. Very often, once a problem is fully stated, automatic methods can be employed in its solution. The designer requires support to design exploration that can grow as the design develops. Knowledge used in the act of designing is dynamic and in order to capture it effectively it must be handled during the process, not as an after thought. Indeed knowledge development is an essential part of the process.

iii) Reformulation of Knowledge : the Emergence of Concepts

A key requirement is the opportunity for being able to respond to change. Designers need an opportunity to come back and reformulate the design in a flexible manner. During the process new concepts emerge. For example the addition of a new parameter or the conversion of a constant into a variable may play an important role in innovation. The tracking of the designer's emerging ideas is a significant research problem [15].

The discussion and scenario that follows takes a number of steps towards meeting these core requirements.

3.2 Knowledge System Support for Conceptual Design

Most conventional CAD models contain geometric modelling information only and do not provide the designer with support for the optimisation of the design: for example, information about design rules and procedures, legal requirements, engineering and quality standards, tooling, costs etc. Because the process of generating a choice of designs and evaluating them against a set of criteria is not given explicit support, the designer is more likely to seek a best fit solution. This can imply a form of "premature closure" where there is a failure to push beyond initial ideas and find more options. There is some evidence that this might explain why many ideas and products do not reach the level of innovation and originality that the effort would lead one to expect [16]. Where the aim is to provide computer-based support for conceptual design, it implies addressing the more creative and exploratory processes in which the designer is engaged. This suggests a need to provide forms of knowledge that may be used during the actual process of design at the appropriate time [17]. To that end, more expressive interaction techniques that do not impair creativity in design are required.

The nature of the design process is solution-oriented in that the designer analyses general problems by way of the generation of solutions to the immediate problem in hand [18]. There will be a set of requirements and a given set of constraints depending upon the initiating events and the organisational situation. During the course of the designing activity, the designer may encounter an unintended implication of the modifications to the existing design made. This issue has to be resolved and, in doing so, there is a need to re-formulate the design problem in hand. The designer is not simply applying well understood knowledge to a well defined problem, development is taking place in the designer's knowledge. Thus, the evolution of a design is an exploratory activity and there will be unforeseen developments as the designer pursues ideas in detail [19].

In order to support this process, a Knowledge Support System [14, 20] enables the designer to analyse the design in progress against a given set of constraints,. The knowledge in the system embodies sets of prototypical designs as well as rules about legal requirements, costs, manufacturing processes and other constraints. The designer thus, has available knowledge about a range of issues that may or may not be useful to the evolution of an individual new design. This may, at any given point, bring to bear considerations in respect of the design that impel the designer towards rethinking the original problem. Indeed, a complete reformulation may then take place and have consequences for the design under consideration.

Where the knowledge is considered by the designer to be inappropriate, he or she may make a conscious decision to ignore some elements of the existing knowledge base. The system reasons on the basis of the new values and provides the designer with a further analysis of the implications of those changes. The designer can continue to change values or to leave them as they are and make the design conform to them. Thus, interaction with the knowledge has two effects:
i) it provides information about constraints and underlying assumptions (of the system) that are immediately accessible to the designer during the process of developing the design. They can either be observed or ignored by the designer as the design proceeds.

ii) it allows the designer to make explicit changes to the formal knowledge in the system i.e. to alter the basic status of the design constraints. This might take the form of changing the ranges of the parameter values or, indeed, adding new parameters or relating existing ones in different ways.

In the approach described below, the requirement to support early design ideas by offering both generative and evaluative capabilities in the system is addressed in the following ways:

i) It provides access to knowledge as a set of constraints in the form of domain specific rules that can be used to test and generate design ideas as they are being developed.

ii) It provides facilities for the consultation of rules in the knowledge base that allows the designer to consider all the elements of the design under consideration and whether these meet existing rules. Where they do not, the designer may alter the rules by adding new constraints or modifying existing rules. In this way the knowledge may be refined and extended.

iii) It provides facilities for capturing knowledge dynamically as the design proceeds. The assumption is that, in conceptual design, the designer's ideas are always developing because the generation of new design solution is a major goal.

These features and their use will be elaborated in the following section in the context of concept design using a vehicle packaging approach.

4 KNOWLEDGE BASED APPROACHES TO CONCEPT VEHICLE PACKAGING

Concept vehicle packaging is an approach to whole vehicle design that incorporates a representation of the key elements or sub-systems such as body structure and style, passenger compartment and drive-line. As such, it is an inter-related set of activities involving the manipulation of the major components and multiple sub-systems. The fundamental requirements of the overall vehicle design are described in terms of the operating characteristics of the whole. Creating a specific design solution to provide the required characteristics is, therefore, very complex, and involves compromise at many points. Changes in order to optimise particular parts of the design demand a very large amount of design effort to rearrange all other parts of the design. The scale of this effort tends to restrict the number of optimisation iterations which can be undertaken without effective automation.

4.1 The Lotus Vehicle Packager System

In the Lotus Vehicle Packager system (LVP) [11] data and rules about vehicle packaging are provided in order that the designer can explore the package by specifying the minimum number of attributes, the system being left to deduce the implementations. The attributes are typically numerical values, such as the required wheel base.

In a typical scenario, the designer approaches the LVP design in a given task order which has been devised specifically for this system. The order is partly determined by the interactions between the vehicle sub-systems and partly by the priorities set by the client organisation. In a typical class of car, for example, the passenger cell might be of primary importance in driving the overall design, whilst, in other cases, the engine and power train might have a higher priority.

Given the required values, which may be defaults, the system generates and displays the vehicle package. Values may then be varied. The process enables and encourages iterations. Because the system *generates* designs according to the rules, there is no need to check, for example, that legal requirements are met. Providing that the legal rules are included, only legal solutions will be generated.

The LVP system has demonstrated that a knowledge-based approach can support concurrent engineering design. Whether or not the type system has the makings of a support tool for conceptual design is not yet proven. The *ease* with which the organisation, and in particular, its design section, can maintain the knowledge, even during the process of investigating a particular design, is an important question. This is because one way in which innovation often occurs is by posing such questions as , "What if we changed that rule?" [21]. Knowledge Support Systems offer a way forward here. In such systems, domain specific interaction languages are used to provide experts with direct access to knowledge within the system and facilities for updating and augmenting the rule base.

4.2 A Knowledge System Approach

A demonstration Knowledge Support System has been developed at the LUTCHI Research Centre that shows, amongst other things, an approach to knowledge acquisition and maintenance in design. The designer is provided with access to historical knowledge, the rules of received wisdom and databases of technical information. A key element of the approach is to enable the designer to review, evaluate and modify the knowledge in the system. A description of the way the KSS supports approaches to the exploratory and analysis activities in conceptual design follows.

4.2.1 Exploration and Analysis in Concept Design. Let us envisage a situation in which a designer wishes to change part of a previous design. This may be to conform to some standard or to evaluate the impact the changes will make on the overall design. The designer displays an existing design represented in the system by an image, possibly even an old sketch.

Graphical objects represent individual features of the new design, for example, in vehicle design, the fuel tank, battery, spare wheel, seating positions of the passengers, etc., which can be manipulated by the designer. The system may also make available graphical objects which the designer can use to represent elements of the knowledge or facts within the system, for example, the rear chair height or the distance between the hips of the front and rear passengers.

At any time, during a design in progress the designer can interrogate the existing knowledge in the system. Here, the designer supplies known parameters and values of the given features and the parameter of the unknown value required. It is assumed that the designer has a good understanding of this kind of knowledge. When the designer confirms the new design decisions, the system then generates a set of parameters and values using the graphical objects which represent elements of the knowledge. The parameter is the identifier of the object and the value may be, for example, the length of the object. The system uses these facts together with a selected knowledge-base to analyse the new design.

For each design parameter the system checks against the existing knowledge base. If no match between the value in the knowledge and the value of the design parameter is found, then it stores the parameter and the value in the knowledge. Once all design parameters and values are analysed the system checks for the completeness of the design. This is done by finding parameters which have been omitted from the new design. Default values may be automatically substituted. In this way, the basis for making the designer aware of the underlying implications of any design change is derived. This supports the drive for a closer relationship between early design and design feasibility.

If, for example, the results of the analysis shows that the value of the parameter has not been matched against the value in the knowledge base, there are two options open to the designer. He or she may change the design to conform to the existing rules, or alternatively, the knowledge may be refined to include the new design attributes. Where the designer chooses to refine the knowledge, he or she may not necessarily know which rule element does not match with the design parameters. A trace facility, which provides a means of interrogating the knowledge in the system to see which rule and elements have not been matched, is provided for that purpose. Once the appropriate knowledge base has been selected, the system provides the designer with a list of all the rules in the knowledge base. The designer selects the rule to be refined and is provided with a rule editor. The trace shows which element did not match against the design parameters, and also the value of that parameter generated from the design. In order to refine the knowledge to conform with the features of the design, the designer may simply change the value in the rule and if confirm later may be then included in the knowledge-base.

In the following section, the Vehicle Packager Knowledge Support System architecture and modules are described. In addition, the approach adopted to graphical interaction is summarised.

5 THE VEHICLE PACKAGER ARCHITECTURE

In this section, we describe an architecture that facilitates the support for conceptual design discussed above. A number of subsidiary facilities, such as the notepad, that are supportive of knowledge development are included in the description for completeness. Each system module, described below, can be used independently or as a whole. For example, the designer may use the analysis module to view the results of a previous design option or he may use the analysis module along with the knowledge base module to view the implications of changing elements in the knowledge. Figure 2 illustrates the architecture of the Vehicle Packager.

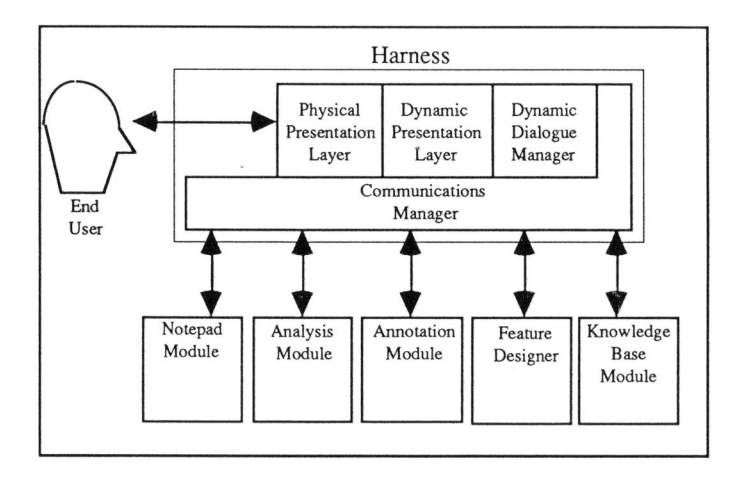

Figure 2 - The architecture of the VPKSS

5.1 System Modules

i) The *Annotation module* provides access to libraries of previous designs, sketches, images and domain specific annotation objects to enable the designer to explore new designs. The designer can select and manipulate these objects to create new designs. The design options are shown in Figure 3 below which illustrates the annotation of a vehicle packaging drawing.

The designer displays an existing design represented by an image in the system. To explore new design decisions the designer places graphical objects over the image. There are two types of graphical objects. The first type of graphical object can be manipulated to represent features of the new design, for example, the seating position of the rear passenger or the position and shape of the fuel tank. The second type of graphical object can not be manipulated by the designer. These objects represent parameters in the knowledge, for example, the rear chair

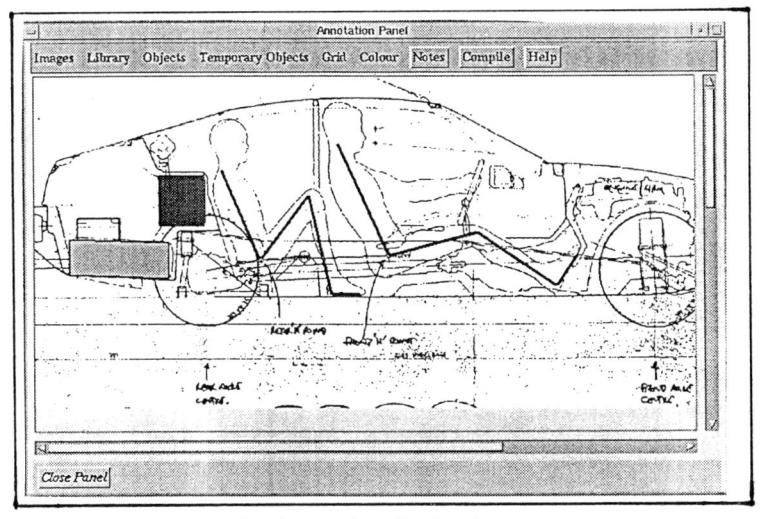

Figure 3 - Annotation of a drawing

height or the distance between the hips of the front and rear passenger, known as the couple distance. This type of object attach to existing objects, and change dynamically as the design evolves. For example, the *couple distance* depends upon the front and rear passengers and will change if either passenger hip points are moved.

During the design process the designer may wish to know values for various features of the new design. The system provides a facility which enables the designer to interrogate the knowledge within the system. If, for example, the designer does not know the value for the percentile of the rear dummy he may provide the system with known values for the vertical displacement, the horizontal displacement and the parameter of the unknown value. Once the designer is ready to confirm the design decisions, the system identifies graphical objects which represent the knowledge within the system and generates values from the information supplied by the objects.

ii) The *Analysis module* uses the specified knowledge base to give feedback to the designer about characteristics of the design. The designer can explore the feedback analysis to identify any mismatches which occurred between the knowledge and the design. The trace facility shows the designer which element of a rule does not match the values from the new design features. The designer may then change the design to conform with the knowledge or refine the knowledge to conform with the new design decisions. The output of a trace identifies items of knowledge not matched during the analysis.

iii) The *Knowledge Base module* provides access to the knowledge allowing the user to browse, edit and add new rules using graphical editors and indexes. The rule structure is dynamic and the user can change the structure of the knowledge using a graphical specification. Once the appropriate knowledge base is selected, the designer is provided with a list of all the rules contained within it. This is the *rule index*. The designer can then add, remove or alter rules and review the whole knowledge base using the rule editor. (See Figure 4).

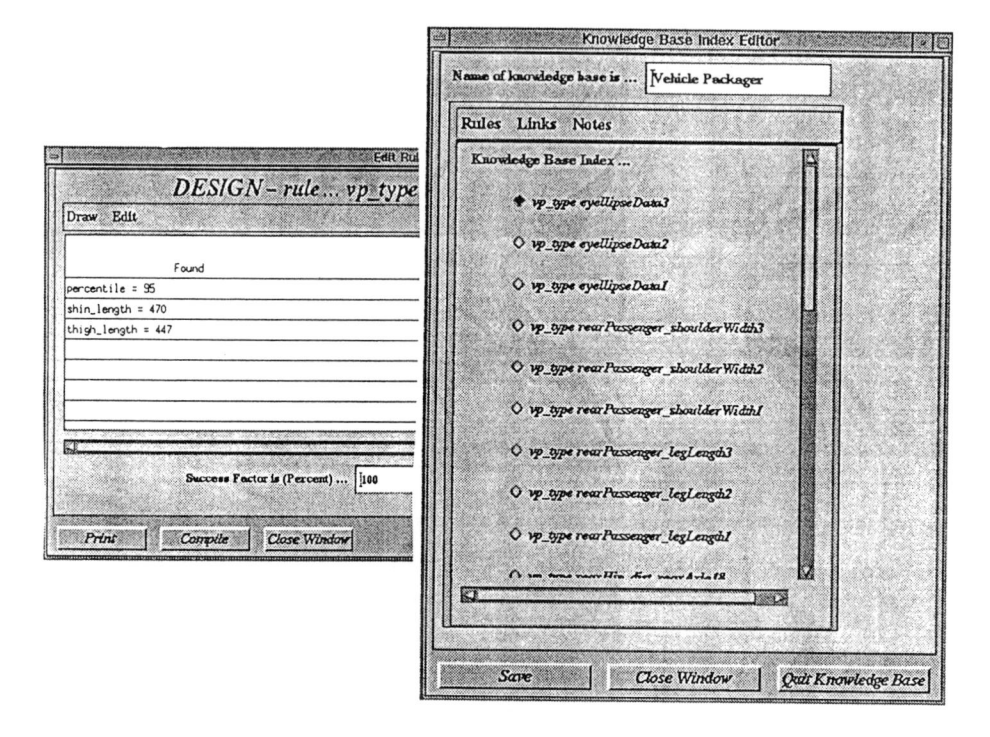

Figure 4 - Rule index and rule editor

The VPKSS provides facilities to support knowledge maintenance. These include the ability to copy entire rule constructs and rule elements between individual rule editors, therefore, the designer can construct rules quickly with fewer errors. The designer may copy rules and rule elements between rule editors. The elements of a rule which are displayed within a rule editor are graphical text objects. They can be manipulated in the same way as the schematic graphical objects. That is, the designer can select, move or delete rule elements at will.

iv) The *Notepad module* enables the user to attach notes about ideas, explanations, sources of reference etc., to any individual item of knowledge in the system. For example, images, and knowledge are indexed contextually with reference to the design knowledge and the favoured representation. The designer can attach notes to elements of the whole system. For example, he or she may attach notes to graphical objects, complete designs and rules in the knowledge base. The system provides a note pad editor which enables the designer to create the documents quickly and create links to other note pad documents. Hypertext documents are created from the text and links which have been entered by the designer and the system allows the user to create links to other hypertext documents [22]. Modules can communicate with the note pad module via the FOCUS messaging system to view the documents at any time (see 5.2 below).

v) The *Feature Designer* enables the designer to graphically create new objects or 'features' for new designs to be used in the Annotation Module. The system enables the designer to apply constraints to graphical objects that define how the they will be drawn and related to existing objects.The designer may add attributes to the primitives of objects and the object itself: for example, line thickness, line style, line colour, text font/size, the objects colour, and how the object will interact with the background. Graphical objects can be defined as being transparent or opaque. Further details about the Feature Designer can be found in Murray et al [23].

5.2 Abstract Interaction Objects

The Vehicle Packager Knowledge Support System was developed using the FOCUS Front End system development software. It consists of several modules running as separate processes that may be located on different machines, connected via a network. Modules can communicate with each other using the FOCUS messaging system. Within the FOCUS distributed architecture, the Harness module is responsible for providing user interface functionality and controlling user dialogues. The graphics functions have been developed as extensions to the Dynamic Presentation Layer and the Physical Presentation Layer [22].

A set of graphics primitives forms the basis of a set of graphical Abstract Interaction Objects (AIOs). An AIO is an object which defines *contents* and *attributes* of an object but does not specify its presentation on screen. Standard AIO definitions represent basic interaction functions, for example, selection, question and hypertext. The AIO approach to interface specification can be seen to break down into three levels of complexity:

- interface primitives - tool kit widgets
- standard AIOs - composed of interface primitives
- interaction object/frame/group - composed of standard AIOs

The final high level interaction object is specified with a simple specification that hides, via the AIO definitions, a considerable degree of complexity at the lowest level of tool kit widgets.

The fundamental design of the FOCUS system envisaged the extension of the library of AIOs stored in the Presentation Knowledge Base. The standard AIOs can be extended by the application developer using an interactive editor.

A set of graphics primitives forms the basis of a set of graphical AIOs. In this way, graphical AIOs are an integral part of the AIO library. They extend the library by adding graphical interaction objects, rather than altering it in any fundamental way. Thus, it is possible to combine lines, circles, ellipses, boxes and text, specify suitable constraints, assign an AIO description to them and allow other modules to reference them in the normal way through the messaging system. This allows the system to create graphical interaction objects on the screen dynamically by sending an appropriate message.

The definition of a graphical interaction object can be updated in real time by the user of the system. A graphical interaction object has a set of points defined by the system developer. These points are the extents of graphical primitives, for example, the start and end points of a line, which can be manipulated by the user of the system. The Presentation Knowledge Base will keep a copy of the original library definition, until the updated object is deleted or saved

6 FUTURE RESEARCH DIRECTIONS

6.1 Implications of User Evaluation

The demonstrator VPKSS has been evaluated by experts in vehicle conceptual design. The evaluation process was informal and based on demonstration and use. The response to the system was, however, very positive and gave rise to a number of specific issues to be addressed in future versions. In particular, the following additional desirable attributes of such a knowledge support system for vehicle packaging were identified :-

i) It was acknowledged the designer might wish, according to the needs of the design under consideration, to "bend the rules" in the knowledge base. However, and conversely, some types of rules, such as legal regulations, should be, in theory, never variable. To impose this rigidly has implications for the boundaries or constraints imposed upon the designer. Breaking rules may be creative in that by moving outside the conventional design space, new solutions may be explored. Thus, a potential conflict may exist here between optimisation and exploratory design activities. Differentiation between fixed and inviolable rules and those open to change would, in a real application in practice, need to be handled within the system explicitly.

ii) It is necessary to index the knowledge from a number of points of view in order to inspect, for example, all of the knowledge associated with a given graphical entity or all of the knowledge relevant to the relationship between the two objects. To assist the designer, explanation documents should be associated with the graphical objects.

iii) A wide range of drawing input is required and, of the physical user interface level, pen-based input for sketching is very important. This form of input device required a high degree of subtlety especially for the expression of tentative ideas.

More generally, vehicle concept design is undertaken by a team comprising designers and engineers with different skills and expertise and those factors must be addressed in any knowledge-based concurrent system approach. This issue is briefly reviewed below.

6.2 Individual and Shared Knowledge

The designer has individual design knowledge as well as the shared knowledge of design practice. The success of innovative designers depends upon the distinguishing contribution that they bring to the field. Hence, successful support systems must tackle both the problem of the acquisition of knowledge of the individual designers and the mechanisms for the transfer of that knowledge to other designers.

Vehicle Packaging, for example, is a team task that requires the participation of people with different skills. Where multi-disciplinary teams are engaged typically in complex design tasks, it is important from the outset that there is suitable exchange of knowledge about the nature of the rules each member is using, modifying or creating. One approach is to provide a documentation facility that automatically creates a hypertext link to a standard format explanation file for each rule. In this way information can be gathered in a standard, agreed format. Thus, as the user creates a new rule, the system will add the rule title to the list and create a hypertext link to a new file, all automatically. The system can create suitable file names, links, etc and maintain the system. The user of such a system would be presented with a text (or hypertext) file to explain the rule and to record any other relevant information about it. Other members of the team, when encountering an unfamiliar rule, could directly access the explanation document. Maintenance is, for example, considerably enhanced by this facility. However, the control of change, familiar in the database world, becomes a significant issue.

To serve the needs of a team effort, the support system must be extended to become a fully fledged multi-user system. Different solutions might be appropriate in different companies depending on the organisational culture. The technical options are well known but the degree of sensitivity with which designers and companies might hold the knowledge in the system poses additional challenges to the already significant technical ones. A private design knowledge base might contain individual design strategies that form part of the design knowledge of the experienced designer. Shared access to standard domain knowledge, organisational constraints, legal regulations, existing product models could be provided as part of the shared design space. Some of the issues implementing such facilities discussed by Jones et al. [25] and in Edmonds et al. [15].

One specific issue that must be addressed is the heterogeneous character of the participants in the design team. The specific languages of interaction used (typically graphical) may well, therefore, vary from team member to team member. The approach being investigated by the authors uses restricted natural language as the commonly available output. Thus, as a designer manipulates the graphical interface and the internal formal knowledge representations are formed, a translator program makes a direct restricted English equivalent of the knowledge available, when required. This description provides an interpretation of designer actions that are not normally readily accessible to other members of the team.

7 CONCLUSIONS

An approach to the provision of support to conceptual design by means of an interactive Knowledge Support System was described in this paper. This was demonstrated in an architecture and a system applied to the task of concept vehicle design. Our aim is to provide knowledge that will support the designer during the design process itself. The knowledge is used to support the exploration of possible design solutions by providing generative and evaluative capabilities in the system. Graphical interactive techniques allow the designer to interact with the domain knowledge in specific concept vehicle design terms. Future developments were discussed, in particular, the extension of the system support to a multi-user platform to support team working. An issue that is being addressed in our current research is how to support different participants in the design team.

ACKNOWLEDGEMENTS

We wish to thank the members of the Vehicle Concepts Department of Lotus Engineering, U.K. for their support and encouragement and to Kelvin Clibbon for his comments on the paper. The work was partly funded by the Engineering and Physical Sciences Research Council, Contract No. GR/J41680.

REFERENCES

[1] Coyne, R. and Snodgrass, A. (1993). Rescuing CAD from rationalism. Design Studies, 14, 2, pp 100-123.

[2] Bowman, D. and Cooper, C. (1994). Conceptual design in an industrial context: An investigation into the use of Schemebuilder within the Aerospace industry.Proceedings of Lancaster International Workshop on Engineering Design CACD'94, April.

[3] Pahl, G. and Beitz, W. (1984). Engineering Design. London: The Design Council.

[4] French, M. (1985). Conceptual Design for Engineers, 2nd ed. Design Council, London.

[5] Blessing, L. T.M. (1994). A Process-Based Approach to Computer-Supported Engineering Design. Black Bear Press Ltd, Cambridge.

[6] Van Dijk, C.G.C. (1994). Evaluation of a surface modeller for conceptual design. Proceedings of the 1994 Lancaster International Workshop on Engineering Design CACD'94, 11th-13th April, pp 149-169.

[7] Gregory, A. (1992). Separating fact from fiction. Manufacturing Breakthrough, November/December, pp 329-333.

[8] Wagner, M.R. (1990). Understanding the ICAD system, ICAD, Inc. Cambridge: MA.

[9] Tovey, M. (1992). Intuitive and objective processes in automotive design. Design Studies, 13, pp 23-41.

[10] Wilson, J. (1992).Knowledge-Based Systems in CAD/CAM. ICAD Engineering Automation Ltd.

[11] Lotus (1993). Knowledge-Based Engineering at Lotus. Internal Document.

[12] Candy, L. and Edmonds, E.A. (1994). Artefacts and the designer's process: implications for computer support to design. Journal of Design Sciences and Technology, Hermes, 3, 1, pp 11-31.

[13] Tovey, M. (1994). Form creation techniques for automotive CAD. Design Studies, 15, 1, pp 85-114.
[14] Edmonds, E.A. and Candy, L. (1993). Knowledge support systems for conceptual design: the amplification of creativity. Salvendy and Smith (eds).Human-Computer Interaction: Software and Hardware Interfaces, HCI International '93, Elsevier, Amsterdam, pp 350-355.

[15] Edmonds, E.A. Candy, L. Jones, R.M. and Soufi, B. (1994). Support for collaborative design : agents and emergence. Communications of the ACM, 37, 7, pp 41-47.

[16] Perkins, D.N. (1981).The Mind's Best Work. Harvard University Press: Cambridge, MA.

[17] Oh, V., Langdon, P. and Sharpe, J. (1994). Schemebuilder: An integrated computer environment for product design. In Computer-Aided Conceptual Design, J. Sharpe and V. Oh (editors), Proceedings of the Lancaster International Workshop on Engineering Design CACD'94, pp 339-362.

[18] Lawson, B. (1980). How Designers Think. The Architectural Press Ltd: London.

[19] Logan, B. and Smithers, T. (1993). Creativity and design as exploration. In Modelling Creativity and Knowledge-Based Creative Design, J.S.Gero and M. L. Maher (editors), Lawrence Erlbaum Associates: Hillsdale, pp 177-210.

[20] Gaines, B. (1990). Knowledge Support Systems. Knowledge-Based Systems, 3, 4, pp 192-201.

[21] Boden, M.A. (1990).The Creative Mind: Myths and Mechanisms. Weidenfeld & Nicolson, London.

[22] McAleese, R. and Green, C. (1990). Hypertext State of the Art. Intellect Ltd: Oxford.

[23] Murray, B.S., Edmonds, E.A., Candy, L. & Foster, T.J. (1993). Constructing Semantic Graphical Objects. Proceeding of Third International Conference on Computational Graphics and Visualization Techniques, Santo (ed), Compugraphics '93, pp 46-57.

[24] Edmonds, E.A., Murray, B.S., Ghazikhanian, J. and Heggie, S.P. (1992). The re-use and integration of existing software: a central role for the intelligent user interface. Monk, Diaper and Harrison (editors), People and Computers VII, Cambridge University Press, Cambridge, pp 415-427.

[25] Jones, R.M. and Edmonds, E.A. (1994). A Framework for Negotiation. In CSCW and Artificial Intelligence Connolly, J.H. and Edmonds, E.A. (editors). Springer-Verlag: London, Chapter 2, pp 13-22.

A SUPPORT SYSTEM FOR BUILDING DESIGN - EXPERIENCES AND CONVICTIONS FROM THE FABEL[1] PROJECT

B Bartsch-Spörl & S Bakhtari

---

[1] This research was supported by the German Ministry for Research and Technology (BMFT) within the joint project FABEL under contract No. 01IW104. Partners in FABEL are German National Research Center for Computer Science (GMD), Sankt Augustin, BSR Consulting GmbH, Muenchen, Technical University of Dresden, HTWK Leipzig, University of Freiburg, and University of Karlruhe.

1 INTRODUCTION TO THE PROJECT

FABEL [12] is a joint research project partially supported by the German Ministry for Research and Technology for four years from 1992 till 1996. Its methodological goal is the seamless integration of case-based and model-based approaches to problem solving. Its application oriented demonstration prototypes are built in the domain of building design where we started with supporting the conceptual phases and are now exploring ways to extend the system's capabilities to detailed design.

In particular, FABEL concentrates its software implementation efforts primarily on the design of the heating, ventilating and air conditioning systems (HVAC) for highly complex buildings and uses a methodological approach for the spatial coordination of the different systems called ARMILLA [14].

FABEL's six partners are the German National Research Center of Computer Science as coordinator and system integrator, the University of Karlsruhe for providing the application knowledge and data, the University of Freiburg with a focus on the cognitive science part, the University of Leipzig with a focus on the machine learning part, the University of Dresden with a focus on the data base part and BSR Consulting with a focus on the case-based reasoning and integration part.

2 PROBLEMS WITH USING AI FOR COMPLEX DESIGN TASKS

During the first half year of the project, we received a lot serious warnings from our friends in the German AI community telling us that they would never have chosen such an application domain which is by far too complex and ill-structured to be tractable with AI methods. We tried to reassure them that we intend to start small and that at least with case-based reasoning techniques we will be able to provide support which is desired and will be accepted.

The main reasons for our friends' worries were grounded in the apparent mismatches between the indispensable prerequisites of today's mainstream AI technology and the characteristic features of complex design domains. The most critical mismatches are discussed in the following [3].

Classical AI methods usually require

- **a complete domain model**
  as a necessary prerequisite for being able to come to reliable and well-founded results

- **a consistent domain model**
  as a necessary prerequisite for using sound logic-based approaches

- **the so-called closed world assumption**
  as a necessary prerequisite for being able to use deduction as a well-established problem solving method

- **problems which can be decomposed and recomposed according to known and effectively applicable algorithms**
  as a necessary prerequisite for using e.g. recursive algorithms for problems of very different sizes

- **solution spaces which can be formally defined and which are not too large to be accessible by search**
  as a necessary prerequisite for using systematic search in cases where no better problem solving approach is available.

Unfortunately, complex real-life design domains like FABEL's building design domain or like software design in the large do not fulfil the prerequisites stated above. On the contrary, in such domains quite different situations are typically discovered. Our corresponding real-life experiences are as follows:

- **It is an illusion to hope for being able to acquire a complete domain model and keep it up-to-date**
  because of the huge amount of information necessary and because of the changes that can happen every day at different places so that it is practically impossible to register all these changes in the moment they occur.

- **It is not feasible to guarantee the logical consistency of a steadily growing and changing domain knowledge base**
  because of the incompleteness of the available knowledge about the (in)compatibility of e.g. the actually collected cases and other chunks of the domain knowledge.

- **All kinds of reasoning have to be grounded on an open-world assumption**
  because of the permanent incompleteness of the available domain knowledge.

- **There are (sub-)problems which are hardly decomposable and maybe not recomposable**
  because of complicated and partially not explicitly known interrelations between the individual parts of the (sub-)problem and its possible solution(s).

- **There are problems with solution spaces which are out of reach for enumeration or search methods**
  because of the complexity and the non-discrete nature of the artefacts.

This leads to the consequence that one can

- either stay on stable AI methodological ground and distort the application domain as far as necessary or

- accept the application domains' characteristics and try to find and apply an appropriate mixture of methods which are suitable to provide adequate design support.

For us it was never questionable that we choose the second alternative. This means that in the following we will not show ways of how to overcome the problems with the described mismatches. Instead we will show how we cope with the mismatches by using design support methods which ground on less ambitious prerequisites and how we use them in an integrated way.

3 APPROACHES HOW TO COPE WITH THESE PROBLEMS

The first and from our experience most important step is to establish an adequate definition of the role of a design support system. From our point of view such a system is not aimed at solving the design problems on its own. Instead it plays the role of a design assistant which is known to be of limited competence but able to improve its competence in time and through being used.

This role definition has the consequence that the human designers are and will always remain fully responsible for the artefact to be designed. They may delegate certain tasks to the system and even ask the system for advice or for the suggestion of alternative solutions - but they always have the last decision which path to follow and which suggestion to take over and carry on with.

As soon as this role definition is established, the problem solving capabilities of a design support system are allowed to possess some weaknesses. Particularly, they need not cover the whole spectrum of problems that might occur because if the computer cannot solve a (sub-)problem there is always the designer who is able to try it without computer support.

This role definition has the following advantages:

- If the designer already works with a computer then there is no necessity for fundamental changes in his/her current working practice. All additional support functions can be tried out and accepted or rejected without causing critical acceptance problems.

- The system's functionality can be built up incrementally. This is a robust approach especially for the exploration of new kinds or new combinations of support functions in a complex real-life domain.

- The approach allows the designers to bring in their own ideas and help improving the usability of the system as early as possible. This ensures that the development efforts remain focused on both useful and desired support functions.

4 TOWARDS BETTER CHARACTERISTICS FOR REAL-LIFE DESIGN TASKS

In the next section we describe essential characteristic features of real-life design tasks which influence the selection and development of design support tools.

Complex real-life design tasks like building or software design

- **usually aim at constructing complex one-of-a-kind artefacts**
  which means that the amount of effort spent for optimising the design of the artefact will be considerably less than in the case of artefacts designed for being produced in large numbers.
  This leads to a design process where a certain amount of risks coming from using innovative solutions has to be carried because it would be too expensive to build prototypes and examine their long term behaviour etc.

- **usually start with a set of initial requirements which are far away from being complete and may turn out to be even contradictory**
  which means that the final solution cannot be derived from the initial requirements by a deterministic and straight-forward process.
  This leads to a design process where the solution space has to be partially explored and elaborated and where conflicting requirements have to be harmonised e.g. through weakening the less important ones to a degree which makes the overall problem solvable. With every design step the designer has to make decisions which also help reworking, completing and refining the initial requirements.

- **usually come along unsolvable subproblems which may lead to significant changes in the initial requirements**
  which means that the paths to a solution may contain dead ends which force to go back and substantially rework the requirements and intermediate solutions from a former stage.
  This leads to a design process where the designer cannot measure his/her progress in terms of the number of parts created and placed because all intermediate results are preliminary and subject to change if conflicts should arise.

- **usually have more than one solution which meets the initial requirements - at least to a certain degree**
  which means that the designer has to examine alternatives and set priorities in order to select the most promising ways to arrive at a final solution which is on the whole satisfactory.

This leads to a design process which has to be guided by a set of criteria which help to select among alternatives. These criteria determine e.g. if short term optimisations like saving money in the artefact's construction or production phase are more important than long term optimisations like saving money over the whole life cycle of the artefact.

From these characteristics follows that the design process in domains like building or software design can be seen as a non-linear stepwise refinement process where conflicts, unsolvable subproblems and other obstacles cannot be foreseen and are therefore unavoidable.

Other important characteristic features of the design process are the following:

- **Design is knowledge-intensive.**
  This follows very clearly from the fact that designers spend around fifty percent of their working time with gathering information. From our experience nearly all of them would really appreciate to have more and better computer support for this task.

- **Design is a group activity.**
  This follows from the necessity that with the increasing complexity of the artefact more and more persons with different professional background and possibly located in different places have to be involved. From our experience the importance of computer support for group aspects grows steadily with the number of group members and increases considerably with their geographical distribution.

- **Design is a quality management challenge.**
  This follows from the fact that the computer cannot provide a fully automated assessment of all the different aspects that determine the quality of the artefact. Moreover the criteria for the assessment of the artefact change with increasing completion. From our experience it is important that assessment functions are activated only on demand because otherwise they might disturb creative phases.

- **Design is a test bed for the integration of heterogeneous approaches.**
  This follows from the necessity to use a great variety of support methods and tools ranging from CAD tools, databases, etc. which can be bought on the market to more specific tools for case-based retrieval, assessment, adaptation etc. which have to be developed and integrated. From our experience it is very important to work out the necessary concepts and a system architecture for these integration needs very early.

5 HOW DO WE BUILD BRIDGES IN THE FABEL PROJECT?

From now on we narrow our focus on the activities which have taken place and are going on in the FABEL project. One very characteristic feature for all these activities is the imagination of building bridges

- between the real world and its computer model,
- between the human designers and the system,
- between approximate and detailed design,
- between non-AI-tools and AI approaches,
- between different knowledge representation formats, etc.

These bridges are possibilities to carry at least a certain amount of information from the one bank to the other. This is usually not possible without losses and compromises but we consider it as important to be able to try it at all and regard it already as successful if at least the relevant parts can be transported in both directions.

6 CASES FOR DEALING WITH THE KNOWLEDGE-INTENSITY

The big advantage of case-based reasoning (CBR) approaches [16, 2] is their applicability even in ill-structured domains. This is due to the fact that case-based reasoning does not require a complete and consistent domain model and does not rely on the closed world assumption as a necessary prerequisite.

Therefore in FABEL we started with the development of case-based retrieval functions in order to deal with the knowledge intensity of the design tasks. The first question to be answered was what do we regard as a case. In principle, only buildings as a whole are real-world cases. But cases of this size are of rather limited usability and moreover very hard to manage. So we asked our experts what are meaningful parts of the building and began to regard all meaningful parts as potential cases. The size of the FABEL cases varies considerably from e.g. small parts of the return air system to parts of e.g. the fresh air system covering a whole storey or the whole building.

In order to grasp the notion of similarity in our domain we both developed classical feature based approaches as well as innovative graphically oriented approaches. Till today the following case retrieval modules can be used either stand-alone or combined:

- a classical distance-based approach which determines the distances between cases from the numbers and types of the objects constituting the cases [17]

- an indexing approach using also the objects constituting the cases which is implemented on the basis of an associative memory [13] and

- another approach based on the graphical appearance of the cases which compares abstracted bitmaps of design pictures [11].

These three modules can get enhanced by a fourth module using so-called gestalt indexes [19]. These gestalts represent characteristic patterns like a comb or a ring and can be used as additional case indexes.

Some other retrieval modules using tree- or graph-structured representations are still in the implementation and integration phase [9, 8].

At the moment the designer can select a part of the artefact to be designed and activate the selection of cases which are similar to the selected part. The most suitable case found can be taken over and adapted by hand.

7 A SHARED ONTOLOGY FOR DEALING WITH THE GROUP ASPECTS AND THE QUALITY MANAGEMENT CHALLENGE

The model-based reasoning (MBR) approaches in FABEL took a little longer because of the rather comprehensive knowledge needs of the model-based design support functions. Similar to the case-based retrieval field, there are again different approaches to knowledge-based assessment and adaptation functions under development [10].

One of these approaches is called DOM [4] which stands for domain ontology modelling. In the following we give a short overview on DOM's main concepts, functions and role.

The DOM ontology establishes the meaning and the permissible use of the design elements and the relations between these elements. Furthermore it includes design maxims and rules relative to a considered design methodology [14].

The ontology [15] plays a central role for

- supporting the group aspects because it lays ground for a common terminology and a shared understanding and use of the design elements
- defining what is the scope of the assessment functions and
- establishing a framework for the adaptation functions.

DOM will provide the following design support functions:

- review a design layout in its current state in terms of the permissible use of the involved design elements
- assess the quality of a case in terms of coherence and conformance relative to the underlying ontology
- provide suggestions for improvement
- draw the users attention to discrepancies and conflicts
- indicate alternatives and evaluate their merits.

The following figure 1 [5] gives an overview of the assessment organisation with and around DOM. It shows that drawn layouts are analysed with the goal of determining whether the design elements and relations contained adhere to the contents of the domain ontology. If this is not the case the system informs the designer about the detected discrepancies like e.g. missing elements or violations of maxims.

It is important to note that also this kind of design support functionality does neither require a complete or consistent domain model nor a closed world assumption. It is sufficient to have models for the criteria to be assessed and if unknown objects are detected the system will tell the user and ask what to with them.

At the moment the DOM functionality is fully specified, partially implemented and not yet integrated into the FABEL system.

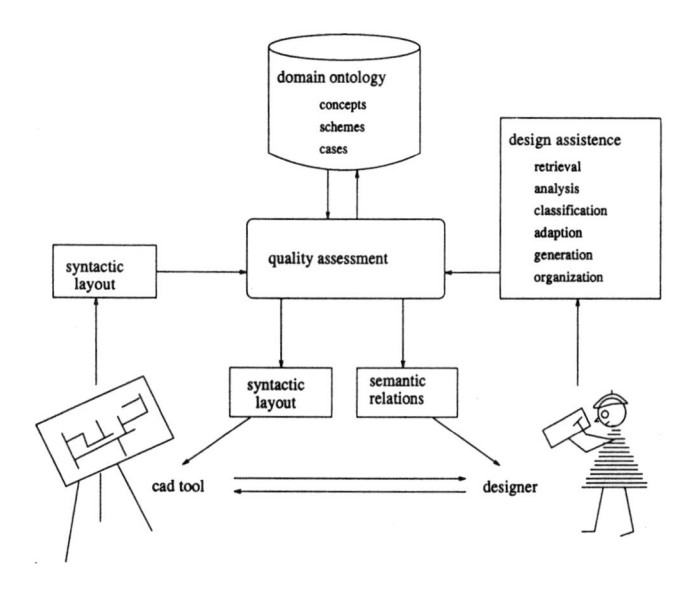

Fig 1 Assessment organisation

8   A SUITABLE SOFTWARE ARCHITECTURE, DEFINED INTERFACES AND SCHEMATA FOR DEALING WITH THE INTEGRATION NEEDS

The necessity to use different sorts of methods and tools, both bought from the market and developed in-house, brings up the topic of a system architecture able to cope with such a variety of software modules.

In the FABEL project the following principles guide the structuring of the system architecture and the development of new modules:

- For the user all design support modules have to operate on the same artefact to be designed and have to exhibit a common look and feel. Graphical input and graphical output are used prior to text communication. This means that e.g. the results of case retrieval, assessment and adaptation functions are shown graphically.

- There is one common data base where the artefact to be designed and the case repositories are located. All modules operate on this data base through a common interface. The individual modules may possess and enhance additional knowledge on their own but all results which are interesting for other modules are exchanged via the common data base.

- In order be able to use different methods alone or in combination for the same or for rather similar tasks there has to be a knowledge representation format for all stages of the artefact which is understood and worked upon by all modules. That means that e.g. different case retrieval modules can work either in parallel or in any sequence. They always get one actual problem description and a set of former cases as input and bring back the problem description and a subset of the former cases.

These three principles sound self-evident and easy to obey to. In the FABEL project they were the result of a discussion process and they are still the topic of ongoing quality assurance work that has to be done during the integration phase of each module.

8.1 Topological Schemata

Up to this point we did not write anything about the integration of different sorts of knowledge chunks like models and cases. Before we go into the details of this line of integration we introduce an intermediate level between models and cases, called schemata.

From our point of view one of the essential insights of the FABEL project [7] is that cases can play a useful role for the design of complex artefacts but a design theory is better expressed with more general terms and with adaptable templates. This has to do with the fact that adaptability plays an important role and that schemata provide convenient ways to combine case knowledge with adaptation knowledge.

The general relations between generic models, episodic cases and schemata are shown in figure 2.

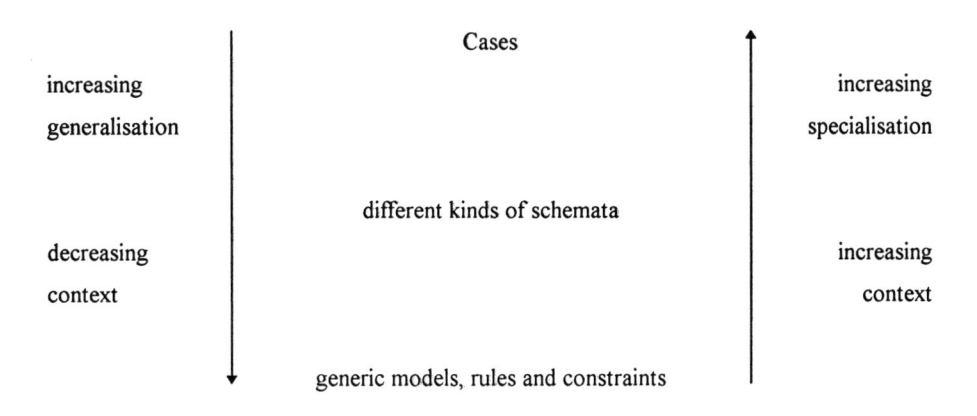

	Cases	
increasing generalisation		increasing specialisation
	different kinds of schemata	
decreasing context		increasing context
	generic models, rules and constraints	

Fig 2 Cases, schemata and generic knowledge

Schemata and schema-based reasoning (SBR) were proposed by Turner [20] for other kinds of application domains:

"Schema-based reasoning is a computer model of adaptive reasoning. It is not a cognitive model, although many of the insights which led to the model and guided its development came from cognitive psychology."

Going one step further, we think that in domains like building or software design it can be shown that schemata evolve from a rational treatment with ways to improve the design process, in particular with the goal of gaining efficiency and ensured quality. As soon as schemata prove to be useful they have a chance to get officially documented in order to enforce their widespread use with the team or the organisation.

Psychological evidence for the use of schemata can also be found in Akin's work on architectural design [1] where Akin uses the term scenario instead of schema and in Pree's work on reusable software components [18] where Pree and others established the term pattern for reusable and recurring pieces of software.

The different kinds of schemata proposed by Turner, namely procedural, contextual and strategic, are not especially well suited for our building design domain. Therefore in this paper we constitute a general view at schemata and call abstract cases in building design 'topological schemata' regardless of the kind of abstraction achieved. So in our terminology topological schemata are reusable design components which carry knowledge useful for the adaptation and integration of cases in actual contexts.

8.2 Representational Issues

In the FABEL domain, we represent all knowledge and data about buildings, cases, generic models and schemata in an object-oriented way [6].

Cases are arrangements of complex design objects with fully instantiated slots. This means that all their attributes have concrete values like a type, a set of features, a location etc. or a reference to another existing design object.

Schemata are stepwise abstractions from cases where
- slots can be omitted or filled with "don't care"
- slots can be filled with variables taken from numerical intervals or from a set of defined values
- slots can contain a formula or an expression to be evaluated
- slots can be filled with references to virtual objects not yet placed etc.

Schemata can be instantiated and thus get specialised to cases. The constraints for the specialisation usually emerge from the concrete situation and from the surrounding context.

8.2.1 An example for a topological schema. In the following we describe an example for the representation and integrated use of ARMILLA design knowledge in the form of cases, schemata and rules.

A very small part of the ARMILLA design knowledge deals with using a grid:
- During the conceptual phases a grid is used.
- The grid size has to be chosen before the design can start.
- All design objects that belong to the conceptual phases have to be placed with their center at a crossing point of the grid.
- The default ARMILLA grid has a side length of 12 units.
- If there are good reasons to deviate from this default, the following adaptations are possible:
    - It is possible to enlarge the grid in steps of exactly one unit.
    - It is recommended to enlarge the grid in steps of 3 units.

A layout pattern for pipes represented as a topological schema which incorporates the recommendation given above is shown in the following diagram:

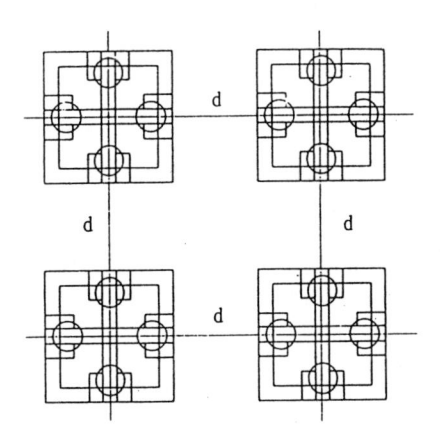

Fig 3 Pipe layout pattern as a topological schema

The variable d has to be taken from the set of possible values {12, 15, 18, 21, 24}. Analogously the rule about the one unit steps can be expressed.

8.2.2 An example for the integrated use. In the following we describe an example for the integrated use of cases and schemata.

In contrast to the rules and recommendations given in the example 8.2.1 there are buildings with a grid size of 10 units. According to the ARMILLA handbook [14] this is possible but not recommended because it leads to a series of modifications which are difficult to carry out. A building with a grid size of 10 units leads to a case which looks similar to the schema given in figure 3 with a distance of d=10.

Now imagine that there is a new building to be designed and there arises the need for a layout pattern for pipes. In this situation it is recommendable to search for layout patterns with a similarity measure that considers the objects contained but abstracts from the exact geometry. With a similarity retrieval method like the one described in [13] the topological schema from example 8.2.1 as well as the case from example 8.2.2 are retrieved. The decision which result to reuse will be determined by the grid size of the actual design. If this grid size is 15, the schema will be instantiated accordingly. If the grid size is 10, the case will be used without adaptation. If the grid size is 11, the architect has to choose which part to take and to adapt it by hand. If the grid size is 9 the adaptation by hand will fail because of forbidden collisions between the conceptual design objects.

These two examples show that generic knowledge and schema representations are more compact and convenient for assessment and adaptation purposes than a large number of concrete cases. But usually it is not clear whether what is searched for will be found in the form of a concept, a schema or a case. Therefore the retrieval, assessment and adaptation procedures have to be able to deal with object oriented knowledge structures on different levels of abstraction.

FABEL's schema-based reasoning part is partially specified and not yet implemented.

9 WHAT DO WE HAVE ACHIEVED TILL NOW?

From our point of view we have learned that we need design support systems which play the role of a smoothly learning to get more competent and cooperative design assistant that takes over primarily the administrative, routine and also partially the innovative tasks and leaves the really creative tasks to the designers. This last point is due both to acceptance and to feasibility reasons.

Besides this role determination we have shown that cases are a good and user-accepted vehicle to a better reuse of former designs. Case-based retrieval is the first step which can give inspiration and support for the reminding process. Case-based adaptation is the second step which is harder to build but can save a lot of manual effort.

Beyond the knowledge incorporated in cases it is very valuable to have a sharable domain ontology which can be used for a variety of purposes. We use the ontology e.g. for quality assessment of design fragments and for abstracting the knowledge incorporated in cases as well.

Last but not least we think that we have been able to demonstrate that schemata are very well suited for the representation of design knowledge and that schema-based reasoning is an important supplement to case-based and model-based approaches. Furthermore SBR is located at an intermediate level of abstraction that supports and facilitates the seamless integration of CBR and MBR.

REFERENCES

[1] Akin, Ö., "Expertise of the Architect", *Expert Systems for Engineering Design* (Rychener, M., Ed.), Academic Press, London 1988, pp 173-196.

[2] Althoff, K.-D., Weß, S., Bartsch-Spörl, B., Janetzko, D., Maurer, F., and Voß, A., "Fallbasiertes Schließen in Expertensystemen: Welche Rolle spielen Fälle für wissensbasierte Systeme?", *KI*, No 2/1992, pp 14-21.

[3] Bakhtari, S., and Bartsch-Spörl, B., "Bridging the Gap between AI Technology and Design Requirements", *Artificial Intelligence in Design'94* (Gero, J.S., and Sudweeks, F., Eds.), Kluwer, Dordrecht 1994, pp 753-768.

[4] Bakhtari, S., Bartsch-Spörl, B., Oertel, W., Eltz, U.: "DOM: An Active Assistance System for Architectural and Engineering Design", FABEL-Report No. ?, GMD, Sankt Augustin 1995, to appear.

[5] Bakhtari, S., and Oertel, W., "Quality Assessment of Design Cases within the DOM Environment", *Proceedings of the 3rd German CBR Workshop*, University of Kaiserslautern 1995, to appear.

[6] Bartsch-Spörl, B., "Aspekte der Integration im Projekt FABEL", FABEL-Report No. ?, GMD, Sankt Augustin 1995, to appear.

[7] Bartsch-Spörl, B., "KI-Methoden für innovative Design-Domänen", *Proceedings of the XPS-95*, Infix, Sankt Augustin 1995, to appear.

[8] Bartsch-Spörl B., and Tammer, E.-C., "Graph-based Approach to Structural Similarity", *Similarity Concepts and Retrieval Methods* (Voß, A., Ed.), FABEL-Report No. 13, GMD, Sankt Augustin 1994, pp 45-58.

[9] Börner, K., "Term-based approach to structural similarity as guidance for adaptation", *Similarity Concepts and Retrieval Methods* (Voß, A., Ed.), FABEL-Report No. 13, GMD, Sankt Augustin 1994, pp 59-72.

[10] Börner, K., (Ed.), "Assessment and Adaptation", FABEL-Report No. ?, GMD, Sankt Augustin 1995, to appear.

[11] Coulon, C.-H., and Steffens, R., "Comparing fragments by their images", *Similarity Concepts and Retrieval Methods* (Voß, A., Ed.), FABEL-Report No. 13, GMD, Sankt Augustin 1994, pp 36-44.

[12] FABEL-Konsortium, "FABEL im Überblick", FABEL-Report No. 1, GMD, Sankt Augustin 1993.

[13] Gräther, W., "Computing distances between attribute-value representations in an associative memory", *Similarity Concepts and Retrieval Methods* (Voß, A., Ed.), FABEL-Report No. 13, GMD, Sankt Augustin 1994, pp 12-25.

[14] Haller, F., "ARMILLA - Ein Installationsmodell", Bericht des Instituts für industrielle Bauproduktion, Universität Karlsruhe 1985.

[15] Hedberg, S.R., "Design of a Lifetime", *BYTE*, No. 10/1994, pp 103-106.

[16] Kolodner, J., *Case-Based Reasoning*, Morgan Kaufmann, San Mateo 1993.

[17] Linowski, B., "Computing distances between attribute-value representations in a flat memory", *Similarity Concepts and Retrieval Methods* (Voß, A., Ed.), FABEL-Report No. 13, GMD, Sankt Augustin 1994, pp 26-35.

[18] Pree, W., *Design Patterns for Object-Oriented Software Development*, ACM Press, New York 1994.

[19]　Schaaf, J.W., "Gestalts in CAD-plans: Analysis of a similarity concept", *Artificial Intelligence in Design'94* (Gero, J.S., and Sudweeks, F., Eds.), Kluwer, Dordrecht 1994, pp 437-446.

[20]　Turner, R.M., *Adaptive Reasoning for Real-World Problems: A Schema-Based Approach*, Lawrence Erlbaum, Hillsdale 1994.

CONFLICT MANAGEMENT IN AN INTERDISCIPLINARY DESIGN ENVIRONMENT

V Oh & J E E Sharpe

1. INTRODUCTION

Engineering design is, of necessity, a complex process that requires the design engineer to exercise initiative and inventiveness as well as deploying a wide range of skills and expertise in attaining a solution. In an interdisciplinary design environment, such as one involving a mechatronics[1] approach to design, the designer is often required to function in a generalist mode with an eye towards employing a wider range of technologies. Where this is not possible, a product design team is often assembled to facilitate and encourage the interaction between different disciplines. The increasing occurrence of interdisciplinary product development has not only removed many of the traditional constraints to design but has now given the designer or design team a much wider freedom of choice as to the best solution to a particular design problem.

1.1 Interdisciplinary design

The interdisciplinary design (sometimes called concurrent design) process is often characterised by the simultaneous consideration and reconciliation of contributions and concerns of expertise from diverse disciplines or of multiple heterogeneous knowledge sources during the evolution of the design. The upshot of this is that potential problems that may be encountered at the later stages of the product life cycle can be addressed early and improvements made at the design stage rather than at the manufacturing stage, which has the beneficial effect of reducing product lead time and overall costs; or more innovative and cheaper means of embodying product functionalities can be accomplished due to the moving of functionality implementations across disciplinary boundaries. Such an approach is taken in the design of mechatronic products. It must be noted that interdisciplinary design does not necessarily imply the presence of a group, although in practice this often is the case and particularly when the product is large and complex in nature. It is possible that as designers and engineers gain in experience and become exposed to areas of expertise beyond their normal regime - hence complementing their existing knowledge - they are able to bring more to bear upon their design problems, and possibly bringing about a more innovative design solution.

Whichever the case may be, the truth of the matter is that conflicts are pervasive and if not handled and managed carefully could make the design process disorderly, uncoordinated and unrationalised. It is common knowledge that current computer-based design environments are much more sophisticated in their level of support and operation than their counterparts of some ten years ago. However, many of these do not provide facilities or

---

[1] Mechatronics may be considered as the integration, at all levels and throughout the design process, of mechanical engineering with electronics and computing technology to form both functional interaction and spatial integration in component, modules, products and systems. Common examples of products that are a result of the a mechatronics approach to design are camcorders and auto-focus cameras.

services to enable and facilitate the resolution and management of conflicts. Even those that purport to support and enable design group activities do not in the main provided enhanced support for conflict management. Most conflict resolution processes encountered in the design office are by and large informal, and institutionalised infrastructures for conflict management are rarely in place.

This paper presents an integrated design environment called Schemebuilder which is a design "workbench" aimed at supporting a designer at the conceptual and embodiment stages of interdisciplinary design, and the strategies used by Schemebuilder to provide enhanced support for conflict management. As a forerunner to the fundamental theme of this paper, a description of Schemebuilder is warranted and the next section, section 2, describes the Schemebuilder project (currently being conducted at the Lancaster University's Engineering Design Centre), the philosophy behind its development and the support tools provided. This is then followed by a section (section 3) delineating the possible sources of conflict in interdisciplinary design. Section 4 then goes on to present the strategies that Schemebuilder employs for the management of design conflicts. The paper then concludes with a summary of the main points.

2. THE SCHEMEBUILDER PROJECT

The field of mechatronics is of increasing industrial importance, providing high value-added products with enhanced performance and flexibility. It provides opportunities for the adoption of innovative solutions to many design problems, which may be missed through a concentration on familiar technologies (Bradley, et al., 1991). To take full advantage of these opportunities requires the ability to generate, from an expression of need, a wide selection of alternative schemes for evaluation. This must be done rapidly, systematically and without any bias towards particular technologies.

To facilitate and assist in the generation of interdisciplinary design solutions and maintenance of their alternatives during the early stages of design, we have identified a need to provide computer support to the decision-making process in a form that is technologically neutral to enable an efficient partitioning of technologies to be obtained. This takes the form of an integrated design support environment called Schemebuilder, which has as its main objective to support and assist the engineering designer in the conceptual and embodiment stages of design. It takes on a role as facilitator for the creation of a model of the system to be designed, and advice provider on the appropriate means to achieve the design. More importantly, it aims to facilitate the exploration of alternative conceptual schemes with an appropriate allocation of function between mechanical, electronic and software elements.

This is achieved by providing a semi-automated, structured and rigorous approach to developing designs from pure statements of functional requirements without any preference to any particular technology (Bracewell and Sharpe, 1994). This has the benefit of removing any bias on the part of the designer and shifting part of the responsibility of generating design solutions onto the computer. Apart from the benefits gained from the reduction of lead time for concept generation and the greater number of potentially good alternative solutions, it partially prevents the imposition of designers' beliefs of what constitutes a good design on each other. We stress that this does not in any respect stifle design creativity. On the contrary, Schemebuilder has been built for that purpose - to allow for the design of innovative design solutions.

2.1 The Schemebuilder environment

The Schemebuilder environment provides highly integrated support for the rapid creation of technologies and enables the comparison of technological alternatives to take place before large commitments are made and irrevocable decisions taken on the basis of partial or biased information. In short, Schemebuilder allows for the exploration of new ideas with minimal risk.

The software tools currently under development in Schemebuilder cover the stages from Problem Analysis to Embodiment Design. Schemebuilder provides research tools to:

- assist in the production of specifications
- provide advice and browsing facilities on available technologies
- aid in the identification of 'technological holes'
- produce system models for dynamic simulation and parameter optimisation
- support embedded software design and development
- monitor system integrity, checking for continuity and matching
- produce preliminary layouts in the form of realistically sized 'solid sketches'
- design and model casings and support structures
- support and enable the management of design conflicts

Figure 1 shows the overall architecture of the Schemebuilder environment, with the arrows indicating information flows. A variety of software systems are used to provide the various supporting functions. These software systems include MetaCard (a hypermedia system), KEE (Knowledge Engineering Environment), Simulink (a commercial simulation package), and AutoCAD. The integration of the heterogeneous software systems is achieved by a combination of Unix pipes and shared file mechanisms.

The man-machine interface, implemented using MetaCard, allows the designer to access the assortment of tools for performing a variety of design tasks such as scheme generation, component selection and design rationale capture. It also provides the designer with advice on what to do and suitable choices to make during designing, access to application notes for reference, and, depending on the current context, intelligent querying of either designer or other ancillary sub-systems until an appropriate advice principle is selected (Langdon, 1993). The user-interfaces for the conflict resolvers are coded in MetaCard (top right hand corner of figure 1), with the underlying mechanisms written in Lisp (in KEE).

Central to Schemebuilder are the knowledge bases, including a repository of stored working principles, that store the functional, behavioural and structural representations of the component models (Oh et al., 1994b) and the mechanisms for scheme generation. With the aid of KEE's Worlds facility, Schemebuilder allows the designer to progress and maintain a set of alternative schemes. All this provide an environment in which competing alternative components or sub-systems can be matched to functional requirements, compared on a wide range of criteria and be tested by means of simulation. The facilities are integrated by sharing the same set of data and by employing cross-disciplinary component descriptions in the form of bond graphs which will be discussed later on in the paper.

As well as modelling systems representing proposed products, the designer is also provided with facilities to perform digital simulations of experimental systems to explore the nature of the components selected. For example, it will be possible to determine the torque-speed curve for a motor and investigate its efficiency on load; and determine its sensitivity to parameter changes and manufacturing tolerances. Furthermore, the mathematical models derived from the bond graphs allow the mathematical optimisation of system parameters and operating conditions. Further details about Schemebuilder can be found in (Oh, et al., 1994a).

3. CONFLICTS IN DESIGN

With the recognition that conflict is a natural occurence in design, particularly in concurrent and interdisciplinary design, several AI researchers have addressed the problem of providing computational support for conflict management in design support systems. Many though assume that conflicts are pervasive only within a group setting. However, when design is performed using an interdisciplinary approach conflicts are just as pervasive in situations involving only a single designer, the reason being that in such an approach many different types of technologies from different disciplines are being brought to bear on the design problem. This is not to say that conflicts do not arise in single domain engineering problems: trade-offs must sometimes be made because of incommensurable criteria imposed on a design solution.

A few notable achievements in this area of computational support for conflict resolution in design include the work of:

• *Sycara* (1991), whose work employed case-based and utility negotiation in which to arrive at a compromise solution. Her proposed negotiation model embraced several key components: knowledge of previous designs, communication of design rationale, justifications and objections to proposed design solutions, constraint propagation and relaxation, and traversal of goal graphs.

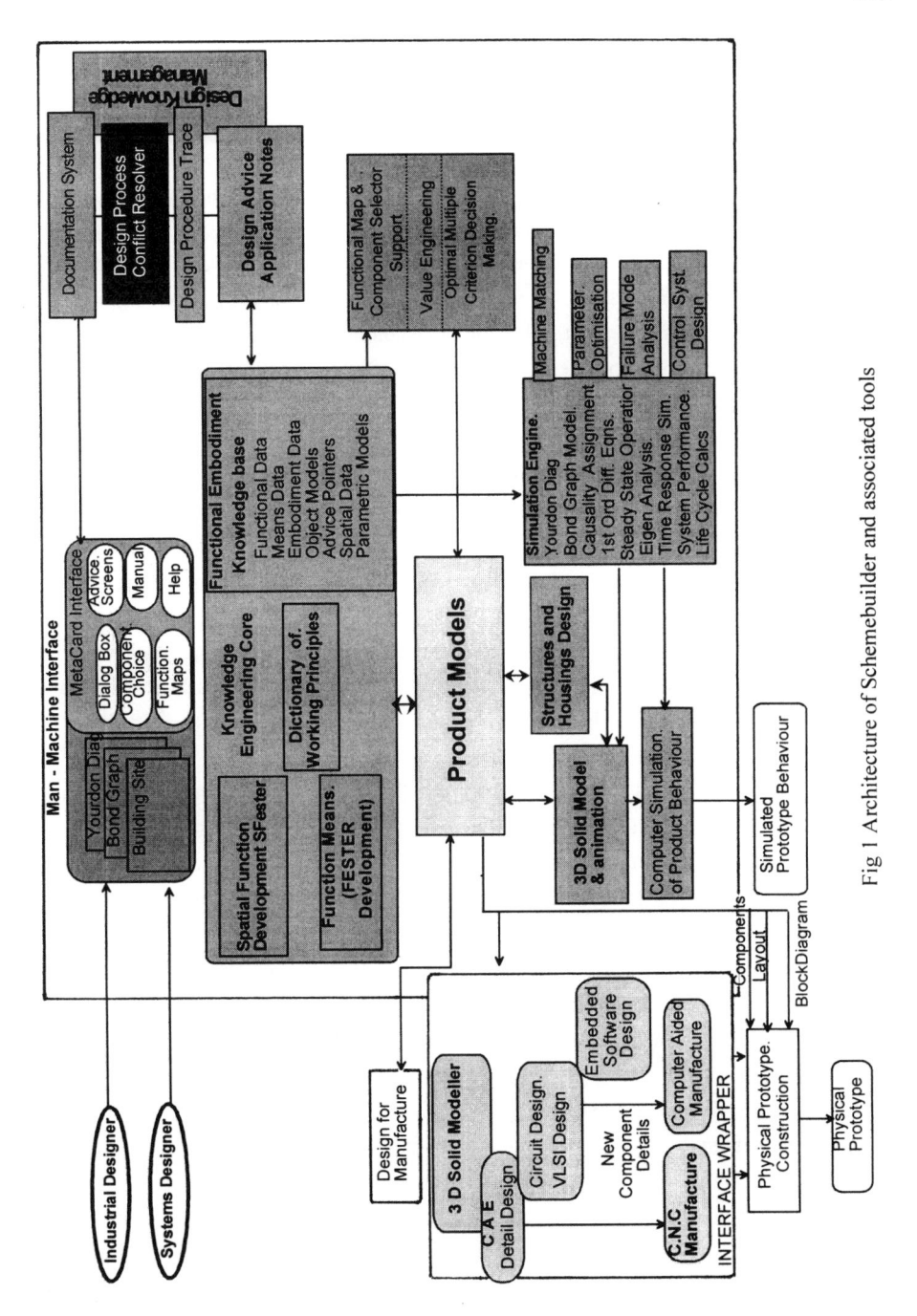

Fig 1 Architecture of Schemebuilder and associated tools

- *Klein* (1993), who developed a heuristic-based conflict resolution shell that contained "first-class" knowledge of conflict resolution strategies acquired from empirical studies of human designers at work. He firmly believed that conflict-resolution expertise should exist separately from domain-level expertise and that it should be made explicit. Klein may possibly be one of the first to articulate the possibility of identifying general conflict resolution expertise that may be applicable to a wide variety of design conflicts (Klein and Lu, 1989). His conflict resolution scheme is based on a taxonomy of conflict classes and contain associated general advice for resolving conflicts in each class. Each agent in his system has different domain expertise but the same conflict resolution expertise. The system has since been extended to include humans, who may be physically distributed, within the conflict resolution process.
- *Bahler, et al.* (1994), who embedded a negotiation protocol within their constraint-based system, Galileo3. Their protocol not only detects potentially conflicting decisions between design agents (clients) but also offers a evaluation procedure for assessing the various design plans proposed by the agents. Rather than opt for a fully automated procedure of resolving conflicts, they adopt a partially automated procedure where properties that the solution must satisfy are "axiomatized" and the degrees of effect on design solution plans and their overall scores are computed. The computer support provided in this approach lies in the provision of feedback to the design agents regarding the impact of their design solution plans. The decision-making as to which direction is to be taken after a conflict has been detected lies solely on the human designer.
- *Lander, et al.* (1991), who proposed a conflict resolution framework in the light of integrated distributed search among a set of heterogeneous and reusable agents. They employ a range of conflict resolution strategies, which are selected on the basis of prevailing conditions, such as "generate random alternatives", "compromise", "generate constrained alternatives", and so on.
- *Kannapan & Marshek* (1992), who applied formal models of group decision-making, particularly the Nash and Kalai-Smorodinsky bargaining theory model, in their implementation of a negotiation schema for a computer-based concurrent engineering environment.
- *Easterbrook, et al.* (1994), whose work focused on managing the consistency - or rather the *inconsistencies* - of viewpoints during capture of requirements for the design of software systems. The strength of their work lies in the fact that their system does not require a central, global database for team members to remain coordinated and consistent in their design specifications. Their implemented system instead allows for distributed design and the tolerance of inconsistencies between specifications. The advantages accruing from this are that design freedom is maximised and that design decisions need not be committed too prematurely for a design to proceed

- *Werkman* (1990), who proposed a knowledge-based model of negotiation that hinges on the notion of *shareable agent perspectives*. Every agent's unique knowledge is supplemented with a shared knowledge base consisting of the domain object knowledge and the history of the negotiation dialog trace. Werkman's model also includes third-party intervention, through the use of an arbitrator, in the negotiation process. The arbitrator stores the agent's perspective and collects the interagent dependencies. Thus, the arbitrator comprises much of the negotiation knowledge and has the responsibility of mediating between conflicting agents when negotiations run into a "deadlock".
- *Haroud, et al.* (1994), who proposed a computational model of conflict management that is based on the dynamic constraint satisfaction paradigm. Their approach is based on the premise that designers ascribe different semantics (types of knowledge) on the assumptions they make during the preliminary stages of design, and this semantic differentiation is used to guide the choice of conflict management strategies
- *Polat, et al.* (1993), whose negotiation platform provides the basis for multiple knowledge-bases to resolve the disparities between solution components. Like the work of Klein, their work is based on the insight that design agents have conflict resolution knowledge that are separate from the domain-level design knowledge. However, unlike in Klein's work, there is no notion of globally-known conflict resolution strategies that may be accessible by all design agents; each design agent encapsulates its own unique store of conflict resolution strategies. The upshot of this approach is that, in principle, knowledge-based agents with their own "brand" of conflict resolvers can be added to the design environment without affecting the conflict management ethos of other agents
- *Oh, et al.* (1993), who implemented a cooperative design model called CIC in a design support tool to aid in the improvement of mechanical assembly designs. A novel feature of their work in the area of computational conflict resolution is the use of a 3-tier conflict resolution scheme that is based on the negotiation metaphor. When a conflict is detected, e.g. different values for the same design variable are proposed, machine-based agents firstly try to reconcile their differences in the light of their own strategies. When a deadlock is encountered the resolution is passed onto another machine-based agent with a special responsibility to arbitrate over the conflict. When all these fail, the conflict is passed on to the third level where the human then decides the next of course of action, supported by available information.
- *Petrie, et al.* (1994), whose Next-Link architecture captures all notions of conflict in their configuration design domain of cable harness design in terms of constraint violations. Instead of providing support for automated reconciliation of conflicts, they have opted for human-guided conflict resolution facilitated by the provision of novel book-keeping facilities that keep track of changes made and their implications.

3.1 Defining conflict

There is currently no singly accepted definition for the term conflict. Scholars who are interested in studying conflict have attempted to define the all inclusive and general definition for conflict but none so far has been satisfactory. For instance March and Simon (1993) considers conflict as a term generally applied to a "*breakdown in the standard mechanisms of decision-making so that an individual or group experiences difficulty in selecting an action alternative*". However, most definitions consider the incompatibility or opposition in goals, activities, or interaction among agents.

Rahim (1986), for one, considers conflict as an "interactive state" manifested in incompatibility, disagreement, or difference within or between social entities. He points out that conflict occurs when agents (1) are required to engage in an activity that is incongruent with their needs or interests, (2) hold behavioural preferences, the satisfaction of which is incompatible with another person's implementation of their preferences, (3) have partially exclusive behavioural preferences regarding their joint action, and (4) are interdependent in the performance of their functions or activities.

3.2 Sources of conflict

Conflict, contrary to traditional beliefs, can be productive rather than destructive or disruptive if their situations are recognised and managed effectively (Galliers, 1990). This is particularly true in design where the occurrence of conflicts can initiate a resolution process that may result in a better design solution.

In engineering there are different types and levels of conflicts that may arise. We ignore conflicts due to socio-psychological factors like human emotions, for example hostility, ill-feelings, anxiety, anger, fear, etc, although we acknowledge that these are important factors to be considered in any conflict situations involving humans. We believe that the management of conflicts arising from or involving such factors is best left to humans to handle. We instead focus on the technological and task-level conflicts.

Interdisciplinary design requires a collaborative style towards design work and conflict is seen as an inherent part of it. Hence, it is constructive to determine these sources of conflict because by identifying and understanding them, it allows us to draw up strategies to manage and exploit them effectively. The various sources of conflict that are inherent within an interdisciplinary design environment are delineated below.

1) *Differences in technical beliefs*: this may lead to technical prejudice on the part of a designer, which may lead him/her to focus on preferred technologies or implementations. Furthermore, it may lead the designer to assume certain facts about the problem. This in itself is not a problem but may become one when the designer has to interact with others

who may have different conceptions as to what constitute a good design solution.

2) *Differences in decisions and assumptions made*: the early phases of design is often characterised as one having incomplete and uncertain information. In order to progress a design, designers often make assumptions or decisions based on such information, or even make assumptions based on individual perceptions of a problem. This is obviously a source for potential conflicts especially when designers are working on separate subsystems that ultimately needs to be interfaced together to complete the final system design.

3) *Differences in technical vocabulary*: potential for conflicts in this area is very real as designers from different functional domains typically use different symbols, technical language and nomenclature to represent and talk about their domain. The problem is further compounded by the fact that sometimes different terms from the different domains are used to mean the same thing, and in other cases the same terms to mean different things. For instance, the term "power" may mean different things to different people and if a common conception or proper understanding of each others conception of the term is not established then there may be a potential for conflicts at some later stage.

4) *Lack of common and shared understanding*: Aggregating expertise from individuals who have very detailed knowledge in their own areas and much weaker understanding of others is characterised by many difficulties: (a) agents are unable to communicate their expertise clearly to non-experts because of differences in technical vocabulary, as in (3) above, and conceptual content; (b) the process is susceptible to incorrect inferences made on the part of an agent on the other's knowledge; (c) some element of uncoordination may arise because of an agent's lack of knowledge about another's needs and requirements

5) *Inconsistencies between design models*: when engineers and designers work on separate design problems, they will typically use modelling tools that they are familiar with; tools that are more appropriate for that particular functional domain; or the same tool but the difference lying in the problem proper. Changes made in one functional domain must be propagated to those that interfaces with the domain that made the change. If this is not done then the design of that domain is inherently in conflict with others and vice versa.

6) *Differences in goals or preferences*: conflicts arise when goals or preferences on design attributes are incommensurable. Individuals, viewing the problem from their own standpoint, often endeavour to maximise the utility in achieving their goals at the expense of others, and/or any preferences they may have. Different functional team members have

goals that inherently conflict; for example, marketing may be interested in maximising product variety to increase sales, while production is interested in minimising product variety to restrict costs. To illustrate further, consider the design of a motorised drug infuser system. One goal is to minimise the cost of the system and hence a decision was made not to use an ASIC (Application Specific Integrated Chip) implementation of the control system that has the function of controlling the motion of the mechanical components, but instead several cheap off-the-shelf chips were used. However, this now conflicts with the goal of achieving a compact system, which imposes a space limitation on the layout of the chips.

7) *Differences between evaluations of criteria*: sources of conflicts arising from these differences are due to different functional team members using different criteria to judge a design. This is often caused by the inclination of individuals from different functional areas to attend only to certain information in a design problem or task, which is due to their exposure to the goals, views, and traditions of their particular functional area. This phenomenon of *selective perception* (Dearborn and Simon, 1958) causes an individual to focus on specific functional and performance attributes of a design, which may not be all that important to another, and evaluate the design based on these - certain attributes of a design is seen to be more important than others depending on the individual's viewpoint. The problem here lies in that typically these criteria cannot be easily satisfied simultaneously and optimally. The design decisions that optimise one set of criteria often conflict with those of that from another set.

8) *Differences between suggested solution components*: conflicts arising from this situation is often due to different specialists specifying different constraints or values on attributes that they share, the interfacing point between two disciplines or functional domain. This means that the proposed value of a design parameter by one specialist makes it infeasible for another specialist to offer a consistent set of values for other parameters. Such conflicts may also be encountered through constraint propagation. Conflicts arising from this nature of incompatibility are common in situations where different team members try to assert different parts of a design into the global product design model database, but which may contain solution components having overlapping areas of concern with others already in the database. This constitutes the global integration problem whereby local solutions emanating from the different heterogeneous agents are integrated to produce a globally integrated solution.

3.3 Individual and group conflicts

As alluded before, conflicts may also arise at the individual level. This would occur as a result of the designer simultaneously considering different competing technologies for the implementation of a functional requirement. This is easily seen in a common situation whereby a control function can either be implemented in software or in hardware. In this case, a conflict can be said to have arisen because of two design assertions stating that a particular function can be implemented in two mutually exclusive technologies. Which of the two that is to be selected as the favoured implementation technology depends on several factors and criteria, which may differ from one product to another, and upon further analysis. Consequently, there is a need for more effective and enhanced decision support for occasions like these.

Conflicts may frequently arise in the light of available alternative schemes for the same functional requirement, particularly when different experts from different functional areas are involved. The reason for this is that these experts view the acceptability and optimality of the alternative schemes from a different viewpoint; multiple experts will yield multiple opinions which may potentially conflict. Furthermore, it is even possible for experts from the same area of discipline to disagree about the optimality of the solution.

4. STRATEGIES FOR MANAGING CONFLICT

There are essentially two ways to manage conflicts: (1) avoid them or minimise their occurrence by designing them away, and (2) resolving them at run-time when they occur. These two approaches are used in Schemebuilder. We employ a mix of formal and informal methods to realise these approaches, the former in the form of analytical methods like multi-criteria decision making tools and the latter in the form of heuristically-driven rules, established protocols or conventions and negotiation techniques.

It must be noted that the outcome of a conflict resolution process does not necessarily result in a "zero-sum" situation or what might be called a win-lose situation, i.e. one party winning at the expense of the other. There are essentially 3 types of outcomes within a conflict situation: win-win, win-lose and lose-lose. Within the human context, there are generally 5 styles of conflict handling: integrating, obliging, dominating, avoiding and compromising (Rahim, 1986). The outcome of a conflict situation depends on the selection of one of these styles for resolving the conflict; the dominating and obliging styles generally lead to a win-lose situation, the integrating style to a win-win situation, and the avoiding and compromising styles to a lose-lose situation. Compromise tends to lead to a lose-lose situation because the parties involved generally disagree about two solutions and settle for something in between.

Such styles are manifested in some form or other within the Schemebuilder environment.

For example, if two specialists propose two conflicting values for a design attribute then a dominating style can be used to resolve the conflict (note that when one dominates the other accommodates). To do this, the system can trace back up the function-decomposition tree or objective tree to determine which between the two proposals is the design attribute considered to be most important, in which case a win-lose outcome results. If the design attribute has the same importance from the two standpoints then a compromising style can be used, in which case a lose-lose outcome results.

Figure 2 illustrates conceptually how Schemebuilder would be used in a group-based design environment. Each copy of Schemebuilder is equipped with an assortment of conflict resolving tools that provides the necessary support to the designer to reconcile conflicts arising during the design process. As the figure shows, each Schemebuilder environment is provided with tools that provide services for local and global conflict resolution activities. However, there is no clear distinction between these tools as some tools that are used within the local workspace can also be used within the global workspace. However, the latter would be supplemented with services that take into account the asynchronous and synchronous nature of design in a virtually co-located design team environment.

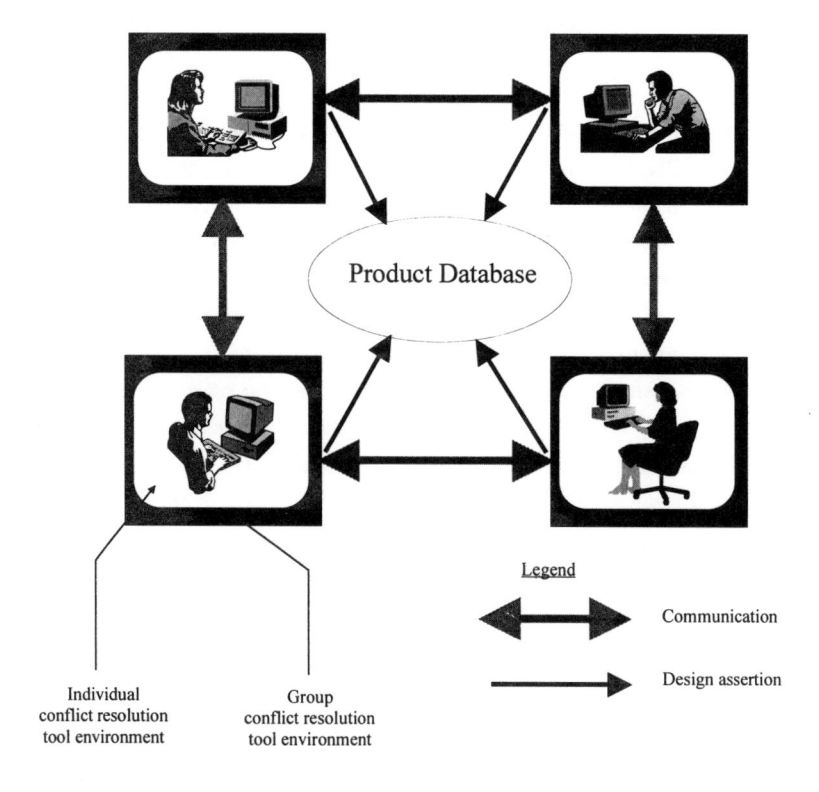

Fig 2 Conceptual view of a multi-Schemebuilder environment

4.1 Conflict management by design

We believe that it is possible to minimise the potential for disruptive conflicts by creating an environment with an infrastructure for effective collaborative design work consisting of institutionalised rules, standards and guidelines on best practices and the use of design support systems augmented by appropriate knowledge bases and on-line documentation, which together helps increase the shared understanding and co-ordination between members of the team. In doing this we are not attempting to avoid or suppress conflicts but to reduce, if not eliminate, occurences of disruptive ones. Conflict management by design attempts to address, to some extent, the conflict situations of *differences in technical beliefs, differences in technical vocabulary, lack of common and shared understanding* and *inconsistencies between design models*.

Knowledge browsing system

Schemebuilder provides a knowledge browser system that allows designers to browse through on-line electronic documentation on technical matters, supporting information and in-house information bases to allow them an opportunity to keep abreast of technologies not normally familiar to them and also to be kept well-informed of developments within the organisation. Formal procedures for best practices and access to group memory design repositories also form part of the facilities provided by the knowledge browser system. The system enables designers to search, browse and become acquainted with accepted company-wide design practices and in the process develop a shared view of design rules and practices. The knowledge browser system takes on the form of an arbitrator system that designers appeal to when "difficult-to-resolve" design conflicts are encountered.

Bond graphs

A common vocabulary system design language is used as the underlying representation for design objects in Schemebuilder. As products become more complex and highly integrated, design teams will find it increasingly necessary to have a common language, independent of traditional engineering disciplines, in which to communicate. The use of *bond graphs* (Karnopp, et. al., 1990), with its common structure and clear rules across engineering domains, for Schemebuilder's representation of the functional and behavioural aspects of energetic systems allows a very natural approach to design and allows the development of an integrated suite of object and rule-based computer aided conceptual design tools. Very briefly, the modelling process is based upon the conservation of energy, with physical processes being linked in a labelled digraph through energy flows. Basic concepts like effort, flow, inertia, and capacitance are used in modelling, and their generality allow the method to be applied across domains like thermodynamics, rotational and translational mechanics, fluid dynamics and electronics. Thus, bond graphs cover essentially all classical macrophysical

Change coordination

Schemebuilder limits inconsistencies between design models by providing mechanisms for integrating heterogeneous software design tools and storing the design on a central blackboard structure. A group data management system will eventually be included in Schemebuilder to manage and integrate the design of sub-systems. This arrangement facilitates co-ordinated updates of design and help reduce inconsistencies between design models but does not handle conflicts that arise when constraints are violated, and incompatibilities between design solution components. These are generally treated as runtime conflicts and are discussed in section 4.2.

Proper design partitioning

Another way to limit the potential for conflicts is to define and partition the design problem in an optimal way so as to facilitate concurrent design. This in practice is not an easy task as this is in itself a design problem. An example will serve to illustrate this problem of poor design problem definition. Consider a facile - but by no means unrealistic - example of two sub-systems to be designed: a drive system **D** and a system to move a body of fluid from one place to another **M** (figure 3a). If the design problem is partitioned in this manner and defined as such then there could be a potential for a design conflict in the form of a causal conflict occurring at the interface **I** between the two subsystems when the two sub-systems are integrated.

The designer (a mechanical engineer perhaps) in sub-system **D** may have designed a drive motor system with control feedback to supply a constant source of effort (torque), assuming that that was what was required of the design. If the designer (a hydraulics engineer perhaps) in sub-system **M** adopts a solution consisting of a piston system then there will not be any causal conflicts arising at their interface when the designs are integrated. However, if designer for **M** had adopted a centrifugal pump solution that demanded a constant source of flow then a causal conflict occurs at their interface. In this case, a strategy of using convention (discussed in next section) such as removing redundant functionality as much as possible is used (the controller in sub-system **D** in this case), or introduce an intermediary component to remove or balance out the causal conflicts.

However, this situation could have been prevented if the design problem had been stated in terms of their interface, i.e. the interface between the two sub-systems constrained. The effect of this is to shift any resolution of conflicts away from the interface between sub-systems into the designer's workspace, where the designer has a greater control over the resolution of conflicts arising within his/her workspace.

Resolution of conflicts can be localised by
defining problem at the interface

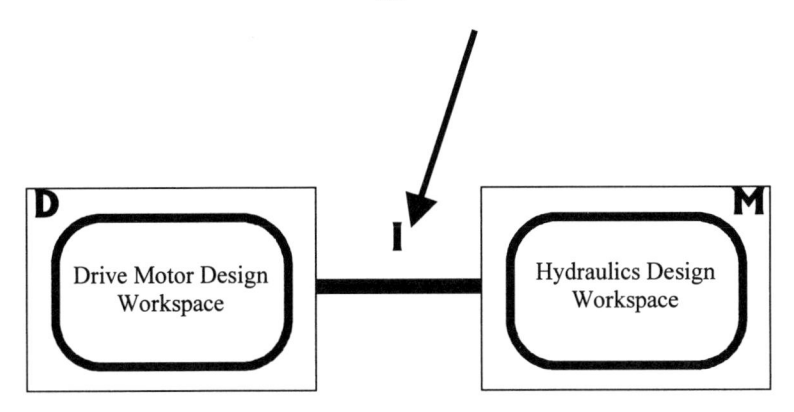

Fig 3 Concurrent design of two sub-systems

There is a limit to which the methods described above are able to manage the occurences of conflict as these only provide the infrastructure and are largely static in nature. Ways in which to manage and resolve conflicts dynamically must be sought and these are discussed below.

4.2 Resolving run-time conflicts
Multi-criteria decision making techniques

Adopting a strategy of employing formal, analytical techniques provides a rigorous, analytically sound procedure for guiding the design process in the light of conflicting preference requirements and incommensurate attributes. Methods like weighted sum, multi-objective decision making (Keeney and Raiffa, 1976), fuzzy set analysis and the analytic hierarchy process (Saaty, 1990) are relevant to our work and we are currently investigating them. We are aware that each method is best suited to particular types of problems, and each offers varying degrees of success, ease of use and sophistication, but each method helps the designer to deal with conflicting attributes by defining the attributes governing the design and addressing their relative importance or the willingness of the designer or design team to make trade-offs between them.

Using MCDM techniques helps to tackle the conflict situations by providing an analytical manner to handle trade-offs between incommensurable criteria and goals. The acquisition of importance weightings on criteria and preferences and their use in an appropriate MCDM procedure implicitly compromises and trade-offs the objectives or attributes of the various specialists in a manner dictated by the MCDM procedure.

Rules and convention

Rules and conventions are somewhat equivalent to arbitration and resort to them typically leads to a win-lose outcome as discussed above. The use of prescriptive rule strategies in Schemebuilder is similar to that of failure handling mechanisms in AIR-CYL (Brown and Chandrasekaran, 1989). This means that for any action there can be attached one or more strategies to perform if a particular action fails. Domain-dependent and problem-specific strategies can be encoded by using this type of prescriptive strategy. Prescriptive strategies are a way to establish a protocol or convention to resolve conflicts in a prescribed manner as soon as the characteristics of the conflict situations are adequately mapped onto the appropriate conflict resolution strategy. Klein's work is another example of such an approach (Klein and Lu, 1989).

Domain-independent strategies are also possible to enforce conflict resolution by convention. For instance, in the example of figure 3 the rules "remove redundant functionality where possible" and "add intermediary component" are general enough to be applied in any engineering domain.

Constraint relaxation and adjustment

Relaxation of conflict is when two parties in conflict decides not to enforce their individual goals on each other. This can only occur if the constraint involved in the conflict is a soft one. Hard constraints are constraints that must be met at all times; they represent constraints imposed by strict regulations or engineering laws. Soft constraints on the other hand represent desirable or preferential constraints on design attributes.

Figure 4 illustrates some of the strategies employed to relax constraints. Assume that there are two types of constraints originating from two input sources on a design attribute: one hard and the other soft (figure 4a). The strategy employed here is to relax the soft constraint and allow the value to be shifted to point C, with the value of the hard constraint also shifted to C since this constraint is still being satisfied. Consider another situation as in figure 4b. In this case, there is a zone of agreement and the strategy employed here is to take a point (N1 and N2 are equal) mid-way between the two original values.

There are cases when a fuzzy utility function can be applied on a design attribute across the range of the constraint. The fuzzy utility function indicates the degree of preference for a particular value on a design attribute. So in the cases in figure 4, a fuzzy utility function can be imposed across the range of the constraint. Decision-theoretic analytical techniques like bargaining, for example the Nash bargaining technique, could be used to find the maximum *joint payoff* for the two conflicting proposed solutions.

In some cases it may not be desirable to relax any of the constraints and an alternative must be selected to resolve the conflicts taking into account this set of constraints. The conflict resolution strategy rule of "add intermediary component" initiates a search in the component library for a component that provides the desired functionality and still satisfy the set of constraints. For example, in the design of an actuator system to provide rotary motion for the rear fin of an unmanned aircraft, a designer may opt for a solution containing a DC motor that directly drives the shaft of fin. However, the requirement for low turning speed (low angular velocity) on the rotation of the fin will constrain the DC motor to run at an efficiency level well below its preferred level. Hence, the system will initiate a search of the component library for a component that can match the two constraints while still providing the required functionality of providing rotary motion. In this case, a reduction gearbox was selected.

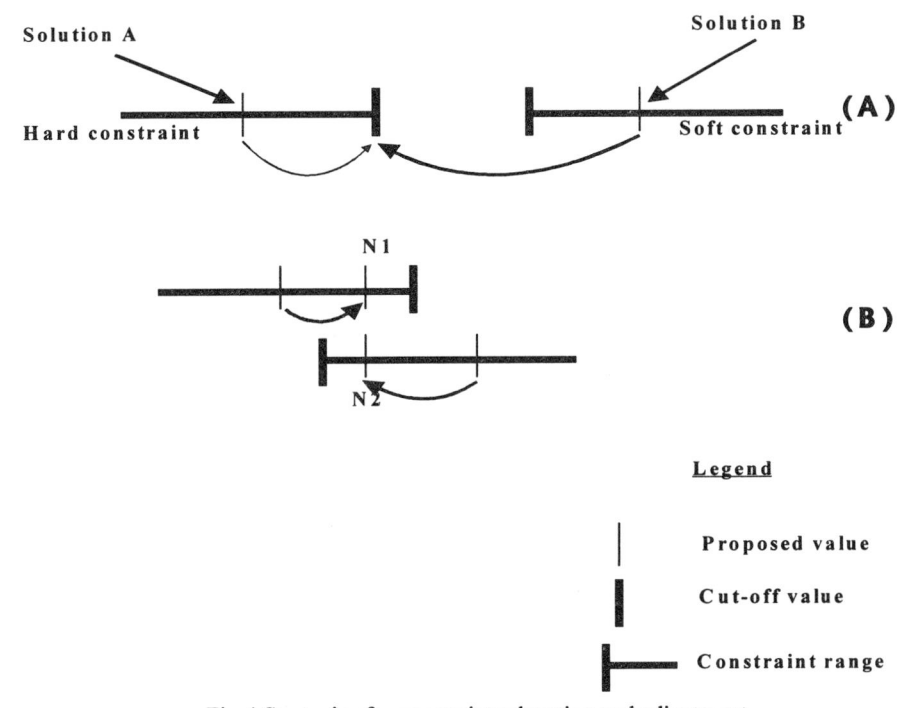

Fig 4 Strategies for constraint relaxation and adjustment

Negotiation

We are currently exploring ways to extend the existing conflict management strategies within Schemebuilder by considering the use of negotiation techniques. Negotiation becomes important when parties in conflict do not have rule-oriented conventions or techniques available to them. It is a general method to resolve conflicts involving at least two parties. The outcome of any negotiation is dependent on both the nature of the conflict problem and the type of strategy used to solve that problem. Information exchange in negotiation serves as the primary medium for justifying one agent's position with another's, and eventually for making concessions.

In general, negotiation can enter the design process at the following points:
- when different relevant specialists have made conflicting recommendations regarding values of attributes of a design
- when an attribute value proposed by one specialist makes it infeasible for another specialist to offer a consistent set of values for other attributes
- when a design decision made by one expert adversely affects the decision other experts consider as optimal
- when alternate approaches can achieve similar functional results. This is particularly true in an interdisciplinary design environment when a function can be realised by several different technologies, each with its own unique implementation characteristics.

5. CONCLUSIONS

From observations and empirical studies of designers at work, it is easy to conclude that conflicts is not only pervasive but also an inherent part of design. From our studies, we have observed that conflicts do occur not only at the group level but also at the individual level. Such conflicts are accentuated when design is interdisciplinary and integrated. To increase the effectiveness of computer-based design support systems, facilities and services must be provided within such systems to support and enable designers to manage and resolve the conflicts that arise in the course of the design process at the individual and group level.

This paper has presented some of our ongoing work on the management of conflicts in an computer-aided interdisciplinary design environment. We have sought to provide an assortment of tools that designers may call upon to aid them in the resolution of a conflict. What we hope to achieve is to provide a design environment that embraces an architecture consisting of techniques and mechanisms to accommodate, recognise and resolve conflicts. We have taken the approach that the tools and mechanisms we provide in Schemebuilder are more supportive in nature than automating; the ultimate decision has to be made by the designer when the situation calls for it. However, owing to the mathematical nature of formal and analytical tools, designers who employ such tools may rely on the final outcome as the optimal resolution.

More work, however, is still needed to consider and explore if these strategies and the manner in which they are performed are effective and do not pose a hindrance to the human designer during the design act. The interface between the human and computer, and between human and human mediated by the electronic medium within the context of conflict management is still a subject for interesting research.

ACKNOWLEDGEMENTS

The authors would like to thank Bob Chaplin, Man Li and Xiu Tian Yan for their valuable input and discussions.

REFERENCES

Bahler, D., Dupont, C., & Bowen, J. (1994). In *Artificial Intelligence in Design '94* (Gero, J., & Sudweeks, F., Eds.), pp 363-379. Kluwer Academic, Dordrecht.

Bracewell, R.H., & Sharpe, J.E.E. (1994). Computer aided methodology for qualitative development of schemes from first principles. In *Computer Aided Conceptual Design* (Sharpe, J.E.E., & Oh, V. Eds.), pp 79-94. Lancaster EDC, Lancaster.

Bradley, D.A., Dawson, D., Burd, N.C., & Loader, A.J. (1991). *Mechatronics: Electronics in Products and Processes*. Chapman and Hall, London.

Brown, D.C., & Chandrasekaran, B. (1989). *Design Problem Solving: Knowledge Structures and Control Strategies*. Pitman, London.

Easterbrook, S., Finkelstein, A., Kramer, J., & Nuseibeh, B. (1994). Co-ordinating conflicting viewpoints by managing inconsistency. In *Artificial Intelligence in Design '94 Workshop on Conflict Management in Design*. Lausanne.

Galliers, J.R. (1990). The positive role of conflict in cooperative multi-agent systems. In *Decentralized AI*, (Demazeau, Y., & Muller, J.P. Eds.), pp 33-46. Elsevier, Amsterdam.

Haroud, D., Boulanger, S., Gelle, E., & Smith, I. (1994). Strategies for conflict management in preliminary engineering design. In *AI Design '94 Workshop on Conflict Management in Design*. Lausanne.

Kannapan, S.M., & Marshek, K.M. (1992). In *Intelligent Computer Aided Design* (Brown, D.C., Waldron, M.B., &Yoshikawa, H., Eds.), pp 1-25. Elsevier, Amsterdam.

Karnopp, D.C., Margolis, D.L., & Rosenberg, R.C. (1990). *System Dynamics: A Unified Approach 2nd ed.* Wiley, Chichester.

Keeney, R.L., & Raiffa, H. (1976). *Decisions with Multiple Objectives: Preferences and Value Tradeoffs*. Wiley, New York.

Klein, M., & Lu, S.C.Y. (1989). Conflict resolution in cooperative design. *Artificial Intelligence in Engineering. 4(4)*, 168-180.

Klein, M. (1993). Supporting conflict management in cooperative design teams. *Group Decision and Negotiation. 2(3)*, 259-278.

Lander, S.E., Lesser, V.R., & Connell, M.E. (1991). Knowledge-based conflict resolution for cooperation among expert agents. In *Computer-Aided Cooperative Product Development* (Sriram, D., Logcher, R., & Fukuda, S., Eds.), pp 269-297. Springer-Verlag, Berlin.

Langdon, P. (1993). Software for Schemebuilder: an aid for design creativity. Technical Report EDC 1993/01. Engineering Design Centre, Lancaster University, Lancaster.

March, J.G., and Simon, H.A. (1993). *Organizations 2nd ed*. Blackwell, Oxford.

Oh, V., Taleb-Bendiab, A., Sommerville, I. & French, M. (1993). Incorporating a cooperative design model in a computer-aided design improvement system. In *Prospects for Artificial Intelligence* (Sloman, A., Hogg, D., Humphreys, G., Ramsay, A., & Partridge, D., Eds.), pp 101-110. IOS Press, Amsterdam.

Oh, V., Langdon, P., & Sharpe, J.E.E. (1994a). Schemebuilder: an integrated environment for product design. In *Computer Aided Conceptual Design* (Sharpe, J.E.E., & Oh, V., Eds.), pp. 339-362. Lancaster EDC, Lancaster.

Oh, V., Chaplin, R.V., Yan, X.T., & Sharpe, J.E.E. (1994b). A generic framework for the description of components in the design & simulation of mechatronic products. In *Mechatronics: The Basis for New Industrial Development* (M. Acar, M., Makra, J., & Penney, E., Eds.), pp 515-520. Computational Mechanics, Southampton.

Petrie, C., Cutkosky, M., Webster, T., Conru, A., & Park, H. (1994). Next-Link: An experiment in coordination of distributed agents. In *Artificial Intelligence in Design '94 Workshop on Conflict Management in Design.* Lausanne.

Rahim, M. A. (1986). *Managing Conflict in Organizations*. Praeger, New York.

Saaty, T. (1990). *The Analytic Hierarchy Process*. McGraw-Hill, New York.

Sycara, K. (1991). Cooperative negotiation in concurrent engineering design. In *Computer-Aided Cooperative Product Development* (Sriram, D., Logcher, R., & Fukuda, S., Eds.), pp 269-297. Springer-Verlag, Berlin.

Werkman, K.J. (1990). *Multiagent Cooperative Problem Solving through Negotiation and Perspective Sharing*. PhD Dissertation, Lehigh University.

A FUZZY THESAURUS FOR SEMANTIC INTEGRATION OF DESIGN SCHEMES

I Mirbel

1 INTRODUCTION

Several designers should have to work together on the design of the same part of the real world. Maybe because of the extent of the work to be done, maybe to take into account knowledge from several designers, from several application domains : all these people do not perceive the real world in the same way. The integration goal is to obtain the richest view of the part of the real world under study. It is often pointed out that there exists two different kinds of scheme integration [2, 21] :

- Existing-databases integration : this process gives a global scheme, which represents a set of several existing databases. It creates a virtual view in which all the different databases are brought together. Then the users see and work on one database, instead of several [10, 18, 20, 25, 15, 1].

- Design-scheme integration : this process gives a global design scheme, from several schemes, each of them depicting the same reality according to different perception ways.

Our work belongs to this second kind of integration.

The first integration tools proposed where based on the relational model [3, 18], on the entity-relationship model [2, 28], on the extended entity-relationship model [21, 11], and on other semantic models. Most recently, we find works on object-oriented models [30, 17]. Our approach belongs to this last kind of works, being based on an object-oriented model defined in [5].

Throughout the scheme integration, there are some problems and conflicts. They are due to the several ways of representing the semantic of knowledge and of structuring knowledge (using the same design model) [2, 7, 22, 26]. When the detection and solution of the structural problems are model dependent, the detection and solution of the semantic problems are not model dependent and require some designer operations. Those requirements are more or less considerable. In J.M. De Souza's works, the solution of the semantic conflicts is entirely done by the designers. They must find the synonyms and homonyms alone [29]. C. Batini, M. Lenzerini and S. Navathe use a name convention (scheme-name.element-name), in order to remove homonym conflicts [2]. In more recent works, there exists a more deeper approach of the semantic integration : inside the process, it appears a distinct structure, allowing the representation of the semantic of design-scheme words [13, 4]. In the same way, we have defined a semantic representation structure, inspired by some linguistic works. This structure, named fuzzy thesaurus, allows to compare pairs of elements from the two schemes that must be integrated. This structure makes it possible to quantify the word likeness with a degree, and to qualify it by a kind of likeness, in order to semantically locate one word from another. Quantifying the likeness allows to choose the possible integration from several others. Qualifying the likeness makes it possible to know what kind of integration must be done between the two elements. We will, first, present the structure of our fuzzy thesaurus, then the way to enrich it. And lastly, we will show how to use it while integrating knowledge.

2 THE FUZZY THESAURUS

To represent the semantic of the words wich are used to designate attributes, methods, classes and links of the schemes, we have defined a model, inspired from a domain dealing especially with meaning of the words : the linguistic [8]. For linguists, a knowledge model is a multidimensional space, in which intersecting axes represent some conceptual primitives. A concept, a knowledge unit, may be represented and identified uniquely by refering to its coordinates on each axis. Listing the values of the concept with respect to each axis is equivalent to defining its position in the knowledge space [23]. In our data structure, we distinguish words (or terms) from concepts (or meanings), following Quillian ; he was the first to make this distinction [12]. Each concept is represented by a sentence, the most explicit one, which can be sometimes a definition. Each concept sentence must explain the meaning of words that are connected to the concept. We make the distinction between the words and the concepts in order to be able to work at a level (the concept level) where knowledge can be defined very precisely, without ambiguity, without some synonym and homonym problems. The concepts are linked together by the conceptual relationships of a semantic kind. The interpreting relationships between the words and the concepts allow the expression of the meanings given to the words. The use of the thesaurus allows the calculation of the ambiguity and nearness coefficients, from the conceptual and interpreting relationships.

3 THE THESAURUS STRUCTURE

We define two kinds of relationships in our thesaurus : the conceptual relationship of a semantic kind, that links the concepts together ; and the interpreting relationship, that links the words and the concepts together, in order to explicit their meanings. Now, we will present each of these kinds of relationships more precisely.

3.1 The conceptual relationships of a semantic kind

We use two kinds of conceptual relationships : the generic relationship and the aggregation relationship. Those relationships are the most often used [27]. There exists a lot of other relationships linking the concepts together [23], [27], [6]. But keeping them inside our thesaurus will bring nothing more, because we cannot exploit them. We must not forget that this thesaurus is only a tool for scheme integration, wich makes it possible to know the proximity between the words in order to know how to integrate them.

3.1.1 The generic relationships. When we link the concept c_a and the concept c_b, by a generic relationship, we want to indicate that c_b belongs to the class of the specific elements of c_a. But in order to capture more semantic, we graduate this membership. c_b can belong more or less strongly to the class of the specific elements of c_a. The set of all this specific elements is a fuzzy subset. So we can represent some categories with ill-defined boundaries, some situations between everything and nothing, like the quasi-generic relationship [23]. We can, for example, express the fact that all dandelions are members of the family of compositae, and that some members of the family of compositae are dandelions (generic relationship), or that some people consider that dandalions are vegetables, and that some vegetables are dandelions (quasi-generic relationship). Let C be the thesaurus-concept set, we define S as the fuzzy relationship which characteristic function assigns to each element of $C \times C$ a likelihood degree (between 0 and 1), f_S : $C \times C \to [0, 1]$, which indicates the membership degree of the concepts pair in the generic relationship. On the figure (see figure 1), this membership degree appears on the arc which represents the relationship between c_a and c_b. Of course, we do not represent arcs with nul degree.

3.1.2 The aggregation relationships. An aggregation relationship links a component and a compound together. We distinguish several kinds of aggregation relationships, following several criterions. Let study the **aggregation relationship** which links a house and a door and a door and a key together. In this relationship, the whole do not preserves the part features. Therefore this relationship is not transitive. The house do not preserves the door features, and saying that the house is composed of some keys has not real meaning. On the contrary, the aggregation relationship which links a cake and a piece, and a piece and a mouthfull together is transitive. It preserves features. Indeed, we can say that a cake is composed of some mouthfulls. Therefore, we distinguish transitive relationships from non-transitive ones. (This feature will be used in the thesaurus-exploitation phase). D. A. Cruse [9] et J. Lyons have shown in their works the existence of the transitive and non-transitive aggregation relationships [16].

In the aggregation example linking a house and a door together, we also can observe that the door is not the only possible component of the house, for a given aggregation ; the house can be seen as an aggregation of some doors and some windows simultaneously. On the contrary, a cake is composed only of some pieces simultaneously. Therefore, we distinguish the transitive relationships with several kinds of simultaneous components, and the transitive relationships with one kind of component. And we do the same discrimination on the non-transitive relationships. So, we differentiate four different kinds of aggregation relationships. They are the following :

- The transitive aggregation relationships :

    - The transitive aggregation relationships with several kinds of simultaneous components, called **set relationships** [16]. An example of this kind of aggregation can be the meal, which is composed of a seafoodboard (heterogeneous set), a cheeseboard (heterogeneous set) and a fruitbasket (heterogeneous set). The three components appear simultaneously in the meal, and the relationship is transitive : if we consider the fruit basket as composed of a red fruit basket (heterogeneous set) and an exotic fruit basket (heterogeneous set), we can illustrate the transitivity of this kind of relationships by describing the meal as composed of a seafoodboard, a cheeseboard, a red fruit basket and an exotic fruit basket.

    - The transitive aggregation relationships with only one kind of component, called **segmentational relationships** [16]. An example of this kind of aggregation is the cake, which is only composed of the pieces, which are only composed of the mouthfulls. Each compound is only composed of one kind of component. We can illustrate the transitivity of this kind of relationships by describing the cake as composed of the mouthfulls.

- The non transitive aggregation relationships :
  - The non transitive aggregation relationships with several kinds of simultaneous components, called functional relationships [16]. An example of this kind of aggregation is the bicycle, which is composed of the wheels and a handlebar. A wheel is composed of the rays, but we cannot say that the relationship which links the bicycle and the rays together has a meaning : this kind of relationships is non transitive.
  - The non transitive aggregation relationships with only one kind of component, called collection relationships (homogeneous sets) [16]. We can illustrate this kind of composition like this : the lump sugar is packed into some boxes containing a lump sugar collection (homogeneous set). The boxes are dispatched by the cardboard crates containing a collection of sugar boxes (homogeneous set). Therefore, we can say that a crate is composed of a collection of boxes, that are composed of a collection of lumps. This relationship is non transitive, and each component is only composed of one kind of components.

D.A. Cruse, in his works [9], explains the exitence of several kinds of aggregations. He show how to detect differences in the used vocabulary representing the aggregation relationship ; the word "part" represents a functional relationship, the word "piece" represents the segmentational relationship. The aggregation relationship appears like a family of four different kinds of relationships. We note A_e the set relationship, A_s the segmentational one, A_f the functional one, and A_c the collectional one. We define them like we have defined the generic relationship, that is to say with fuzzy relationships, in order to indicate if the component is optional, or not, [23], and how much so. Our model support several different ways of composing the same compound, several alternatives [23]. Let C be the thesaurus-concept set, for each aggregation of the concept c of C, in a general way, we define A_k like a fuzzy relationship which characteristic function, for the k aggregation alternative, assigns to each element c of C, the degree $f_{A_k}(c)$, varying between 0 and 1, with which c belongs to A_k, $f_{A_k} : C \rightarrow [0,1]$. So, we obtain, for each aggregation alternative of a concept, a fuzzy subset of the concepts that are its components. On the figure (see figure 1), this necessity degree of the component for the compound appears on the arc representing the aggregation relationship between the concepts. Of course, we do not represent arcs with nul degree. We assign the notation symbols $> -$ and $- <$ [27], in order to differentiate aggregation relationships from generic relationships.

3.2 The interpreting relationships

The interpreting relationships allow to link the words and the concepts together. As was done previously, in order to have a better representation of the real world, we define them as fuzzy relationships. Let M be the thesaurus word set, and let C be the concept set. We define Cpt like a fuzzy relationship which characteristic function assigns to each element of $M \times C$, a degree, between 0 and 1, $f_{Cpt} : M \times C \to [0,1]$, which indicates the membership degree of the word-concept pair to the interpreting relationship. This relationship will be named "have-meaning" relationship. We obtain for each word of the thesaurus a fuzzy subset of the concepts with which it is linked. On the figure (see figure 1), this probability for the concept to be a meaning of the word appears on the arc representing the interpreting relationship. Of course, we do not represent arcs with nul degree. Thanks to this kind of relationships, we can know the semantic nearness of the words and the ambigous words. Two semantically-near words are the words linked to the concepts linked together. An ambigous word is a word with several interpreting relationships, that is to say several meanings.

The figure 1 summarises all the notions described above.

Now we will show how to put knowledge in the thesaurus, and how to exploit it. The building phase consists in entering the words and the concepts in the thesaurus, and in linking them together, with the different kinds of links. The exploitation phase allows to know, from knowledge entered throughout the building phase, the ambiguity of a word, its nearness with other words, in order to integrate the schemes from a semantic point of view.

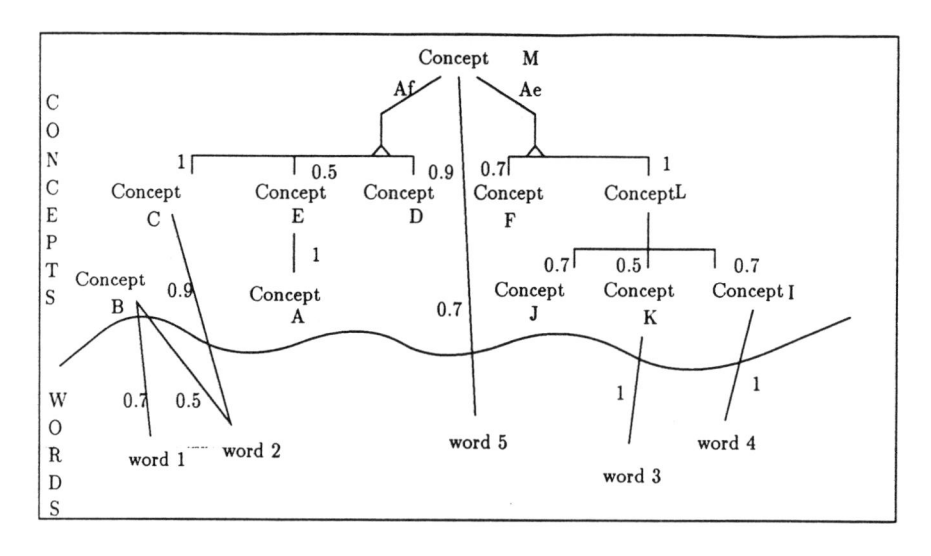

Figure 1: The thesaurus structure

4 THE THESAURUS CONSTRUCTION

The different experts will put their knowledge in the thesaurus, and link them together with the relationships that we have defined.

4.1 The conceptual relationships of a semantic kind

Each expert must structure the concepts that he wants to add to the thesaurus, thanks to conceptual relationships. He must examine each concept that he wants to use to qualify words he uses in his scheme. If the concept he wants to use does not exist in the thesaurus, he must add it, and link it to one or more concepts of the thesaurus, by using the conceptual relationships. He will qualify each of those links by a fuzzy coefficient : a likelihood degree for the generic relationships, a necessity degree for the aggregation relationships. Each expert will also be able to give his point of view about values of coefficients of all conceptual arcs of the thesaurus.Then, we will do the wheigted average of all the given knowledge about degrees from all experts, in order to know globally the linkeness degrees of the concepts. Let $f(c_a, c_b)$ be the semantic relationship (generic or aggregation) which links c_a and c_b together, $V = \{v_1, v_2, ..., v_n\}$ the set of expert-given values for this relationship (n experts), and q_i the credibility coefficient assigned to the expert i, then $f(c_a, c_b)$ is defined by :

$$f(c_a, c_b) = \frac{1}{\sum_{i=n}^{i=1} q_i}(\sum_{i=n}^{i=1} q_i v_i)$$

4.2 The interpreting relationships

To add a word in the thesaurus, an expert will have to link it to one or several concepts, with the interpreting relationships. For each new arc, the expert will give a value, varying between 0 and 1, indicating the probability of use of this meaning for this word by the expert. The sum of all the meanings for one word for one expert will be 1. And we will also compute the weighted average of all the probabilities from all the experts. Let $C = \{c_1, c_2, ..., c_m\}$ be the concept set assigned to m by the different experts, and $E_i = \{p_{i1}, p_{i2}, ..., p_{im}\}$ the set of use probability of the elements of C that are meanings for m for the expert i, with $i \in [1, n]$ and q_i the credibility coefficient assigned to the expert i, then we define f_{Cpt} as :

$$f_{Cpt}(m, c_k) = \frac{1}{\sum_{i=n}^{i=1} q_i}(\sum_{i=n}^{i=1} q_i p_{ik})$$

with $k \in [1, m]$

This way of proceeding has two advantages :

- each expert can put his knowledge in the thesaurus, without really worry about knowledge of the other experts : we do the global computing of f_S, f_A, f_{Cpt} ;

- it allows the incremental enrichment of the thesaurus : adding new knowledge of new experts requires only to compute the weighted averages again.

Moreover, we can weight expertise significance when they are done simultaneously or in an incremental way. We can be more favorable with some schemes, for example, the more recent.

Now, we will show how we compute the ambiguity of a word, and how we find its nearness.

5 THE THESAURUS EXPLOITATION

The thesaurus exploitation allows to know what words are ambigous, when they can be used with different meanings. It also makes it possible to detect similar words in the schemes under study.

5.1 The semantic nearnesses

Two lexical relationships often appear in the linguistic papers : the synonym relationship and the homonym relationship [24]. G. Hirst presents them among the three kinds of semantic ambiguities (the last one is about grammar, and useless for us) [14]. In our structure, the lexical relationships do not appear explicitly, because they can be deduced from the interpreting relationships [19]. Moreover, in order to capture more semantic, we speak more globally about the semantic nearness and the ambiguity. We deduce weighted knowledge from the conceptual and interpreting relationships. From this knowledge, we can represent gradually the word membership to the fuzzy sets of semantically-near words or semantic-ambigous words, for a given word. We define D_{V_o} as the nearness degree (varying between 0 and 1) which links m_a and m_b. And we define D_{Am} as the ambiguity degree of the word m (varying between 0 and 1). We will now show how we compute D_{V_o} and D_{Am}.

5.2 The ambigous-word treatment

To know if a word is likely to be ambigous, we must see if it is connected to several concepts, that is to say to several meanings. In this case, it can be considered as ambigous. But, we think it will be more interesting to qualify this ambiguity. A word is linked to concepts by the f_{Cpt} relationship which have a probability coefficient, varying between 0 and 1, and indicating how much the word is linked to the concept. We can use the information provided by these coefficients to detect the words likely to be ambigous. Therefore, we compare the probability coefficients of all the interpreting relationships connected to the word. If the coefficients have near values, the different meanings are as often used. If the values are very different, some meanings are more used than the others. In this last case, the word is less ambigous than in the first one. Therefore, we compute the ambiguity degree of a word D_m in this way : let m be a word, and $C = \{c_1, ..., c_m\}$ the set of its concepts (meanings), sorted by decreasing values. Then D_{Am} is defined by the formula written by Shanon in information theory :

$$D_{Am}(m) = \Sigma_{k=1}^{k=m}(f_{Cpt}(m, c_k))log(f_{Cpt}(m, c_k))$$

So, when we find the same word in the several design schemes, we can know its ambiguity degree, and make it easier to carry out the scheme integration. It will avoid to integrate a word used in different meanings in the different schemes.

5.3 The word-semantic-nearness treatement

Knowing if two words have a semantic-nearness is more difficult than determining some ambiguity of one of them. It is more difficult, because we must consider the concepts linked to the two words under study. And they are several cases :

- case 1 : the two words have some common concepts,

- case 2 : the first word has some concepts directly linked to some concepts, themselves directly linked to the second word

- case 3 : the first word has some concepts not directly linked to the concepts linked to the second word.

Case 1 : in this case, the two words are synonyms (noted Sy) and one must determine how much they are synonyms. Let m_a and m_b be the two words under study, and C be the set of common concepts to m_a and m_b, we define the semantic-nearness degree D_{Vo} between m_a and m_b by :

$$D_{Vo}(m_a, m_b) = max_{c \in C} min(f_{Cpt}(m_a, c), f_{Cpt}(m_b, c))$$

Case 2 : In this case, like in the first one, one must determine the nearness degree between the two words. And we also can specify the kind of nearness :

- the first word may be more generic than the second one (G),

- the first word may be more specific than the second one (S),

- the first word may be composed of the second one (C^{ed}),

- the first word may be a component of the second one (C^{ent}),

Let m_a and m_b be the two words under study. Let $C_a = \{c_{a1}, c_{a2}, ..., c_{am}\}$ be the set of the concepts linked to m_a, and $C_b = \{c_{b1}, c_{b2}, ..., c_{bm}\}$ the set of concepts linked to m_b. We consider in C_a and C_b that concepts from m_a and m_b are directly linked two by two ($\forall k \in [1, m]$, $\exists f = f_S(c_{ak}, c_{bk})$ or $\exists f = f_S^{-1}(c_{ak}, c_{bk})$ or $\exists f = f_A(c_{ak}, c_{bk})$ or $\exists f = f_A^{-1}(c_{ak}, c_{bk})$). We define the semantic nearness degree D_{Vo} between m_a and m_b by :

$$D_{Vo}(m_a, m_b) = max_{c_{ak} \in C_a, c_{bk} \in C_b} min(f_{Cpt}(m_a, c_{ak}), f(c_{ak}, c_{bk}), f_{Cpt}(m_b, c_{bk}))$$

In this case, the kind of nearness depends on f :

- if $f = f_S(c_{ak}, c_{bk})$, then it is of the S kind,

- if $f = f_S^{-1}(c_{ak}, c_{bk})$, then it is of the G kind,

- if $f = f_A(c_{ak}, c_{bk})$, then it is of the C^{ed} kind,

- si $f = f_A^{-1}(c_{ai}, c_{bi})$, then it is of the C^{ent} kind,

Case 3 : In this case, we must, first, value the proximity between concepts linked to the words under study, in order to be, in a second time, in a situation like the one of the case 2. That is why, we will first present the cases where transitivity is admitted. An important point is the transitivity of the aggregation relationship. We have described it like a set of four kinds of relationships, and we have explained (see paragraph 3.1.2) that the transitivity property is relationship-type dependent. In the next table (table 1), we recapitulate the different cases of the succession of the different relationships : Let c_a, c_b et c_c be the concepts under study, such as $\exists f_l(c_a, c_b)$ and $f_j(c_b, c_c)$. In lines, we find the different kinds of relationships that can link c_a to c_b. In columns, we find the different kinds of relationships that can link c_b and c_c.

		generic relationship		aggregation relationships			
		$f_{j,S^{-1}}$	$f_{j,S}$	$f_{j,A_{\{f,c\}}}^{-1}$	$f_{j,A_{\{e,s\}}}^{-1}$	$f_{j,A_{\{f,c\}}}$	$f_{j,A_{\{e,s\}}}$
generic relationship	$f_{i,S^{-1}}$	f_S^{-1}	•	•	•	$f_{A_{\{f,c\}}}$	$f_{A_{\{f,c\}}}$
	$f_{i,S}$	•	f_S	$f_{A_{\{f,c\}}}^{-1}$	$f_{A_{\{f,c\}}}^{-1}$	•	•
aggregation relationships	$f_{i,A_{\{f,c\}}}^{-1}$	•	$f_{A_{\{f,c\}}}^{-1}$	•	•	•	•
	$f_{i,A_{\{e,s\}}}^{-1}$	•	$f_{A_{\{e,s\}}}^{-1}$	$f_{A_{\{f,c\}}}^{-1}$	$f_{A_{\{e,s\}}}^{-1}$	•	•
	$f_{A_{\{f,c\}}}$	$f_{A_{\{f,c\}}}$	•	•	•	•	•
	$f_{A_{\{e,s\}}}$	$f_{A_{\{e,s\}}}$	•	•	•	$f_{A_{\{f,c\}}}$	$f_{A_{\{e,s\}}}$

Table 1: transitivities of relationships

When we substitute, like in the precedent table, the aggregation relationship (f_A^{-1}) followed by the generic relationship (f_S), by an aggregation relationship (f_A^{-1}), one must assign to it a fuzzy coefficient, computed from the two initial ones. Therefore, we have differentiate several cases : let c_a and c_c be the two concepts under study, and c_b be the concept linked to c_a and c_c.

- If the three concepts are linked together with the same kind of relationship, then the semantic distance between the two concepts is :

$$dist = f(c_a, c_b) \times f(c_b, c_c),$$

with $f = f_A$ or $f = f_A^{-1}$ or $f = f_S$ or $f = f_S^{-1}$, it depends on the case.

- If the two relationships are transitives, we see that the aggregation relationship provides more semantic than the generic one. Therefore, the semantic distance depends on the aggregation coefficient. For example, if a car (c_a) is composed of a petrol engine (c_b), which is a specialization of an engine (c_c), then we can consider that there exists a semantic link between a car (c_a) and an engine (c_c) and that the kind of this relationship is aggregation. And the coefficient of the link from c_a to c_c, must not be the same that the one which links c_a to c_b (even if it is right to say that a petrol engine is linked to a car as well as an engine), because in the case where the engine and the petrol engine appears together in a scheme, we must integrate the car and the petrol engine, rather than the car and the engine. Therefore, the semantic distance between c_a and c_c is defined by :

$$dist = 0.9 \times f(x, y),$$

where $f(x, y)$ represents the aggregation relationship.

Finally, the last step of our semantic-nearness treatment consists in keeping as result of the exam of the semantic nearness of one word the best result from all those obtained in cases 1 and/or 2.

At the end of this step, we obtain :

- Either a result : we have found some links between the two concepts, and we have assigned a coefficient to this semantic distance ; then we are in a situation like in the case 2;

- Or no result : there is no way to link this concepts together. Then the words are not semantically near.

Now we are able to know if two words are semantically near, following two criterions :

- a kind of nearness, which could be :

  - synonym (it corresponds to the case 1), named Sy,
  - generalization (G),
  - specialization (S),
  - composed (C^{ed}),
  - component (C_{ent}),

- a nearness degree which consists in a fuzzy coefficient varying between 0 and 1.

6 CONCLUSION

We have define a thesaurus structure which allows to take into account several expertises of several designers working simultaneously and/or in an incremental way. Moreover, we affect the relative weights to the knowledge of the different experts, while acquiring several perceptions, at different periods, in order for example to favour the most recent ones. Lastly, the knowledge exploitation makes it possible to integrate the schemes on a semantic point of view, because it allows to compute semantic-nearness degree, and to specify the kind of nearness. It also makes it possible to compute the ambiguity degrees of the words in the schemes, and it avoid the integration of a word used in several schemes, but used in different meanings.

REFERENCES

[1] R. Ahmed, P. De Smedt, W. Du, W. Kent, M.A. Ketabchi, W. A. Litwin, A. Rafii, and M.C. Shan. The pegasus heterogeneous multidatabase system. *Computer*, 24(12):19–27, December 1991.

[2] C. Batini, M. Lenzerini, and S. Navathe. A comparative analysis of methodologies for database schema integration. *ACM Computing Surveys*, 18(4), December 1986.

[3] J. Biskup and B. Convent. A formal view integration method. In *International conference on the management of data*, Washingtown, 28-30 May 1986. ACM.

[4] M. Bonjour and G. Falquet. Concept bases : a support to information systems integration. In *Advanced information systems engineering : proceedings of the 6th international conference CAISE'94, June 6-10, 1994*, volume 811 of *Lecture Notes in Computer Sciences*, pages 242–255. Springer Verlag, 1994.

[5] A. Cavarero and E. Vittori. COD, un système de définition de classes d'objets. In *Inforsid*, Lille, France, Mai 1993.

[6] R. Chaffin and D. J. Herrmann. The nature of semantic relations : a comparison of two approaches. In M. W Evens, editor, *Relational models of the lexicons*, pages 289–334. Cambridge university press, 1988.

[7] I. Comyn-Wattiau. *L'intégration de vues dans le système SECSI*. PhD thesis, Université Paris VI, France, Octobre 1990.

[8] A. D. Cruse. Aspects of lexical relations. In *premier séminaire de sémantique lexicale*, pages 1–13, Toulouse, Janvier 1991. Université Paul Sabatier.

[9] D.A. Cruse. *Lexical semantics*. Cambridge textbooks in linguistics. Cambridge university press, 1986.

[10] U. Dayal and H. Hwang. View definition and generalization for database integration in multidatabase system. *IEEE Transaction on software engineering SE*, 10(6), November 1984.

[11] J. Diet and F.H. Lochovsky. Interactive specification and integration of userviews using forms. In *8th international conference on entity-relationship approach*, Toronto, 18-20 October 1989.

[12] M. W. Evens. Relational models of the lexicon : introduction. In M. W Evens, editor, *Relational models of the lexicons*, pages 1–37. Cambridge university press, 1988.

[13] P. Fankhauser, M. Kracker, and E.J. Neuhold. Semantic vs. structural resemblance of classes. *Sigmod record*, 20(4):59–63, December 1991.

[14] G. Hirst. *Semantic interpretation and the resolution of ambiguity*. Cambridge University Press, 1987.

[15] H.C. Howard and D.R. Rehak. KADBASE interfacing expert systems with databases. *IEEE Expert*, 4:65–76, 1989.

[16] M. A. Iris, B. E. Litowitz, and M. W. Evens. Problems of the part-whole relation. In M. W Evens, editor, *Relational models of the lexicons*, pages 261–288. Cambridge university press, 1988.

[17] P. Johannesson. Schema transformations as an aid in view integration. In *Advanced information systems engineering : proceedings of the 5th international conference CAISE'93, June 8-11, 1993*, volume 685 of *Lecture Notes in Computer Sciences*, pages 71–92. Springer Verlag, 1993.

[18] M.V. Mannino and W. Effelsberg. Matching techniques in global schema design. In *IEEE Compdec conference*, Los Angeles, Californie, 1984.

[19] G. A. Miller, R. Beckwith, C. Fellbaum, D. Gross, and K. J. Miller. Introduction to wordnet : an on-line lexical database. *International journal of lexicography*, 3(4):235–244, 1990.

[20] A. Motro. Vague : a user interface to relational databases that permits vague queries. *ACM transactions on office information systems*, 6(3):187–214, July 1988.

[21] S. Navathe, R. Elsmari, and J. Larson. Integrating user views in database design. In *IEEE Computer*, January 1986.

[22] C. Parent and S. Spaccapietra. Intégration de vue et relativisme sémantique. In *VI journées bases de données avancées*, Montpellier, France, Septembre 1990.

[23] J.C. Sager. *A practical course in terminology processing*. John Benjamins, 1990.

[24] P. Saint-Dizier. A denotational semantics for synonyms and opposites. In *premier séminaire de sémantique lexicale*, pages 70–75, Toulouse, Janvier 1991. Université Paul Sabatier.

[25] A.P. Sheth and J.A. Larson. A tool for integrating conceptual schemas and user views. In *4th international conference on data engineering*, 1988.

[26] P. Shoval and S. Zohn. Binary-relationship integration methodology. In *Data and knowledge engineering*. Elsevier science, North-Holland, 1991.

[27] G. Van Slype. *Les langages d'indexation : conception, construction et utilisation dans les systèmes documentaires*. les éditions d'organisation, 1987.

[28] W.W. Song, P. Johannesson, and J.A. Bubenko. Semantic similarity relations in schema integration. In *Entity relationship approach, ER'92, 10th international conference on the entity relationship approach*, volume 645 of *Lecture Notes in Computer Sciences*, pages 97–120. Springer Verlag, 1992.

[29] J.M. De Souza. SIS - a schema integration system. In *BNDODJ conference*, 1986.

[30] C.J.E. Thieme and A.P.J.M. Siebes. Schema refinement and schema integration in object-oriented databases. Technical Report CS-R9354, Centrum voor Wiskunde en Informatica, P.O. Box 4078, 1009 AB Amsterdam, The Netherlands, August 1993.

MANAGING DESIGN AND MANUFACTURING CONSTRAINTS IN A DISTRIBUTED INDUSTRIAL ENVIRONMENT: THE CREATION OF A MANAGED ENVIRONMENT FOR ENGINEERING DESIGN

A Medland

1 BACKGROUND

Considerable effort is currently being directed in industry toward the reduction of time in the development of new products. This has led to a great interest in the principles of concurrent engineering [1]. Here the aim is to draw as many of the processes required into parallel, or as a minimum, overlapping operations. Such processes, whilst in theory providing a great reduction in lead times, result in greater complexities in the organisation and management of the whole operation.

There are advantages in having many operations occurring concurrently, as long as errors found in one do not cause any of the other activities to be reworked or abandoned. If so all the advantages may be lost.

Within the concurrent approach, research has been undertaken into the definition and construction of complete product data bases [2]. Here all aspects of the design, manufacturing and function of the proposed product are recorded. These are then accessed throughout the product development and newly created data added. Such an approach assumes that the original definitions are well founded, not in conflict or leading, in hindsight, to the creation of unreal solutions.

The nature of practical engineering design is that it evolves throughout the designing process, it is by nature based upon an incomplete definition and will develop in an ill-structured manner [3].

2 MANAGEMENT REQUIREMENTS

The employment of a common data base structure necessitates that all design activities have access to that common information. This then needs to be controlled and decisions made as to who has the rights to change individual data elements and who has only the rights to use. Whilst this can be handled by any owner/reader management procedure, there are many cases in which conflict occurs and higher levels of decision making need to be undertaken. It may be insufficient for a designer to simply have control over a variable and to force all other processes to comply with that demand. For example the designer may choose the smallest blend radius which will satisfy a bending stress condition, whilst being unaware that this is smaller than the standard range of tools available in the workshop. Although he may be happy to accept the slightly larger radius (giving a greater margin of safety), under the authoritarian structure the machine shop is required to invest in the creation of a new unnecessary tool.

A management structure is thus required to provide not only control but also a decision making and reporting structure.

3 A CONSTRAINT RESOLUTION APPROACH TO HANDLING DESIGN PROBLEMS

Work has been undertaken, initially within the Brunel research group for over ten years and now continued at Bath, into the solution of design and manufacturing problems by constraint resolution techniques [4]. A constraint modelling language has been created, called "RASOR", that allows the constraint rules of a problem to be defined and resolved. Within this process a number of rules are created and clustered together. Free variables are declared for the group that is to be manipulated, within a direct search procedure, to seek a true solution.

Such techniques have been successfully applied to resolve a wide range of engineering and design problems [5][6]. These extend from conceptual design through to the determination of errors in manufacturing from inspection data. The approach has been used to provide an intelligent interface between CAD and coordinate measuring machines and is currently being used in industry to design and optimise packaging machinery.

As larger problems are addressed the collection of constraint rules into one large function that needs to be resolved, presents problems. Firstly the search time increases considerably. In addition, it is found that the level of control over the rules and their definition needs to be refined. A greater degree of control is found to be necessary to stop the search moving away from the simple (and perhaps more obvious) solutions into more complex relationships. It is then necessary to impose additional rules to ensure that the global searching of all the variables does not disturb the design solution into an entirely new and unexpected form.

These problems can also be considered as ones of managing large and complex problems (as well as ones of mathematically searching large and sparse matrices). In practice, these problems are only rarely solved in large problem groups with complex rules. More often the problem is resolved by decomposing it into a number of sub-problems that are then defined and addressed in a specified order. It is often a management decision as to which sub-problem is given the greatest priority and resolved first, and hence the form of the final solution. If again conflict is found when addressing the subsequent sub-problems, it is a management decision to re-visit previous decisions, with the new knowledge, and to re-evaluate the problem.

4 ASSIGNMENT OF VARIABLES

A new management approach had thus to be conceived that would allow all of the above problems to be addressed within a constraint resolving structure.

The approach currently under investigation, allows all variables to be assigned to a specified sub-group or user that can itself be managed within a higher level token-based resolving system.

Any problem can be broken down into a chosen set of hierarchical relationships. These may reflect any chosen "view-point" on that design but in the main will be used to define the relationships between the parts of the functional assembly. This set of relationships can be redefined at any stage to create a different view, such as when the design is frozen and authority is moved from design to manufacturing, and further when the product is in full production and under marketing control.

Within such a system the user is associated with an activity or responsibility and, within an assembly hierarchy, will thus enter by first specifying the assembly or part to be investigated. All the free variables of that group are then available to the user and can be manipulated to change the style and relationships, as required.

The system must then be able to search all rules associated with the complete design to establish which of these are affected by any variable parameters of the chosen group. These rules are thus isolated and formed into a sub-problem that has to be made true or remain true after all manipulations have been carried out. The operator can thus make any changes that are considered necessary and then have the truth of the sub-problem evaluated.

If the sub-problem remains true then the changes in the design are acceptable and the whole design problem remains true (as these variables effect only the truth of the rules assembled in the created sub-problem).

If however the sub-problem is untrue, then the rules of the sub-problem need to be further investigated. Each rule is investigated in turn and its truth determined. The one with the largest untruth can then be investigated further and other variables (not assigned to the user) can be manipulated on a "temporary loan" basis. Whilst these values can be changed, in an attempt to re-establish the truth of the sub-problem, the final value of these cannot be stored in the global data base. Only the user with the correct assignment can carry out such an action.

The chosen approach thus allows the user to investigate possible changes and establish what negotiations need to take place with the assignee of the disturbed "borrowed" variables.

If the sub-problem created around the group, containing the modified variables, remains true with the disturbed "borrowed" values, the final design can be re-established as true and is therefore declared to be acceptable.

Should the new values result in a conflict within their own sub-problem, the changed values can be rejected by the assignee. This then leaves a conflict existing in a given set of rules. The problem is then reported up the hierarchical management system to be resolved by the first sub-system in which both sets of variables are inherited. Here a strategic decision can be made to accept either of the chosen parameters and force the other sub-problem to seek an alternative but true state. Alternatively the two sub-problems can be combined into a higher cluster of constraints and an optimum solution sought.

5 TOKEN MANAGER

The proposed management approach has thus been created around the principle of dividing the design into a number of assigned sub-problems. These can then be resolved by normal constraint resolution techniques.

A higher level system has been created to allow the relationships between these sub-problems to be handled. This has been built around a token handling structure, that is an extension of a Petri net form and incorporates constraint net techniques that have previously been created to handle manufacturing and design problems [7].

Here all sub-problems are considered as separate processes. The entries and conditions required to run each are set up as a series of tokens. Once the required tokens are in place, for a particular process, the activity is undertaken. The "firing" of a process will result in the "truth" of the outcome of that activity being reported to the higher level decision making process (incorporating rules if necessary). If the outcome results in a true or acceptable state, then this, in its turn, can result in tokens being inserted that, with others, will result in further processes being triggered. The designing process thus proceeds as a cascade of firings and re-firings that settles down as a completed state, or an intermediate one, to await the implanting of other tokens of information before further firings can commence.

The role of the higher level management system is to seek a state in which all necessary processes have been activated and a "true" outcome achieved from all simultaneously. Logical rules can be implanted within its structure to allow different actions to be taken when different (or inappropriate) combinations of tokens arrive. This may result in certain combinations of tokens being reset and processes refired (or even new processes initiated).

6 SYSTEM APPLICATION

The proposed approach is currently being incorporated within a constraint based environment for the design of mechanisms and assemblies. A series of test programs has been created to allow the approach to be investigated. These have been employed within the following test case example.

The test case has been built around the design and assembly of a pen. This was defined as being comprised of four separated parts; a cap, a body, an ink refill and an end bung. Within the adopted hierarchical management structure the cap and body were directly associated with the "pen", whilst the ink refill and bung were directly associated with the body. Through such a structure higher to lower level relationships could be explored, as with the relationships existing between cap and ink refill. Also within this structure, the relationships between body, ink refill and bung could be addressed as individual parts or clustered together under the single "body" node.

Whilst this problem is hierarchically and geometrically simple, many aspects of the proposed management approach could be investigated, by initially allowing all four parts to be modelled independently. In the examples shown these parts are constructed from primitive solid elements, to provide a clear and simple model structure. More complex styled components can however be created and used. Here each part was modelled to provide an acceptable geometric form, whilst no account was taken of scale or fit necessary between parts.

Rules were then created, through discussion, to ensure that the final pen design met all the desired objectives. These rules ranged from external conditions (such as "the distance from the underside of the clip to the end of the pen must be less than the depth of the pocket"), through constraints on geometry to ensure that parts fit (such as "inside diameter of cap same as outside diameter of top of body"), to constraints on style (such as "what proportion of pen can be seen above the pocket"). Throughout this test case twenty six rules in total were developed and applied to the resolution of the combined problem.

These rules were initially written into a text file that was read into the main management program. Through the management hierarchy the problem could be addressed in different ways to allow many designs to be created that conformed to the set of rules (it should however be noted that before a complete engineering design could be completed further rules would need to be added to this original set).

Figure 1 shows the result of firstly inserting the original parts together in the same work space and assembling. No rules were applied, resulting in a number of regions of interference existing (show cross-hatched) and styling proportions not yet satisfied.

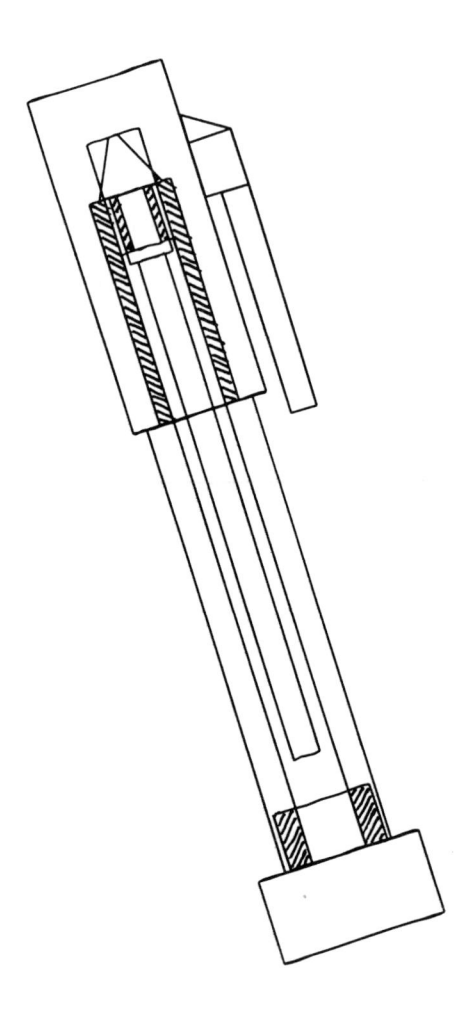

Figure 1. Assembly of individual piece part designs for a pen.

By entering the design at "body" level the ink refill and bung can be automatically optimised for a solution conforming to the given rules and made to fit the cap part. A solution found is shown in figures 2a and 2b. Here, in order to fit within the small hole in the original cap design (shown in figure 2a), the pen body has been severely reduced, leaving only a minimum of material but still conforms to the design rules. This can be seen clearly on the body assembly (figure 2b).

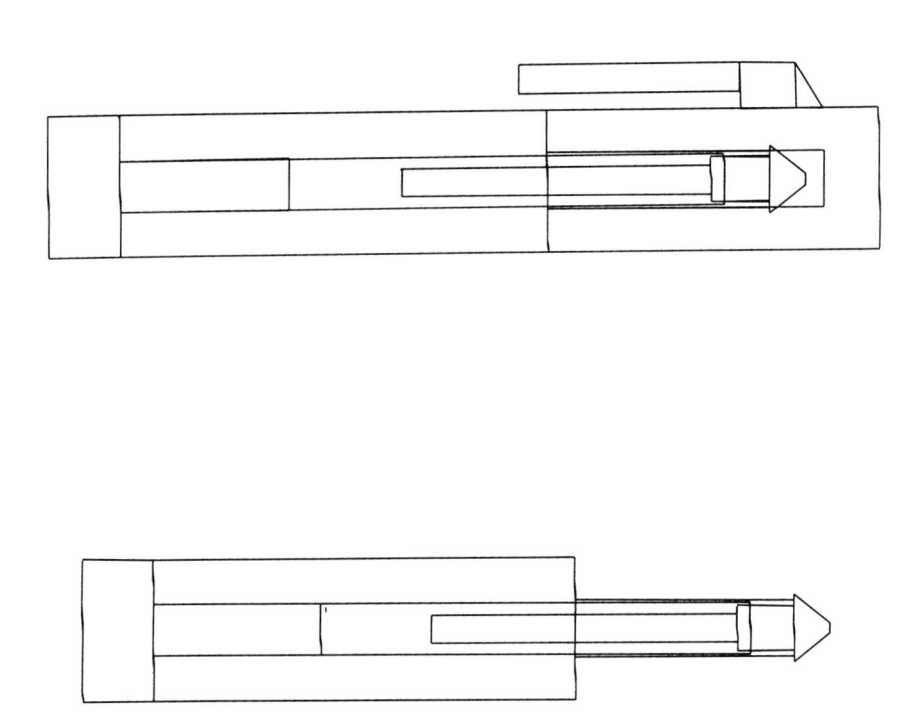

Figure 2. Pen body, ink refill and bung designed to fit existing poor cap design.
                a) body and cap assembled.
                b) body only.

Such a design would be considered as unacceptable. The management system was then used to allow a new style of cap to be produced. Proportions and wall thicknesses were chosen and the rules invoked to ensure that the design conditions could be met. A preferred design of cap thus emerged (as shown in figure 3) but failed to fit to the original body design.

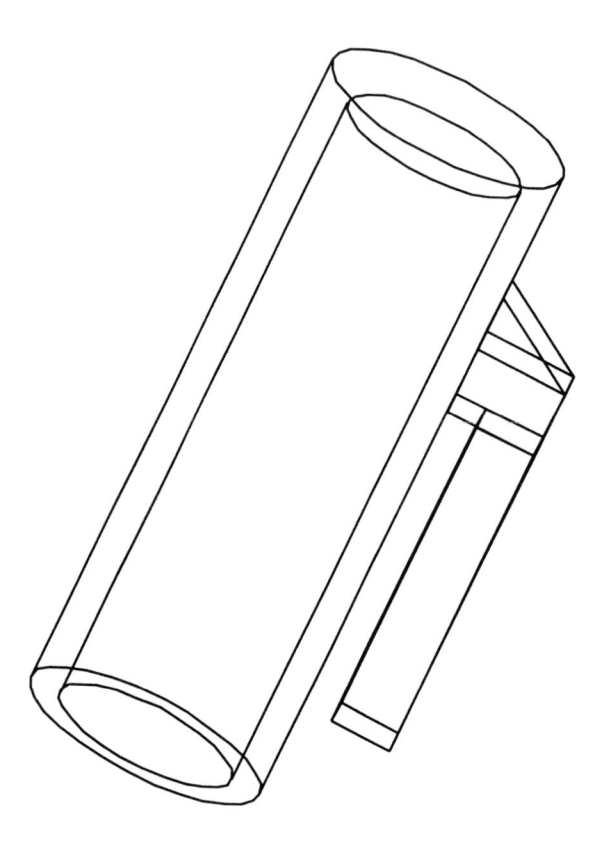

Figure 3. Preferred cap design.

By re-running the body design a new configuration of body, refill and bung was created that satisfied the rules (as shown in figure 4). This design was considered to be acceptable but for the chosen lengths of the bung and refill. Here the bung to refill relationship is satisfied by the bung being large and the refill small. For practical reasons it is desirable to have the refill as long as possible.

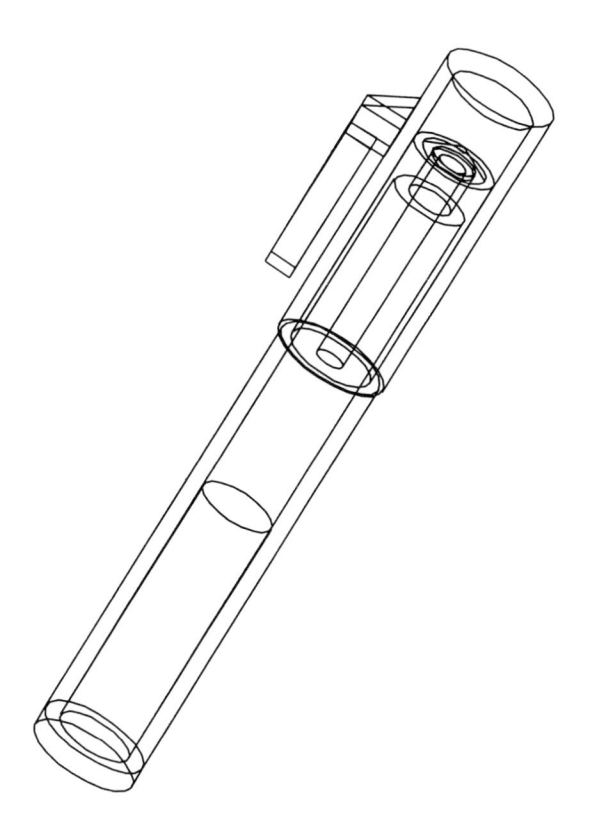

Figure 4. Body designed to fit new cap, showing short ink refill and long bung.

A new length was then chosen for the refill, making all parts true and acceptable but for the bung. This was then re-evaluated though the management system and a new bung designed to meet the new condition. This then resulted in a finally accepted pen design as shown solid-shaded in figure 5.

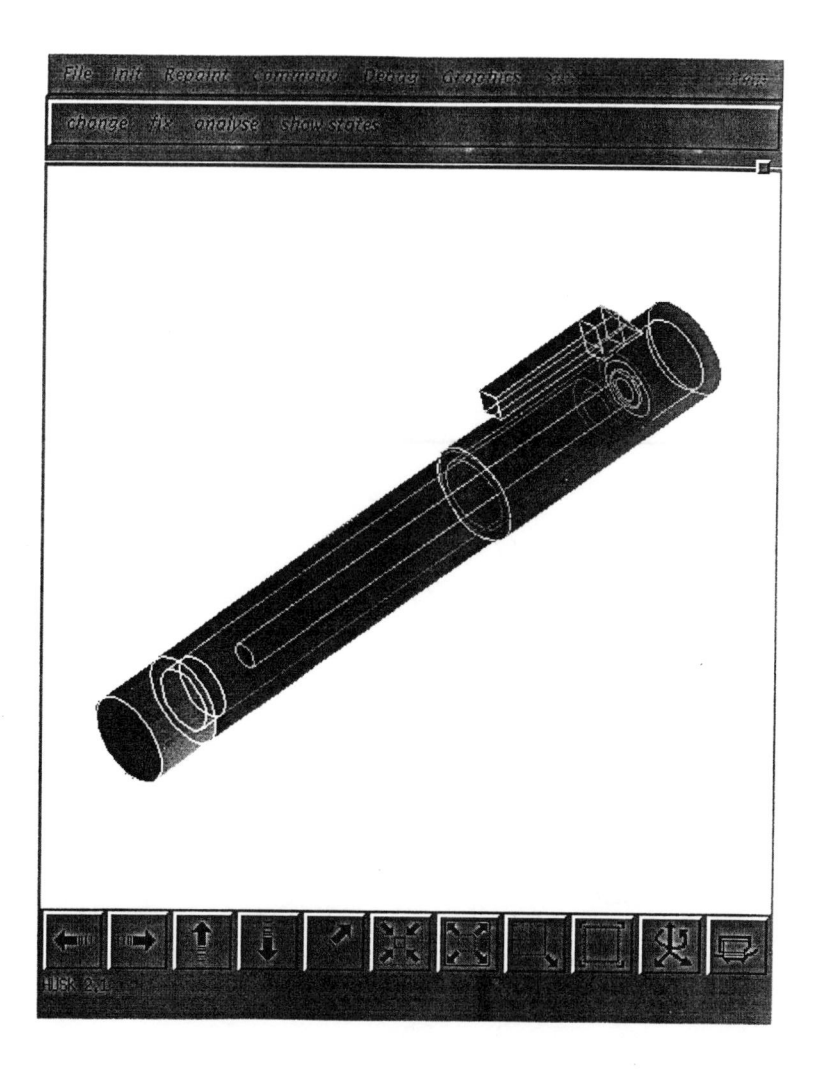

Figure 5. Final design with extended ink refill and bung to fit.

7 CONCLUSIONS

The problems associated with handling complex design problems within a managed environment have been investigated.

The approach adopted for the problem solving system has been to allow constraint rules to be generated for all aspects of the design and for these to be solved within a constraint resolution approach, through direct search of the free variables of the individual part geometry.

A management system was constructed and evaluated, based upon the approach of assigning these variables within a hierarchical management structure that could be formed to represent the part assembly problem, or any other desired view-point on the design. The individual elements of the design were resolved independently and the truth of the sub-problem reflected up the management structure as a set of tokens. Different sub-problems were fired or re-visited dependent upon the state of the tokens returned.

The approach, together with the token-based management structure, investigated in this study is now being implemented in detail within a general designing environment employing constraint resolving procedures. Currently this system is being populated with analysis, information and libraries for the design and optimisation of mechanisms.

ACKNOWLEDGEMENTS

The design study reported in this paper was conducted as part of an investigation into the creation of an environment for machinery design, being supported by a EPSRC/DTI LINK project and a group of collaborating companies. The author wishes to gratefully acknowledge the encouragement and support of these organisations throughout this research programme.

REFERENCES

1. *Williams D.J., Manufacturing Systems*, Halsted, New York, 1988.

2. Ellis T.I.A., Young R.I.M. and Bell R., *Modelling manufacturing process information to support simultaneous engineering*, ICED '93, The Hague, August 1993.

3. Medland A.J., *The Computer-based Design Process* - 2nd edition, Chapman-Hall, London, 1992.

4. Leigh R., Medland A.J., Mullineux G. and Potts I.R.B., *Model spaces and their use in mechanism simulation*, Proc. IMechE, J. of Eng. Manf., **203**, 1989, pp 167-174.

5. Medland A.J., Mullineux G., Butler C. and Jones B.E., *The integration of coordinate measuring machines within a design and manufacturing environment*, Proc. IMechE Part B: J. of Eng. Manf., **207**, 1993, pp 91-98.

6. McGarva J.R., *Rapid search and selection of path generating mechanisms from a library*, Mech. Mach. Theory, **29**, 2, 1994, pp 223-235.

7. Gonikhin O. and Medland A.J., *Use of networks in describing the design to manufacturing process*, Computer-Integrated Manufacturing Systems, **3**, 1999, pp 171-177.

COMPUTER SUPPORT FOR DESIGN TEAM DECISIONS

D G Ullman & D Herling

1. INTRODUCTION

Design of even simple devices requires decision making based on incomplete and conflicting information. Decisions made early in the design process, when the information is the least refined, can have far reaching consequences as the device matures. To date, there have been few methods to help engineers make these critical decisions. In this paper, a new method, called the Engineering Decision Support System, EDSS, will be described[1]. The method has been programmed on a PC Windows system that is still in development.

This presentation begins with an example, typical of problems encountered in design. In this problem, a team of engineers is developing a concept for a mechanical device. Characteristics of the problem that make decision support difficult are identified in Section 3. In Section 4, a system which can support the characteristics, the Engineering Decision Support System, EDSS, is demonstrated through an example. The underlying mathematics, a probabilistic method for supporting decision making, are introduced in Section 5. Finally, in Section 6, the plans for testing the system in practice and extensions to more difficult problems are discussed.

2. A TYPICAL DESIGN PROBLEM

To demonstrate the complexity of design problems encountered in early product development, a simple and rather typical problem is used as an example. This problem is taken from one used extensively in European research on design, [1], [3] and [4]. It has been solved by many individual designers and some of their solutions are used as a basis for this fictitious team example.

The problem, in abbreviated form, is:

> *Design a mechanism to mount an optical device to a wall. The mechanism should be able to carry the weight of the device (2kg with center of gravity 100mm from mounting base) and allow the device to be adjusted ±15° parallel to the plane of the wall and 0-15° out of the plane of the wall. Any angle in the range should be possible with an accuracy of ±0.5°. The movements should be smooth and continuous, and the position set must be held. Only one such mechanism will be manufactured.*

---

[1]This research has been supported by the National Science Foundation under grant DDM-9312996. The opinions in this paper are the authors' and do not reflect the position of the NSF or Oregon State University.

The sub-problem used as an example here concerns only the side-to-side adjustment and locking mechanism. In this example, the problem is being solved by a team composed of: Johan, a mechanical engineer specializing in small mechanisms; Hans, a manufacturing engineer; and Iris, an industrial engineer specializing in human factors.

During a 2 hour meeting to develop ideas for the positioning and locking mechanism the three alternatives shown in Figures 1-3, were developed.

The first alternative (taken from Kurt's design in [3]) utilizes a slot and a thumbscrew. This idea was proposed by Johan who liked its simplicity. However, he was concerned that it would be hard to position accurately and might not hold its setting. Hans, from manufacturing, liked its simplicity, thought that the screw wouldn't be too difficult to position and would easily hold the position, and thought it would be easy to manufacture. Iris agreed with Johan that it would be hard to position accurately.

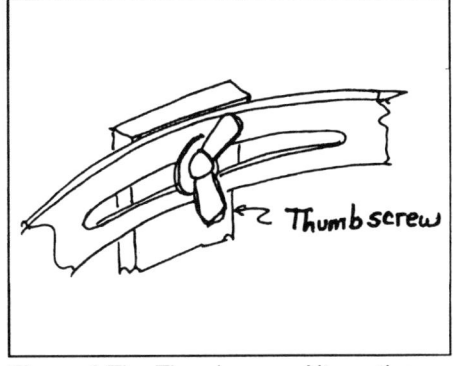

Figure 1 The Thumbscrew Alternative

The second alternative (Taken from Hans' design in [3]) utilizes cranks and locks. This alternative was proposed by Hans who likened the device to a milling machine. He argued that even though it would not be easy to make, it would be easy to adjust and to lock in position. Iris also liked it for its strong human factors. Johan was concerned with its lack of simplicity.

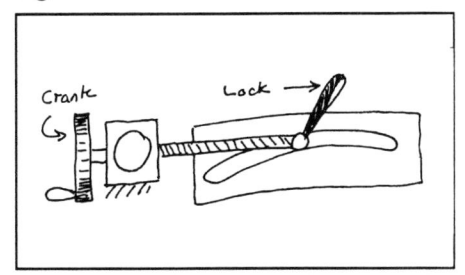

Figure 2 The Cranks & Locks Alternative

The third alternative (Taken from Ingo's design in [3]) utilizes a threaded rod, a sliding block and two lock nuts. This idea generated heated discussion concerning

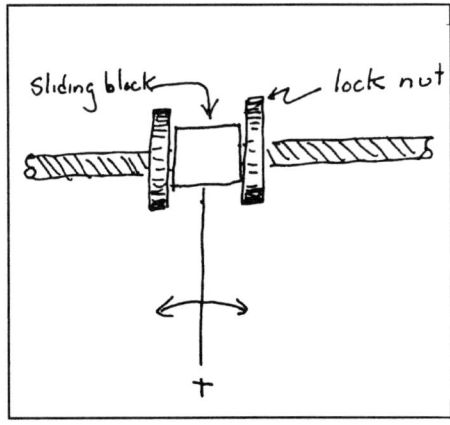

Figure 3 The Threaded Rod Alternative

whether or not it could meet the design requirements. When originally proposed by Iris, Hans raised the issue that, as the block slider was moved off center the device would jam. Iris added the pivot to offset this problem and there was still disagreement. Johan liked the simplicity.

At the end of the meeting the team now had to decide where to put their time and effort. They need to determine which of the alternatives they should spend time refining. Before showing how the Engineering Decision Support System could assist them, characteristics of the problem that make a decision difficult are identified.

3. CHARACTERISTICS OF THE DECISION PROBLEM
This simple example shows a number of characteristics typical of design problems, especially those requiring concept development.
1. The problem is incomplete, there are more criteria and alternatives than the team discussed. Most decision support techniques require complete problem information. It would be useful if a technique could support incomplete problems.
2. The alternatives are qualitative, simple sketches. Thus, there is no numerical analysis possible to support the decision making process.
3. There is no optimum solution. The decision must be made based on the judgement of the engineers.
4. The judgement of the engineers is inconsistent. Sometimes, what one team member favors, another will have low confidence in its potential to work at all.
5. The judgement of the designers is based on differences in knowledge and confidence in the proposed solutions.
6. The evaluation is not complete by any of the team members, meaning that each engineer evaluates each alternative relative to each criteria. This is the case in the example as shown in Table 1, where the alternative-criteria combinations discussed by the team are listed.
7. The issue addressed, the side-to-side adjustment and locking function of the device, is not independent of other issues or sub-problems in the design effort. The results of the work on this issue may affect other aspects of the device.

8. A decision made at the end of the design session may be changed later. The team may reconvene on this issue after refining one or more of the alternatives or the results of another issue may force reconsidering this issue.

ALTERNATIVES CRITERIA	Thumbscrew	Cranks and Locks	Threaded Rod
Position Accuracy	Johan, Hans and Iris	Hans and Iris	
Locking Sureness	Johan and Hans	Hans and Iris	
Simplicity	Johan and Hans	Johan	Johan
Manufacturing Ease	Hans	Hans	Johan
Functionality			Hans, Iris and Johan

Table 1: Summary of the Evaluations made by the Design Team

Ideally, a decision support system should be able to address all eight of the characteristics. In the next section the EDSS is used to demonstrate the potential of such a system and how it can help the team decide which alternative to refine.

4. A TOOL TO SUPPORT TEAM DECISION MAKING

The Engineering Design Support System is still in the development stage. The windows shown in the figures are from an operational prototype system. The underlying mathematics for this system are briefly discussed in the next section of the paper. Plans for its testing and extension are discussed in the final section of the paper.

EDSS is written in Paradox, a Windows database system. This program allows for easy development of a user's interface, for manipulating data and interfacing with other Window's applications. A data base was chosen for this system as there are 19 records needed to record the information in Table 1 (the number of alternative, criteria, evaluator triads). If any of the engineers adds more alternative or criteria, or revises their evaluation of any of the nineteen existing evaluations, new records will be generated. Thus, for a large problem, the amount of information to be stored and queried warrants the use of a database.

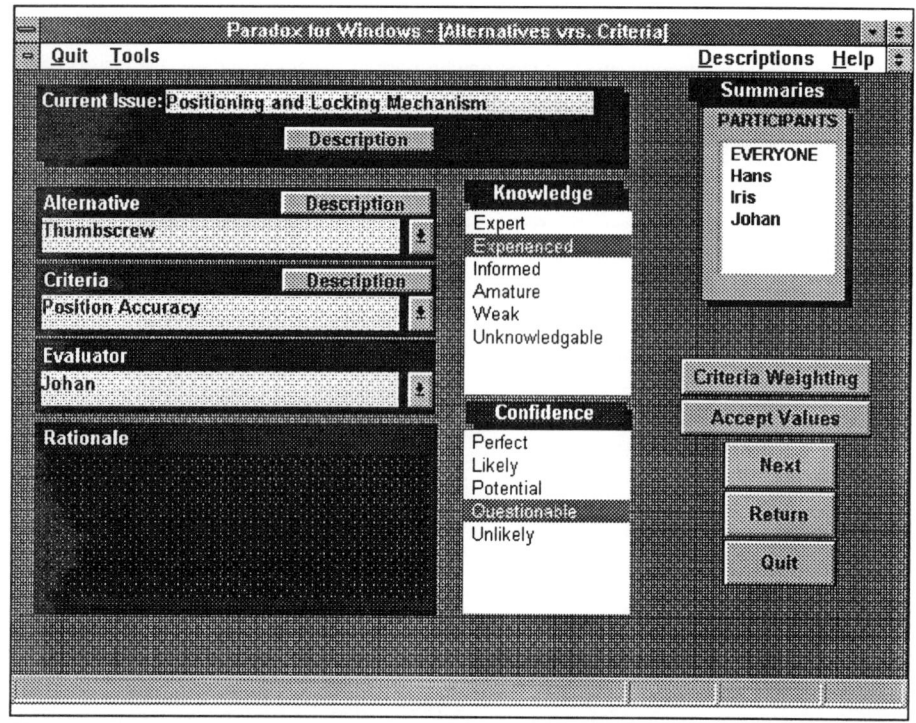

Figure 4 Example of Alternative versus Criteria Window

EDSS allows the users to input the alternatives. Currently, alternatives are represented as word strings, and sketches, as shown in Figures 1-3, will be supported as the program is refined. The alternatives can be input anytime during the use of the tool. Each entry is time stamped for later historical use in a design history or design intent system [9]. The same holds true for the input of criteria information. Either type of information may be entered on the "Alternatives vrs. Criteria" screen shown in Figure 4 or on screens available for adding details about input information (not shown). Figure 4 is the Alternative vrs. Criteria window for Johan's input of his evaluation of the "Thumbscrew" alternative versus the criteria for "Position Accuracy."

In the upper left of the window in Figure 4, the name of the current issue is listed, "Positioning and Locking Mechanism." A detailed description is available (if input by the user) by selecting the description button just below the name of the current issue. Beneath the current issue box is the area for identifying the alternative, the criteria used to evaluate the alternative and the team member performing the evaluation. Evaluation data is input for each alternative/criteria/evaluator combination listed in Table 1. For each of the three pieces of information there is a pull down menu of previously input options and a button for detailed description. New alternatives, criteria or evaluators can be added here also. In the example shown, the evaluation of the "Thumbscrew" alternative is being evaluated against the "Position Accuracy" criteria by "Johan."

In the middle of the window, there are two areas labeled "Knowledge" and "Confidence." Evaluation of the alternative relative to the criteria is expressed to the EDSS in terms of these two measures. Where most decision support tools have a single measure of utility [10], the EDSS uses two: knowledge and confidence. For each alternative/criteria combination, each evaluator can input his/her knowledge about the combination and their confidence in the ability of the alternative to meet the criteria. Confidence is a measure of the chance the alternative has in actually meeting the intent of the criteria. In the example shown, Johan is "Experienced" in the use of thumbscrews for accurately positioning mechanisms and his confidence for the use of thumbscrews in this application is "Questionable."

The Summaries section on the top right of the window of Figure 4 gives data similar to that shown in Table 1. This table is available for all evaluators, "EVERYONE", or for each individual listed. Thus the completeness of the evaluation for the team or any individual can be readily reviewed. Note that there is no need in EDSS for complete evaluation by any or all of the team members.

Another piece of information needed by EDSS as a basis for decision support is the relative importance of the criteria. This information is entered in the Criteria Weighting Window, Figure 5. Here the relative importance of the criteria can easily be input for each team member doing the evaluation. On the left side of the screen, the evaluators who have input alternatives, criteria or alternative versus criteria information are listed. Any one of them can be selected, then the weights are set by using the arrows in the center of the screen. The values are relative and can be changed at any time, facilitating sensitivity analysis. In the example, Johan's name is

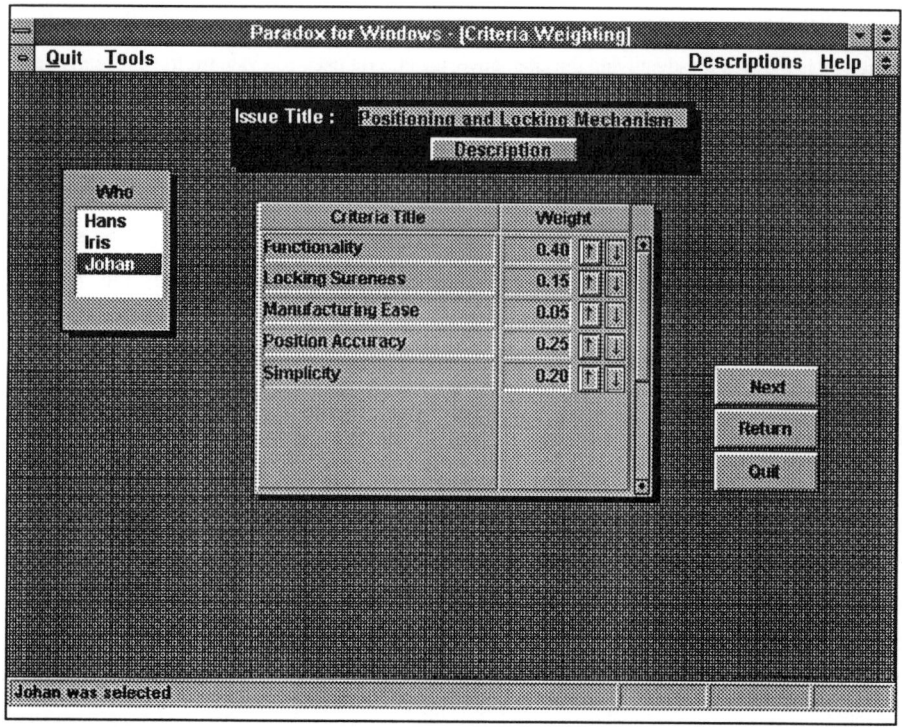

Figure 5 Criteria Weighting Window

highlighted so the values shown are for him only. Here, he considers "Functionality" most important and "Manufacturing Ease" least important.

The evaluation results for this example are shown in Figure 6. These results may be different depending on which evaluator's criteria weightings are used. Thus, in the upper left hand corner, the evaluator can be selected and easily changed. The output is in terms of relative satisfaction. As can be seen in the Figure, with Johan's weightings, the Crank and Locks alternative ranks highest, .45, with the thumbscrew second and the threaded rod last. This result does not tell the team what decision to make. It only gives them a single number based on the sum of all the information input about the problem. These satisfaction scores are based on all the available alternative versus criteria evaluations and Johan's weighting of importance of the criteria. It is easy to change the weightings, update evaluations as new information is learned and add alternatives and criteria to the model of the problem. As each change is made, new records are

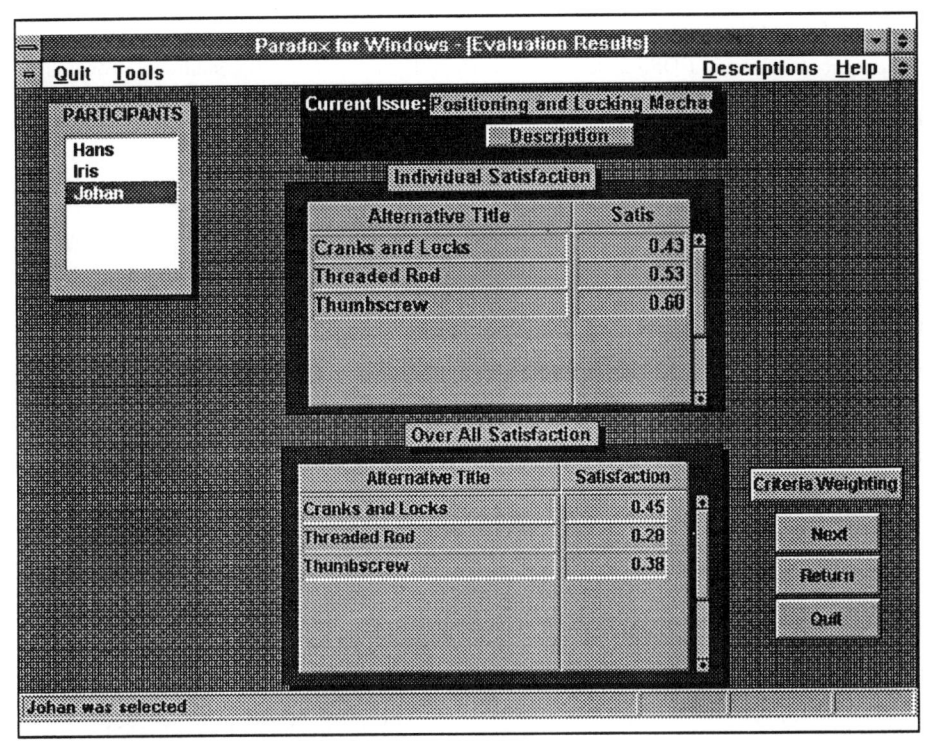

Figure 6 The Satisfaction Window

developed in the database resulting in a complete record of the evolution of the device. Thus, the system supports exploration of what might happen if more is learned about an alternative or criteria, if one evaluator changes his/her evaluation, or if another team member is added.

The use of EDSS supports the eight design problem characteristics listed in the previous section. Since EDSS uses whatever information is available it can support incomplete problem formulations (item 1) and incomplete evaluations (item 6). The alternatives considered can be qualitative (item 2). The judgement of these alternatives (item 3) is based on the evaluator's knowledge and confidence (item 5) which may be inconsistent (item 4). Finally, changes are easily supported (item 8). Currently, the system only addresses a single, independent issue and thus does not support interdependent problems (item 7). However, issue interdependence is a planned extension of the system.

5. MATHEMATICAL BASIS

The decision analysis in EDSS is based on the use of Bayes equation of conditional probability to develop a utility for each alternative/criteria pair. The method uses probability to quantify each evaluator's knowledge of the alternatives and criteria, and his/her confidence in an alternative's ability to meet the criteria. Knowledge is a measure of how much the evaluator knows about the alternative/criteria space. During design activities, knowledge is generally increased by building prototypes, doing simulations (analytical and physical) and finding additional sources of information (e.g. books, experts, consultants). Confidence is a measure of how well the evaluator believes the alternative actually meets the criteria. In the example, Johan is experienced in the use of thumbscrews for position accuracy (his knowledge); he thinks it is questionable that a thumbscrew is good for position accuracy (his confidence).

The explanation of how EDSS mathematically uses this information to develop the satisfaction scores (Figure 6) is detailed elsewhere [5]. The following is only a brief overview.

In EDSS, the words describing knowledge and confidence are converted into numerical probabilities. "Experienced" is equivalent to a numerical value of a probability of 91%. This is based on a scale in which 100% is perfect knowledge and 50% is total ignorance. In the following, the probability of perfect knowledge is noted by, $p(k)$. Thus, $p(\text{"experienced"}) = .91$ implies that an experienced evaluator would give the correct answer 91 times out of 100 questions asked about the alternative evaluated relative to the criteria. The relationship of the term "experienced" to a value of .91 was determined through surveys of subjects [5].

To measure confidence, $p(c)$ is the probability that the alternative fully meets the criteria. The scale ranges from 100% for an alternative that perfectly meets the criteria, to 0% for one that does not meet it at all. In the example, Johan thinks the alternative is questionable. EDSS translates this into $p(\text{"questionable"}) = .42$ implying that there is only a 42% odds that the alternative meets the criteria.

These values for knowledge and confidence along with similar values for the other 18 evaluations listed in Table 1, form the base data for the analysis. For each evaluation (i.e. each alternative-criteria pair evaluated by a single individual), the satisfaction of the criteria by the alternative is calculated by:

$$p(criteria|alternative) = p(k) \times p(c) + (1 - p(k)) \times (1 - p(c)).$$

This calculation gives the likelihood of the confidence given the knowledge and is based on Bayes rule. The combined satisfaction for each alternative-criteria pair is found by taking the product of all the p(criteria|alternative) values. If there was no evaluation for a pair, then a value of .5 is assumed as the probability of the alternative meeting the criteria is 50/50. For example, in Table 1, the thumbscrew concept was evaluated by all three team members and so its overall utility is the product of the satisfaction for each. The threaded rod, on the other hand was never evaluated for position accuracy and thus the utility is set at .5 representing a 50/50 chance that the alternative meets the criteria.

The overall satisfaction of the alternative is found by summing the utility for each criteria, weighted by the values input by the evaluators. Thus, the satisfaction values shown in Figure 6 are based on Johan's weightings as input in Figure 5.

EDSS is programmed so that changes in any of the input values can rapidly be reflected in the satisfaction values. Thus, it is easy to update the system as new information is learned, arguments are made to change other's confidence in alternatives or new evaluators are added to the team.

6. CONCLUSIONS AND PLANNED FUTURE WORK

The Engineering Decision Support System described in this paper is work in progress. The windows shown in Figures 4-6 are from an operational prototype which is currently being tested. Early testing consists of using the system on a laptop computer in actual industrial design situations. As the system is refined, balanced laboratory based tests are planned. It is hypothesized that, beyond giving support for decision making, use of the system will also encourage the development of more alternatives and criteria and a more thorough evaluation of the information available. These hypothesis need to be scientifically tested.

The use of Paradox as a platform for this prototype has proved both helpful and frustrating. There are three components to the system: the user interface, the analytical code and the database. Paradox is a strong database with good user interface development capability. However, the ability to program the analysis within Paradox has proved difficult and limiting.

Thus, before extending the system, the analysis will be moved to C++ which can easily be extended and can, within the windows environment, be linked to Paradox.

Planned extensions to EDSS fall into four main areas. First, the current capture of information is exclusively textual. Since designers often work graphically, sketches and drawings, need to be supported. Sketch capture has been studied, [6] and [8], and is being integrated into the system. This will enable information like that shown in Figures 1-3 to be part of the database record of the product evolution. Second, as noted earlier, the database is a history of the information considered in the development of the product. The concept of a design history is well studied, [2], [7] and [9], although the use of the data in EDSS to give a history is not refined. Third, EDSS currently only supports a single design issue. Most design problems have multiple, interdependent issues. The mathematical theory and the computer program must both be extended to support multiple issues. Finally, even if the satisfaction values generated by EDSS are ignored by the design team, it has been demonstrated that the information organization and display helps support team decision making [11]. Thus, refinement of the user's interface to further enhance the capture and organization of the design information is expected to add value and this too will be tested.

7. REFERENCES

[1]   Blessing, Lucciënne: A Process- Based Approach to Computer- Supported Engineering Design, Thesis, University of Twente, Enschede, the Netherlands, 1994.

[2]   Chen, Aihua: A Computer-Based Design History Tool, Thesis, Oregon State University, Corvallis, USA, 1991.

[3]   Dylla, Norbert: Denk- und Handlungsabläufe beim Konstruieren, Dissertation, Technische Universität Munchen, 1990.

[4]   Fricke, Gerd: Konstruieren als flexibler Problemlöseprozess - Empirische Untersuchung über erfolgreiche Strategien und methodische Vorgehensweisen beim Konstruieren, Dissertation, Technischen Hochschule Darstadt, 1993.

[5] Herling, Derald and David G. Ullman: Engineering Decision Support, poster version accepted for presentation at ICED95, Prague, the Czech Republic, August 1995. Full version in review for Design Theory an Methodology Conference, Boston, Sept 1995 and a journal.

[6] Hwang, Teng-Sheng and David. G. Ullman: The Design Capture System: Capturing Back-of-the-Envelope Sketches, Journal of Engineering Design, Vol. 1, No. 4, 1990, pp. 339-353.

[7] McGinnis, Brian and David G. Ullman: The Evolution of Commitments in the Design of a Component, Journal of Mechanical Design, Vol. 144, March 1992, pp. 1-7.

[8] Ullman, David G., Steven Wood and David Craig, The Importance of Drawing in the Mechanical Design Process, Computers and Graphics, Special Issue on Features and Geometric Reasoning, Vol. 14, No. 2, 1990, pp. 263-274.

[9] Ullman, David G. and Robert Paasch: Issues Critical to the Development of Design History, Design Rationale and Design Intent Systems, DTM94, Minneapolis, Sept 1994, and in Journal review.

[10] Ullman, David G. and Bruce D'Ambrosio, A Taxonomy for Engineering Decision Support Systems, short version accepted by ICED95, Prague Czech, August 95, full version in journal review.

[11] Yakemovic K.C.B. and J. Conklin, Report on a Development Project Use of an Issue-Based KNOWLEDGE System, MCC Technical Report Number STP-247-90, June 1990.

USE OF VISUALISATION AND QUALITATIVE REASONING IN CONFIGURING MECHANICAL FASTENERS

G Zhong & M Dooner

1. INTRODUCTION

Fasteners products, used extensively throughout manufacturing industry for a diversity of applications (typically 'inserts' for joining together plastic assemblies), are supplied by an industry rich in terms of design variety and applications knowledge. In the context of fasteners, conceptual design means the forming of configurations of engineering features (such as knurl or barb), producing designs which satisfy users' specifications. When presented with a particular requirement, the task for an application engineer is to choose the configuration that best matches the requirement. The task is generally aided by company-specific data (geometry, performance graphs) and supported by a wealth of experience, residing in the heads of the application engineer. Such knowledge is generally complex and difficult to express in conventional means (i.e. language description) as there are many interrelationships between engineering features, type of material, and method of installation, all of which have a determining effect on function and performance.

In practice, engineers work with (and think in terms of) engineering features - as opposed to basic geometrical entities - when choosing or forming a configuration. Features are meaningful constructs that have associated properties of performance. In forming a fastener configuration, an experienced engineer knows the respective performance properties of the individual features he is working with, and as an aid to design will sketch the outline shape of a proposed configuration. It is assumed, therefore, that visualisation supports and stimulates the engineer's ability to carry out configurational design. In the prototype system being developed, the outline (or approximate) shape of features is provided by 'intelligent icons', which the engineer can modify. Visualisation is also provided to show (at an abstract level) the cause-effect relationships between features and performance. In the prototype this representation is referred to as a nodal model.

Further support for configurational design is provided by the ability to evaluate proposed or 'candidate' configurations. In practice, while sketching and forming fastener configurations the applications engineer will make an informed judgement on the suitability of a proposed design, based on a knowledge of the features with which he is working and their associated applications. It is a requirement that the system offer the applications engineer more formal support for evaluation, by allowing the system to reason with the qualitative knowledge describing a proposed configuration. A fastener configuration is evaluated against its ability to satisfy the specification.

In summary, it is believed that configuring fasteners requires more advanced forms of support in addition to conventional CAD. Techniques are required which allow applications knowledge to be represented in a computable framework, enabling proposed configurations to be qualitatively evaluated; provided that the use of such techniques do not become laborious and inhibit the engineer's natural creativity.

The paper describes work-in-progress on developing a prototype support system that provides evaluation but maintains visualisation and a high degree of interactivity. Other aspects of the work have been described elsewhere, e.g. [1]. The work is supported by a local manufacturer of fastener products who has provided design and applications knowledge and feedback on the practical aspects of the prototype. The paper is structured as follows. Section 2, which describes the application, is followed by a general discussion (section 3) on system requirements which support fastener configuring. Section 4 describes the system structure and the procedure of forming configurations used by the prototype. Section 5 outlines the implementation. Section 6 concludes the paper.

2. APPLICATION

2.1 Fastener products and practice

In practice, user requirements are generally met from a company's standard family of products (fig. 1a). If a user's requirements cannot be satisfied by a standard design, a special or variant is developed (fig. 1b). A special may have an unusual hole size, or it may be an unusual configuration of features (with respect to the company) which meets a user's particular requirements. A special may also include new features of which the company has little experience. In a typical company an application engineer will work largely with a set of predefined (or known) features, adding special features to the set where necessary, and forming configurations by combining feature members selected from the total set.

Figure 1a. Standard family of inserts

Figure 1b. Specials or variants of standard designs

In general, user requirements are expressed in the following criteria: dimensions (e.g. hole size for parent component, thread size); function and performance (e.g. pull out resistance, resistance to vibration); and method of insertion (e.g. cold, ultrasonic, heat insertion). Satisfying a user's requirements is a complex process, since the above criteria are often conflicting. For example, a knurl feature may only have a high jack-out resistance if ultrasonically inserted, and not if inserted by an alternative method. Identification of an appropriate set of features therefore requires a great deal of experience and skill to balance conflicting requirements.

2.2 Applications knowledge

Types of feature, material, installation and performance are interrelated (fig. 2). In forming a configuration to match a given application, an engineer uses common-sense and incomplete knowledge of these interrelationships. For example, if both high pull-out and high jack-out resistance are required, the engineer knows that a knurl possesses such behaviour provided it is ultrasonically inserted into the parent component. Such knowledge is qualitative and incomplete rather than mathematically precise and detailed. But based on this knowledge the engineer configures and informally assesses possible designs. Clearly, the process is knowledge intensive in terms of cause-effect relationships (e.g. between features and performance) which inexperienced engineers are unlikely to possess.

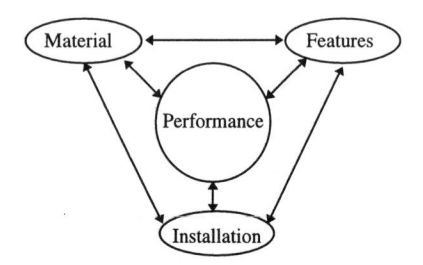

Figure 2. Interrelationships between cause and effect

3. SYSTEM REQUIREMENTS

3.1 Visualisation in conceptual design

One characteristic of conceptual design is that various possible design ideas or candidate solutions may emerge before finding the most appropriate one. Although these ideas or solutions are abstract, qualitative descriptions of the design, visualisation is desirable because it allows designer to see directly the approximate appearance, structure and spatial relationships of a proposed design, visual feedback which helps to judge and modify the design.

Although computer graphics techniques, e.g. geometrical modelling, are highly visual, they are limited in terms of their ability to represent and reason about the physical behaviour of designs. A further weakness of geometrical modelling as an aid to conceptual design is that because geometrical models require precise data input, they cannot handle conceptual designs which are by definition incomplete. The effort involved in geometrical definition can be restrictive with respect to an engineer's natural inclination to explore freely.

The approach taken by some intelligent CAD research in conceptual design is aimed at generating design output driven by functional requirements input [2]. With this 'black box' approach the engineer is excluded from the more creative processes involved in conceptual design, as there is a lack of visual feedback during generation. Further, there is the problem of interpreting textual output, i.e. several different geometric designs may be derived from the same textual specification design output. Figure 3, for example, shows different conceptual designs implied from the following textual specification:

'the main body of the insert is a cylinder, surrounded by a diamond knurl, and a barb at one end'

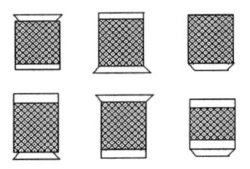

Figure 3.   Implied designs from textual specification

During conceptual design, it is necessary to provide the engineer with a window to observe his visual thinking, to allow him to modify ideas and make quick judgements on the feasibility of engineering concepts. This window need only show the approximate appearance of the concept, and not show a precise geometrical representation. For fastener design, visualisation allows the engineer to see the concepts, forms, proportions and interrelationships of the features that are configured into a fastener solution.

3.2   Visualising applications knowledge

Another, more abstract, form of visualisation can show design and applications knowledge, knowledge which is difficult to express clearly in natural language. Provided with a suitable knowledge framework, corresponding closer to engineer's mental models, engineers will find it beneficial to record and view the complex relationships and interactions between features, type of material, method of installation, and function and performance. A requirement therefore is to provide knowledge structures that can make explicit these relationships. In the prototype (described in the paper), a nodal model is constructed for this purpose, and was used to acquire applications knowledge from fastener engineers.

3.3 Reasoning with applications knowledge

A basic requirement is to provide more formal support for the evaluation of proposed configurations. The system should provide a representation scheme which describes abstract and qualitative concepts about the physical properties of features. Based on both common sense and expert knowledge, the system should generate corresponding qualitative models of configurations of features and reason about their inter-relationships. From the qualitative models, the system should be able to simulate the qualitative behaviour of candidate configurations, in order to evaluate their performance against the specification.

4. SYSTEM STRUCTURE AND PROCEDURE

4.1 Structure

The structure and procedure of the prototype system is illustrated in Figure 4.

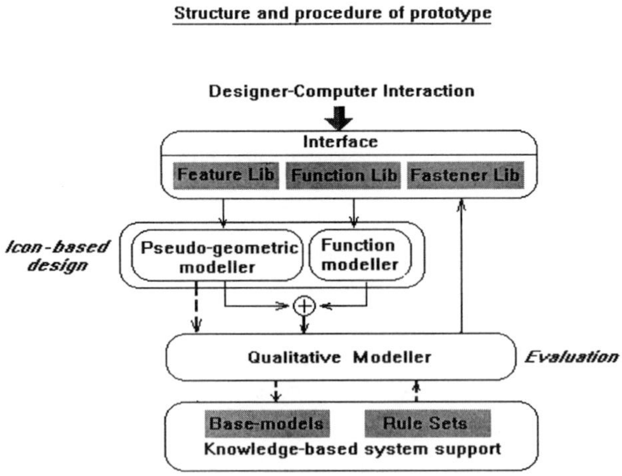

Figure 4. Structure and procedure of prototype

The figure shows the principal structures: the interface (feature, function and fastener icons libraries), the icon based and qualitative modellers, and the knowledge base (base-models, rule sets). It also shows (dashed lines) the information relationship between the functions of iconic design and qualitative evaluation and the knowledge base, which stores base-models and rule sets. Every predefined feature in the library has a corresponding base-model in the knowledge base. Once a feature icon is selected the corresponding base-model will be activated, and any change of shape of a feature icon will be directly mapped on the corresponding base-model, the result of which is called a derived model. Parameters that describe a derived model are generated dynamically, based on stored qualitative knowledge, given as rule sets in the knowledge base.

The figure also shows the procedure of forming fastener configurations (solid lines). It begins with the engineer selecting feature icons (pseudo-geometric or function), after which a candidate configuration is passed to the qualitative modeller for evaluation. A fastener candidate model, which may consist of base-models and derived models, is stored in a temporary buffer while it is evaluated. Qualitative reasoning is used to predict the behaviour of a fastener candidate from the behaviours of base-models and derived models. Satisfactory candidates are added to the fastener library.

4.2 Iconic design

The icons in the feature library represent the approximate shape and size of predefined features. The shape and size of these feature icons can be changed or modified using the edit tool to configure a fastener or generate new features according the engineer's intention. The iconic models are therefore more than symbols, they are approximate or pseudo-geometrical models of features. The feature library also contains a library of feature elements icons (fig. 5). Normally, candidate configurations are formed from features not feature elements. Instead, feature elements are used to configure new features.

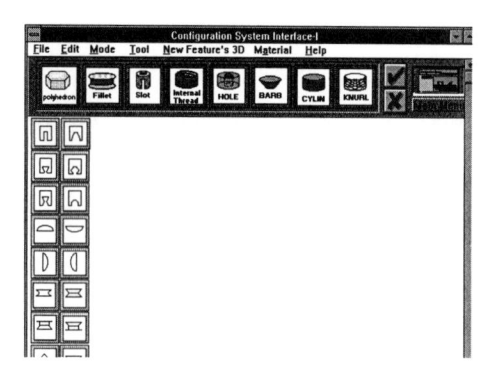

Figure 5. Feature library and feature elements

Aided by an edit facility, the engineer is able to manipulate feature icons, irrespective of physical constraints. This configuring procedure, in some sense, is like using flexible toy blocks to build an artefact. And it enables the engineer to see the appearance of the candidate solution and the spatial relationships between the features. Such visualisation lets the engineer make a quick initial judgement on the feasibility and adequacy of a design prior to qualitative evaluation by the system. The system evaluates candidate configurations and feeds back the results to guide a designer to make modifications until a satisfactory candidate is found (fig. 6).

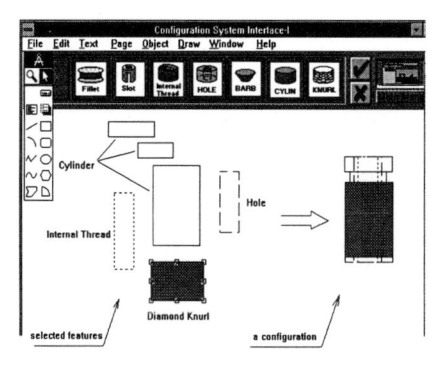

Figure 6. Forming a candidate configuration

The function library in the interface provides a set of icons which visually represent function and performance (fig. 7). Relationships among the functions can be specified by the engineer using three operators: 'And', 'Or', and 'Xor'. The performance can also be specified. The function icon models are linked to base-models (or derived models) by rules. In addition to function information, other user requirements need to be specified, e.g. type of material and method of installation.

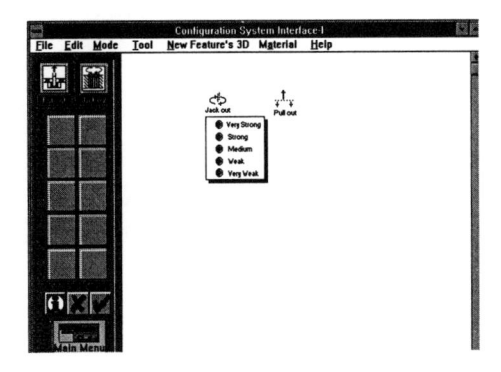

Figure 7. Representing function and performance

4.3 Evaluation

Once a given fastener candidate has been configured using the iconic modeller, its corresponding qualitative model is generated following activation of the 'evaluation' command. A qualitative model of a feature is derived dynamically from the base-model. As the icon models in the interface are mapped to the base-models in the knowledge base, any geometrical change occurring in the interface will directly result in a corresponding change of variables or attributes of the base-model in the knowledge base, resulting in the derived models. Figure 8 illustrates the 'barb' base-model and its derived models.

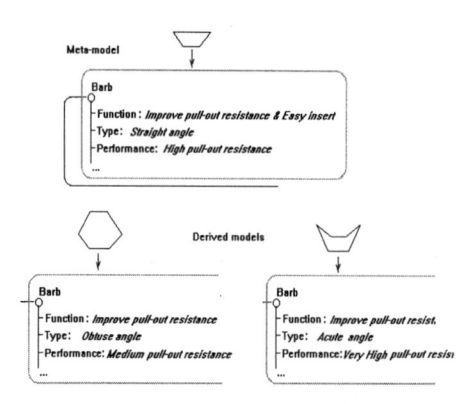

Figure 8. 'Barb' base-model and its derived models

In the process of evaluating fasteners, composite behaviour is inferred (or predicted) from the behaviours of individual features based on composition rules. The composite behaviour of a fastener is not the simple sum of the behaviours of individual features. It depends on the relationships among the features and on the kind of interaction with the parent material in which they are placed. For example, if an insert is a combination of a tapered cylinder and a knurl, when inserted in a parent material, the insert's composite pull-out resistance will not be a simple addition of the knurl's (strong) pull-out resistance and the tapered cylinder's (weak) pull-out resistance (fig. 9a). Method of insertion also affects the performance. For example, if the insert has a reverse taper, and if cold inserted, the insert's composite pull-out resistance will be negligible (fig. 9b).

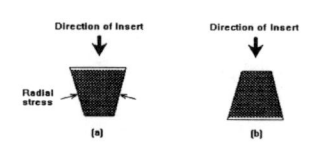

Figure 9. Effect of spatial relation between features

4.4 Viewing underlying structure

A nodal model allows the engineer to visualise the underlying structure of a fastener configuration. It is also used by the engineer to observe, interrogate and update the knowledge base and rule sets. Figure 10 shows a nodal model of a fastener. The four nodes (shown as circles) have a tree structure representation of attributes or variables. These attributes or variables have qualitative states (shown in italic) which are assigned a value within a range of possible (allowable) values. The solid lines and qualitative descriptors show the structural relationships. The dashed lines show casual relationships. The large circle shows influences and functional dependencies. From the figure it can be seen that there is an influence between the taper variable of cylinder and the performance of knurl.

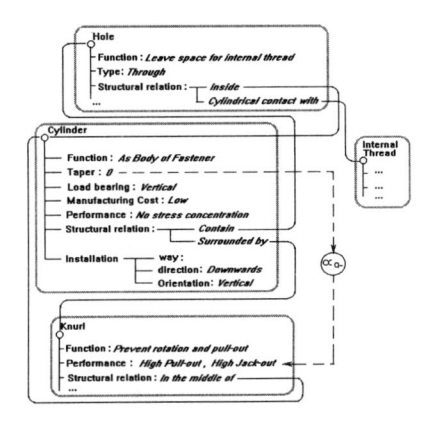

Figure 10. Nodal model (a part) of a fastener

4.5 Specific Example

Figure 11 illustrates a simple example of how the system supports the design process, which is described as follows. First, the engineer selects a cylinder as a main body for the candidate, bearing in mind the functional requirements. The selection is identified by the 'monitor' (a specific program in the interface, its function being to check and analyse the iconic design, and inform the results to the modeller), and is then be registered with the modeller in the knowledge base. Then, the engineer changes the shape of the feature cylinder into a tapered one by utilising the edit tool in the system. The change occurring at the interface is analysed by the monitor to note the type of change happening with the feature. In particular, it identifies the type of taper taking place to the cylinder so that the modeller can map this change to the corresponding attribute of the base model, and activate the process of generating a derived model in the knowledge base. As a consequence of the generated cylinder derived model, the cylinder base model is excluded from the registered list contained in the modeller. Subsequently, the engineer selects a feature knurl as a part of the candidate in order to improve its pull-out and jack-out resistance performance. Once the feature cylinder and knurl are configured in the interface to form an insert candidate, the spatial or relative orientation relationships between the two features are identified by the monitor as well as the generation of the knurl derived model. After the spatial relationships are mapped to the corresponding attributes of the derived models, the modeller may change the value of performance attribute of the knurl model from 'very high pull-out resistance' to 'medium pull-out resistance', based on knowledge about causal relationships stored in the knowledge base. Once the engineer is satisfied with the configuration (i.e. it does not require more features), he can begin qualitative simulation on the candidate. This results in the system checking whether the behaviour satisfies the functional and performance requirements, which is feedback to the interface with an explanation enabling the engineer to decide whether to accept or modify the candidate, or start an entirely different configuration.

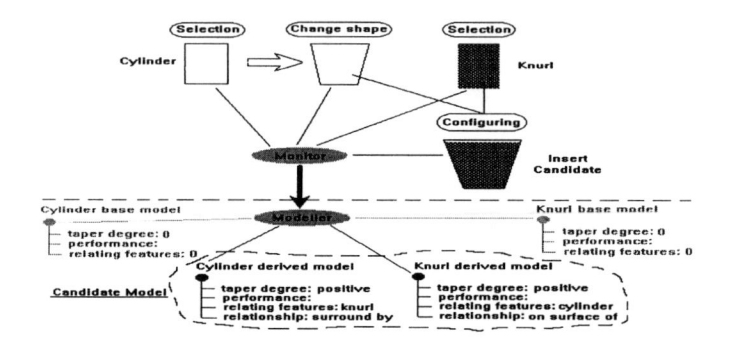

Figure 11. Design process

5. IMPLEMENTATION

5.1 Development system

The prototype is implemented on a PC computer using ToolBook [3] and Kappa-PC [4]. ToolBook is used to develop the design interface, e.g. the iconic models. In the interface all the icons can be manipulated with the mouse. The feature icons are picked out and manipulated (e.g. to move an icon's position or change its shape and size) by clicking the mouse. Selection or changes occurring to the feature icons are mapped to corresponding base-models in the knowledge base by means of Dynamic Data Exchange (DDE) and Dynamic Linked Libraries (DLL). Function icons are also passed down in the same means to the qualitative modeller for evaluation.

Linked Libraries (DLL). Function icons are also passed down in the same means to the qualitative modeller for evaluation.

5.2 Industrial support

The prototype is being developed with support from a local manufacturer of fasteners. This support includes design and applications knowledge, and feedback on the practical aspects of the research. Comment received (to date) as a result of demonstrations indicates that the principles are sound but in order for the benefits to become more apparent, the library will require expansion to cover more realistically the company's range of fastener products. A more formal evaluation phase is planned to commence shortly, during which engineers located at the company will themselves use the prototype.

6. CONCLUSIONS

The paper has described a research project aimed at developing a system for supporting the configuring of mechanical fasteners. A prototype system supports configuring by offering visualisation and manipulation of intelligent icons, and evaluation of candidate designs by qualitative reasoning techniques. Visualisation is also provided to enable engineers to see the underlying structure of fastener configurations, in the form of a nodal model. In general, the development approach has been to provide support which reflects and complements configurational activity, as practiced by fastener engineers. Preliminary feedback from engineers indicates that the functionality included in the prototype offers practical advantages over conventional CAD systems.

ACKNOWLEDGEMENT

The authors would like to thank the support given by the company Armstrong Fastening Systems of Hull, in the way of technical assistance and feedback.

REFERENCES

[1] G Zhong & M Dooner, Multi-level Model for Supporting the Configuration of Mechanical Fasteners, Proc. Expert Systems 1994, Cambridge, December 1994.
[2] Kusiak, A et al (1993), An Intelligent System for Conceptual Design, Expert Systems, Vol. 3, no. 2, pp. 35-44.
[3] Using ToolBook^{TM}, Asymetrix Corporation, 1994.
[4] Kappa-PC User's Guide 2.0, IntelliCorp, June 1992.

CONCEPTUAL DESIGN OF POLYMER COMPOSITE ASSEMBLIES

J K McDowell, T J Lenz, J Sticklen & M C Hawley

1 INTRODUCTION

Polymer composites are advanced materials in which a polymer matrix surrounds a fiber reinforcing structure. The matrix can include thermoset or thermoplastic polymers. The fiber types can include glass, carbon or aramid fibers in continuous and sometimes chopped forms. Design is critical to the successful application of polymer composites. The design activity in composites can be viewed along three dimensions: material design, assembly design and process design. These dimensions are not independent and their interactions give rise to critical design issues that substantially impact feasibility and affordability. Rarely is a product made entirely of composites; in most cases the product is an assembly of composite and non-composite segments. The term *composite assembly* is used to emphasize this hybrid nature.

Material design involves the mapping from requirements (mechanical, thermal, optical, electrical and chemical) to choices of fiber type, fiber length, polymer matrix and any necessary chemical agents. In assembly design, the requirements include partial geometric information, tolerances and loadings (mechanical, thermal and vibrational). Assembly design determines the overall geometry as well as the fiber orientation, ply stacking, integration of other materials (honeycomb and/or foam) and issues of bonding and fastening. Process design involves an array of complex problem solving activities, some of which can be viewed as process technology selection. While consideration of geometry is important in assembly design, the material and process dimensions can provide critical design constraints on shape and configuration.

Much of the assembly design activity involves determining whether a given three dimensional description with the relevant material properties will adequately support the various loadings (e.g. mechanical). In this detailed design phase such verifications are often performed using finite element analysis studies. It is well documented in the design literature that many of the critical cost drivers are determined before detailed design, in the earlier conceptual design phase. Common numbers cited are that 80% of a product's cost are determined in the first 10% of the design process. Providing intelligent decision support for the conceptual design of composite assemblies therefore has considerable potential for controlling downstream costs.

Conceptual design for composite assemblies includes determining which segments of the assembly will be composites as well as determining the overall configuration and relationship(s) between the segments in the assembly. Design criticism and advice is possible before the complete detailed geometry is determined. This paper will focus on the task issues (knowledge representation and inference strategies) of conceptual design for polymer composite assemblies in the context of intelligent decision support systems in an integrated design environment. The specific domain application will be the replacement of an existing metal assembly using polymer composite materials. From a task analysis and a knowledge requirements viewpoint this problem solving scenario is less complex than beginning the design problem from scratch. Additionally, this is a high leverage area for the use of polymer composite materials. Decision support in this area can facilitate the penetration of this advanced material into expanded applications both in the durable goods markets and high performance arenas.

2 INTEGRATED DESIGN AND INTELLIGENT DECISION SUPPORT

Design in polymer composites is a complex activity, rich in knowledge-based problem solving. Success requires an integrated view of materials, assembly and process. Figure 1 illustrates this view and shows the interactions and overlays between the dimensions. As stated previously, the interactions among these dimensions impact affordability. The work described here in conceptual design of polymer composite assemblies is in the context of a wider effort in developing intelligent decision support tools for integrated design. Our view of integrated design begins at the center of Figure 1 and spirals outward. This spiral moves outward through conceptual design into detailed design, with each cycle revisiting the issues of material, assembly and process. Second generation and third generation prototypes are already in place for process design and material design, respectively. These systems complement this current work in assembly design both at the conceptual and detailed phases. The following summaries demonstrate the interactive nature of these dimensions, stressing an integrated approach.

2.1 Material Design

COMADE (COmposite MAterial DEsigner) [5] is a third generation prototype for polymer composite material design. COMADE considers the performance requirements and the environmental conditions an assembly may face and generates multiple designs [4] for the matrix material, any necessary chemical agents, the fiber type and the fiber length. COMADE's coverage of the domain can generate over 1000 material designs. Specific to assembly loadings, COMADE considers the required tensile strength and the required flexural modulus as well as the use temperature.

2.2 Process Selection

COFATE (COmposite FAbrication Technology Evaluator) [9] is a second generation prototype for polymer composite fabrication technology selection. COFATE evaluates sixteen different fabrication technologies based on qualitative economics, abstractions concerning the part/ assembly geometry and other processing issues. Specifics issues involving the assembly include, the size, the geometric complexity, existence of inserts.

2.3 COMADE and COFATE in Assembly Design

COMADE and COFATE compliment the activities in assembly design. COMADE provides a space of choices for materials and COFATE provides screenings on how the composite assem-

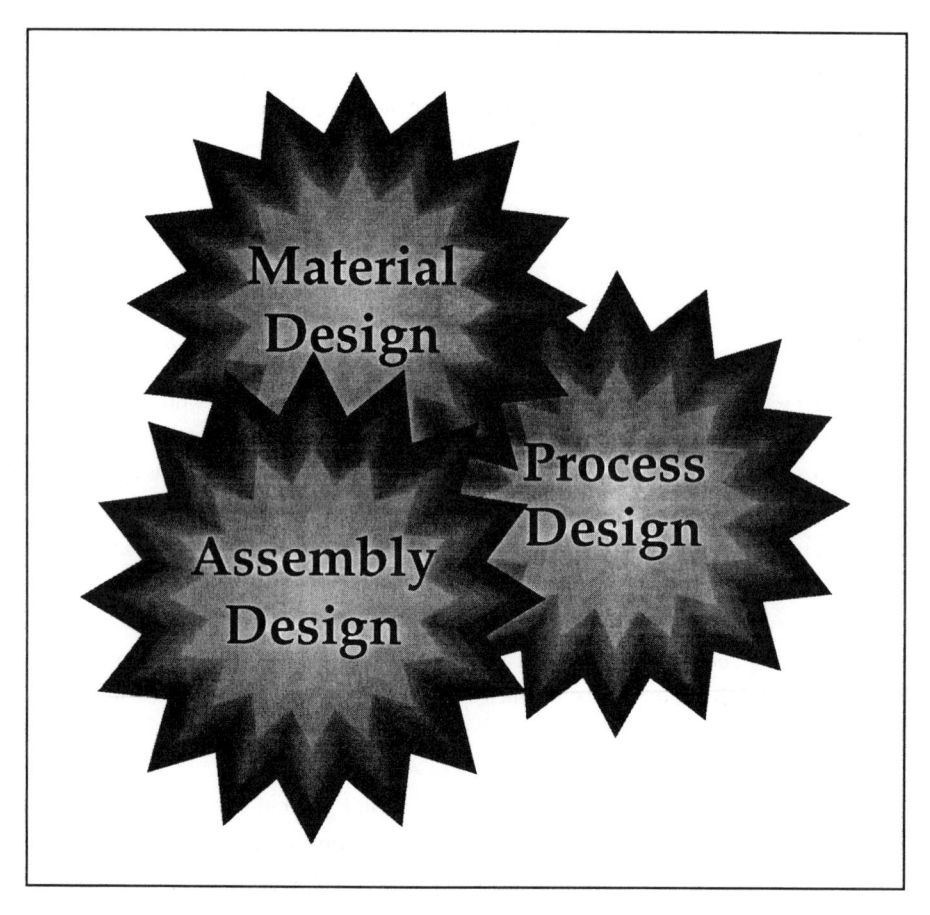

Figure 1. Integrated Design in Polymer Composites.

bly or portions of the assembly might be manufactured. In conceptual design the exact nature of the composite assembly is not established. However much of the needed input for COMADE and COFATE can be abstracted from the existing metal assembly. The performance and environmental conditions for the assembly are known and aspects of the gross geometry are known. For inputs that are not known, the possible values should be explored to appropriately generate the design space. Another important aspect to note is that different portions of the assembly may face different performance and environmental conditions. This expands and particularizes the possibilities concerning materials. Many of these ideas have been played out in MADEFAST, a collaboration design experiment, where COMADE and CTechSel [7, 8] (the first generation process selector) were used in the design of a polymer composite seeker casing [6]. MADEFAST did not involve any intelligent decision support for assembly design, either at the conceptual or detailed phases.

3 TASK STRUCTURE FOR CONCEPTUAL DESIGN

3.1 Overall View of Problem Solving in Conceptual Design

Establishing a task structure for the problem solving activity of conceptual design for composite assemblies will identify the major problem solving modules and how they tie together to perform the overall task. Figure 2 provides an overall view of the conceptual design problem solving activity. The process begins with the existing metal assembly, its geometry, the materials used as well as knowledge about how the segments and features of the metal assembly provide the functionality to achieve various goals. A process of reverse engineering abstracts the knowledge about the metal assembly into two related knowledge models. The first knowledge model represents the functional aspects of the metal assembly. Using functional representation techniques [2, 10, 11] the metal assembly is decomposed into its segments and features which are annotated with functional knowledge about how these segments and features contribute to achieving the overall goals of the metal assembly. The second knowledge model represents the configuration of the metal assembly and shows the relationships among the segments and features in the metal assembly. The configuration knowledge model also shows how the assembly connects to the larger system that surrounds it. These knowledge models are critical to the design of the composite assembly. The composite assembly must reproduce the functionality of the existing metal assembly and it must fit properly in the existing superstructure. The next task is to replace portions of the metal assembly with components made of composite. The goals of this process (and the overall redesign process) are to reduce the cost of manufacturing the assembly and/or reduce its weight, which also translates into a cost reduction. In the context of the load paths in the assembly and the number of fasteners used, the assembly is retrofitted with composite and where possible segments are consolidated to eliminate fasteners. This process results in functional and configuration models of the composite assembly. The assembly then goes through a two part process of criticism and modification in an effort to eliminate incompatible configurations, insure that the loadings are qualitatively satisfied and identify areas contributing to down stream costs. This process is iterative and arrives at an improved design for the composite assembly. The final problem solving module recovers important, relevant geometry information from the original metal assembly and adds this to the functional and configuration knowledge models. The result of the conceptual design of the composite assembly includes partial geometric information, materials information, as well as functional and configuration knowledge models. Given its partial geometric nature and the fact that in detailed design a full solid model of the assembly is produced, the product of the conceptual design phase is referred to as a near-solid knowledge model.

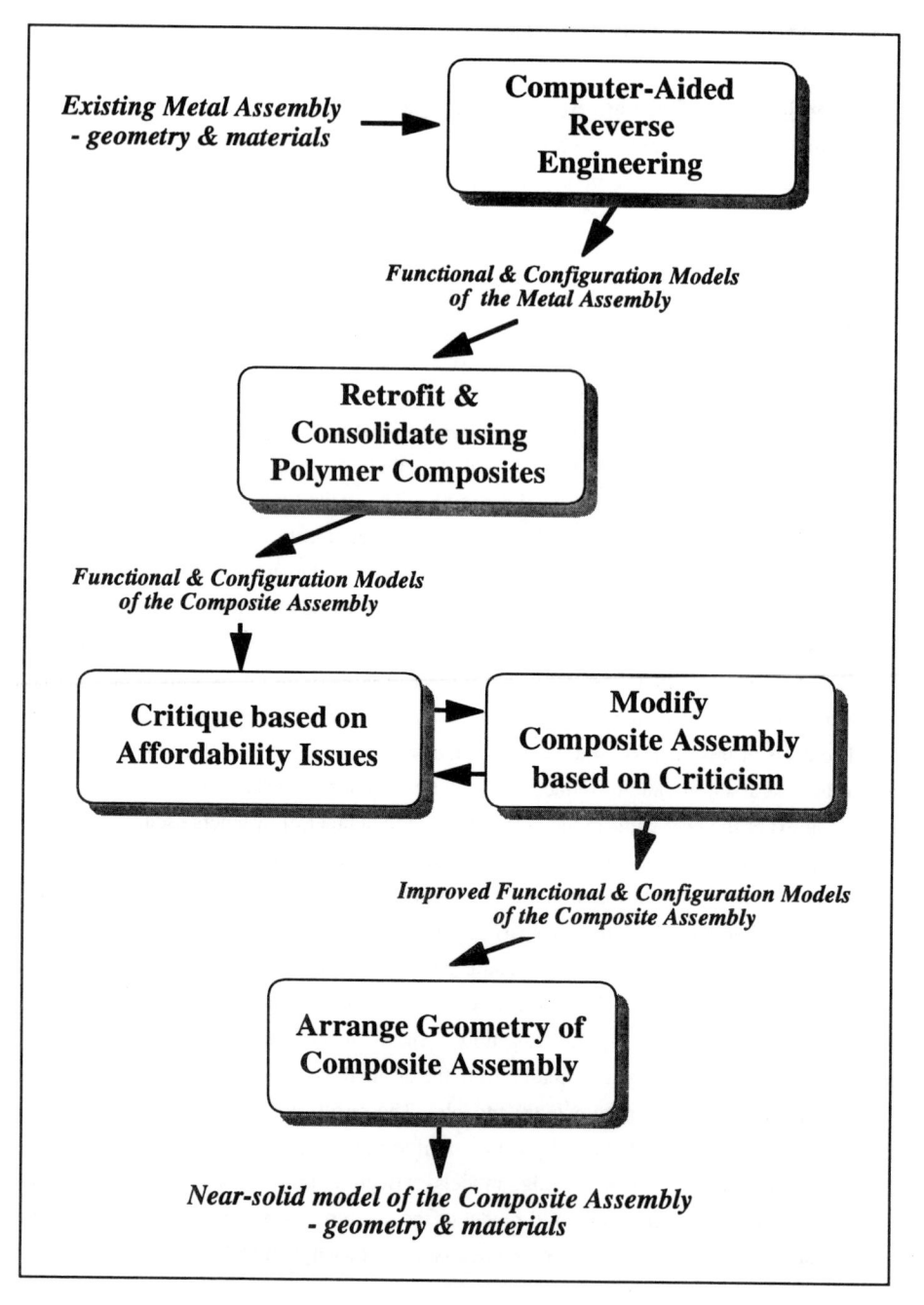

Figure 2. Conceptual Design for Composite Assemblies. Problem Solving Modules and Flow.

3.2 Reverse Engineering of the Metal Assembly

This problem solving phase is critical to the remaining phases of conceptual design of the composite assembly. In many ways this phase can be viewed as a knowledge acquisition phase where functional and configurational information is captured in a form that can be reasoned about in the later phases of retrofit, critique and modify. The success of the composite assembly in actual use will only be as good as the knowledge captured about the original metal assembly.

3.2.1 Functional Knowledge Models. As mentioned previously, the outputs of this phase are knowledge models that express the functionality and configuration of the metal assembly. The functional knowledge model is based on previous work in representing and reasoning about engineering systems like a fighter jet fuel system [11] or the thermal control system of a space station [2]. Functional knowledge augments knowledge about system behavior and system structure, aiding the organization of knowledge for problem solving activities like diagnosis [2], redundancy checking [11] or conceptual design [10]. Function-based Reasoning (FBR) is an emerging subfield in artificial intelligence, for more information readers should see recent special issues of Applied AI, Vol. 8 No. 2 and Vol. 9 No. 1. Other important works in FBR include [1, 3].

Much of the work in functional reasoning and functional modeling has focused on a system schematic and state change behavior viewpoint [2, 10, 11]. The advances in this work include the modeling of physical homogeneous components with specific geometric features that achieve the desired goals. The needed extensions to the representation support later phases on conceptual design namely retrofit and consolidate. It is not clear that these extensions in representation can support the previous types of inference [2, 10, 11].

3.2.2 Configuration Knowledge Models. The configuration knowledge model is composed of segment-feature-connection information about the metal assembly. This includes knowledge about how the segments that make up the assembly are joined together as well as how the metal assembly is joined to the overall superstructure. Much of this joining information forms preconditions for many of the goals the assembly can achieve as represented in the functional knowledge model.

3.2.3 Example Functional and Configuration Knowledge Models. To illustrate the expressiveness of the knowledge models described above, examples using a mock metal assembly will be used. The metal assembly used in Figures 3 and 4 is composed of four segments (numbered 1 through 4), three are mechanically fastened and one segment is welded. The metal assembly has a number of functional roles for example to transfer loads, or to separate other portions of the superstructure. Figure 3 shows a portion of the functional representation for the metal assembly and expands one of the functions, TransLoad, down to the funda-

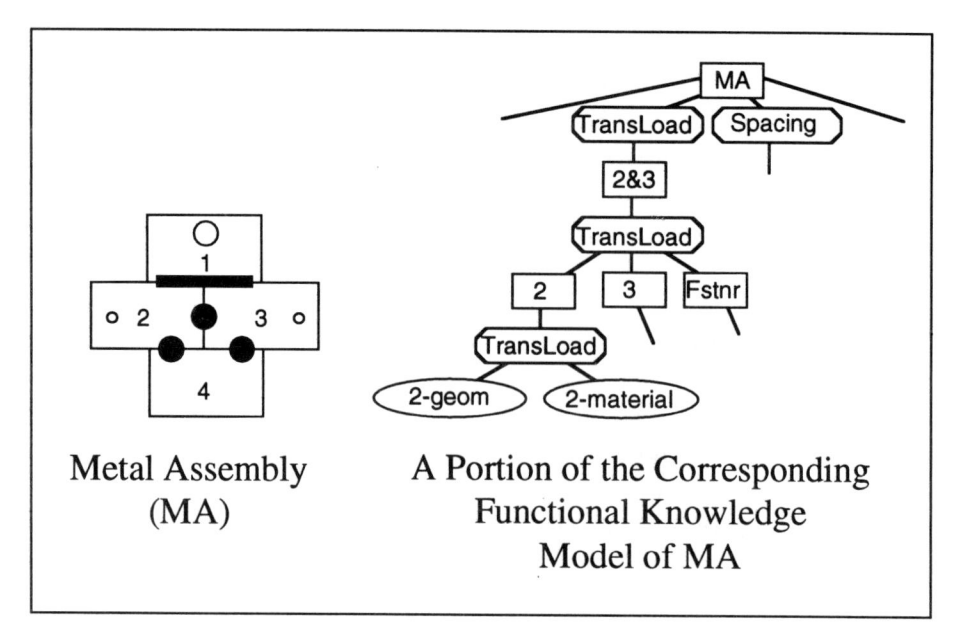

Figure 3. A Metal Assembly and its Functional Knowledge Model

mental aspects of segment 2. Viewing the metal assembly, groupings of segments and individual segments as devices that achieve functions, these are represented as rectangles in the functional diagram in Figure 3. The various functions that hang off of the devices are represented in the diagram as bezels. Fundamental aspects that contribute to achieving a function are represented as ellipses. In Figure 3, the metal assembly has a function TransLoad which is achieved by the grouping 2&3. Expanding further, 2&3's function is supported by segments 2 and 3 and the fastener that joins them. Expanding segment 2 we find that it is able to achieve its function by the nature of the material it is made of and by its geometry. All the functions for the metal assembly (and its groupings and segments) are expanded in this manner.

The configuration model for the metal assembly is shown in Figure 4. The four segments that make up the metal assembly are shown as is their respective configuration and joining (the connecting lines). Also shown is the connections to the surrounding superstructure (connections to the solid squares). This is especially important and not always obvious from fastener holes or other joining points. Note segment 4 in Figure 4. It has two connections to the superstructure because segment 4 is a spacer between segments of the superstructure, there is no fastening or bonding, but there is contact. The links between the squares in Figure 4 contain knowledge about the nature of the joining or connection.

It is important to note that this phase of conceptual design is not automated with intelligent decision support tools. The design engineer is responsible for constructing these knowledge models. Though assisted with modeling tools and enabling graphical user interfaces, the

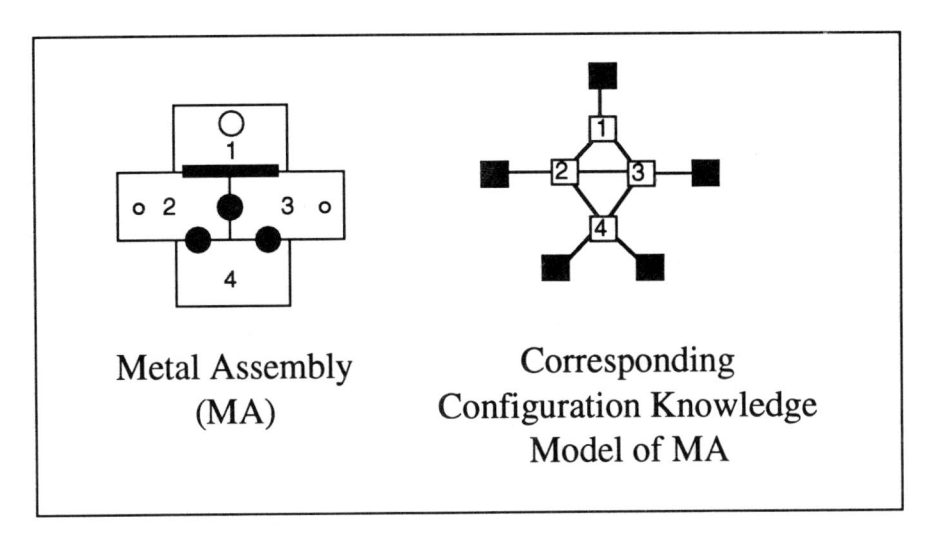

Figure 4. A Metal Assembly and its Configuration Knowledge Model

design engineer makes the decisions concerning correctness and completeness of the models as well as level of detail.

3.3 Metal to Composites - Retrofit and Consolidate

These knowledge models (functional and configurational) of the metal assembly are the starting points for redesigning the assembly using polymer composites. The goals are not to just retrofit the metal assembly with composites on a segment by segment basis but where possible consolidate segments, replacing multiple metal segments with a single composite segment. Reducing the number of segments improves affordability and also reduces the number of fasteners thus further reducing weight.

The functional model is examined from the bottom up looking for opportunities to consolidate segments of the metal assembly and leveraging the use of composites. One leverage of composites is to align the fibers in the direction of the load path and to note that continuous fibers work best under tension. The goal is replacement and consolidation while at the same time reproducing the functionality of the existing metal assembly. This is why the functional model of the metal assembly is so important. The retrofit and consolidation process works based on knowledge about the best use of composites under various conditions. This knowledge is based on examples of successful redesign efforts and on the expertise of composites engineers.

Both the functional knowledge model and the configuration knowledge model include the changes made by this retrofit and consolidation process. The final form of the models might include an assembly of all composite segments or a combination of composite and metal segments.

3.4 Critique and Modify

With a proposed conceptual design of the composite assembly, an iterative process of criticism and modification begins. The goal of this process is to improve feasibility and affordability. Both the functional and the configuration models are examined using knowledge about common trouble areas for composites. Issues of tooling, processing and final manufacturing can be brought to bear early in conceptual design.

A domain example where incompatibility detection that impacts service of the assembly is when carbon fiber composites are in direct contact with aluminum. This configuration can cause accelerated corrosion in the aluminum portion of the assembly.

Taking this incompatibility to the modification stage the design engineer is presented with options for correcting the contact problem of carbon fiber and aluminum. In situations where bearing stresses might be high, metal inserts common. The modification module might suggest materials and proper geometries for the inserts. Again this knowledge comes from successful design examples and expertise of experienced engineers.

On completion of this cycle of problem solving, the conceptual design of the composite assembly is improved over the original proposal based on only retrofitting and consolidating. Carrying on from the example used in Figures 3 and 4, Figure 5 show the metal assembly and a potential composite assembly. Note that there is both retrofitting and consolidation. Segments 2, 3 and 4 have been replaced by two composite segments. Also note that only the certain geometric features have been reproduced, for example the position and size of bolt holes or passages. Three composite materials are used. The composite replacing segment 1 is a high temperature material because the hole in segment 1 allowed passage of a heat pipe. This composite segment is adhesively bonded to the rest of the composite assembly. A cross ply woven material is used to replace segments 2,3 and 4, providing the stiffness to achieve the spacing function of segment 4. Note the width of segment 4 is preserved. To support the tensile load path across segments 2 and 3, unidirectional composite is used between the bolt holes. Because of the possible bearing stresses on the composite, metal inserts are used in place of simply drilling two holes in the composite.

3.5 Capture Relevant Geometry

The functional and configuration models of the composite assembly include geometric information. However at this conceptual design phase, not all the geometry for the composite assembly is determined. In detailed design the complete geometric description is determined and a solid model of the geometry is possible. At the conceptual design phase only partial geometric or pre-geometric (configuration) information is available. Using this information a

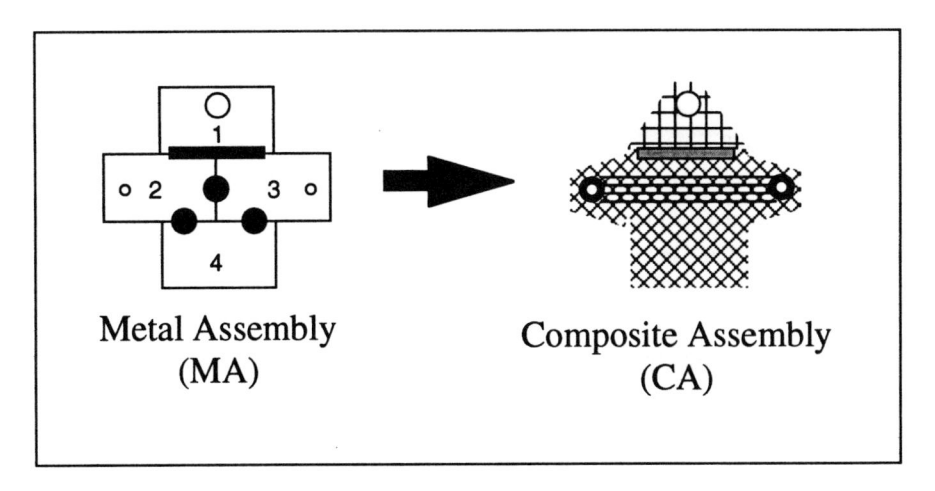

Figure 5. Transformation of a metal assembly into a composite assembly.

"near-solid model" of the composite assembly can be produced. This near solid model is a mix of 3d geometric and configurational information. It is the starting point for detailed design where the complete geometry is determined.

4 CROSSING INTO DETAILED DESIGN

In a comprehensive and integrated design environment the knowledge representations supporting conceptual design would also support detailed geometric design. On completion of conceptual design phase, what could be viewed as a "near-solid model" of the composite assembly is available, including partial geometric information and material information. This near solid model which includes the knowledge representations that supported conceptual design, would serve detailed design and analysis where the assembly's complete geometric description would be defined. Intelligent decision support in this phase could include criticism agents to insure that the simulated performance meets the requirements and that the context of the simulation is realistic. Other criticism agents could suggest changes in dimensions or fiber architectures to meet performance requirements. This is an arena of continued opportunity for intelligent decision support.

5 CONCLUSIONS

A task structure for conceptual design of polymer composite assemblies is presented. The specific application involves the redesign of existing metal assemblies using composites in an effort to improve affordability and/or reduce weight. This activity involves several distinct problem solving stages including activities performed by the designer and intelligent decision support modules. Central to this problem solving activity are knowledge models of both the metal assembly and the composite assembly representing the functional and configurational aspects of each. The completion of the conceptual design activity leads naturally to the phase of detailed design and the knowledge representation models of the assembly should also support the problem solving in that activity as well.

6 ACKNOWLEDGEMENTS

This research is supported in part by ARPA's MADE Program under grant #8673, by the NSF Center for Low-Cost, High-Speed Polymer Composites Processing and by the State of Michigan's Research Excellence Fund. Apple Computer has also supported this research through generous equipment donations. This paper has also benefitted from the numerous discussions and interactions with John Scanlon and Gary Wigell of AutoAir Composites, Inc.

7 REFERENCES

[1] Franke, D., "Deriving and Using Descriptions of Purpose," *IEEE Expert*, Vol 6, No 2, 1992, pp 41-47.

[2] Hawkins, R., Sticklen, J., McDowell, J.K., Hill, T., and Boyer, R., "Function-Based Modelling and Troubleshooting," *Applied Artificial Intelligence*, Vol 8, No 2, 1994, pp 285-302.

[3] Hodges, J., "Naive Mechanics-A Computational Model of Device Use and Function in Design Improvisation," *IEEE Expert*, Vol 7, No 1, 1992, pp 14-27.

[4] Kamel, A., McDowell, J.K., and Sticklen, J., "Multiple Design: An Extension of Routine Design for Generating Multiple Design Alternatives," *Proceedings of the Third International Conference on Artificial Intelligence in Design* (Gero, J.S. and Sudweeks, F., Eds.), Kluwer Academic Publishers, Lausanne, Switzerland, Aug 1994, pp 275-292.

[5] Lenz, T., McDowell, J.K., Moy, B., Sticklen, J., and Hawley, M.C., "Intelligent Decision Support for Polymer Composite Material Design in an Integrated Design Environment," *Proceedings of the American Society of Composites 9th Technical Conference*, Newark, Delaware, Sept 1994, pp 685-691.

[6] Lenz, T., McDowell, J.K., Sticklen, J., and Hawley, M.C., "An Integrated Approach to the Design and Fabrication of a Composite Seeker Casing," MSU Processor, NSF S/I/U Center for Low Cost, High Speed Polymer Composites Processing, Michigan State University, Vol 1, Issue 5, 1994.

[7] McDowell, J.K., Lenz, T., Moy, B., Hawley, M.C., Kamel, A., and Sticklen, J., "Expert Systems for Integrated Material/Part/Process Design in Polymer Composites. A Task Specific Problem Solving Approach," *Proceedings of the Fourth International Conference on Computer Aided Design in Composites Material Technology, CADCOMP 94* (Blain, W.R., and DeWilde, W.P., Eds.), Computational Mechanics Publications, Southampton, UK, 1994.

[8] McDowell, J.K., Kamel, A., Sticklen, J., and Hawley, M.C., "Integrating Material/Part/Process Design for Polymer Composites. A Knowledge-Based Problem Solving Approach," *Proceedings of the American Society of Composites 8h Technical Conference*, Cleveland, Ohio, Oct 1993.

[9] Moy, B., McDowell, J.K., Lenz, T., Sticklen, J., and Hawley, M.C., "Expansion of an Intelligent Decision Support System for Process Selection and Process Design in Polymer Composites," *Proceedings of the 26th International SAMPE Technical Conference*, Atlanta, Georgia, Oct 1994, pp 162-171.

[10] Pegah, M., Hawkins, R., McDowell, J., and Sticklen, J., "Functional Modeling using Standard Parts to Support Conceptual Design," Working Notes of the AAAI'94 Workshop on Representing and Reasoning about Device Function, July 1994, pp 123-131.

[11] Pegah, M., Sticklen, J., and Bond, W., "Functional Representation and Reasoning About the F/A-18 Aircraft Fuel System," *IEEE Expert*, Vol 8, No 2, 1993, pp 65-71.

CONCEPTUAL DESIGN FOR MECHATRONICS

H P Hildre & K Aasland

1 WHAT IS MECHATRONICS?

A universally accepted definition of mechatronics does not yet exist. Many individual researchers, national boards and engineering associations have given the field their attention, so number of proposals of definitions are available. The term has been a catch word and are used in many context.

> Mechatronic is a technology which combines mechanics with electronics and information technology to room both functional integration and spatial integration in components, modules, products and systems.
> Buur [1]

Since the 1970's, the Japanese have been using the term "mechatronics" to describe a combination of mechanics, electronics and software, that is, mechanics controlled by microprocessors. The term has gradually spread to the USA and Europe, thus showing increasing interest in special methods and tools for the design of systems combining these technologies. Mechatronics is now a catchword covering almost anything. I will therefore limit the scope of this paper to: design of mechatronic products where moving parts or mechanisms plays an important role.

It is necessary to understand the essential differences between mechatronic systems and other systems before a new methodology, methods, and models can be made. There are significant differences between design of mechanics, electronics and software. Not only different technical skills are required, but the very nature of design problems differ in terms of functions to be realised, types of solution available, and realisation of the intended function. Buur [1] made a systematic comparison of some basic methodical characteristics of mechanical, electrical and software design. In the following the differences between design of mechatronic and mechanics and design of mechatronic and IT[1] are discussed.

1.1 What is the difference between mechatronic and pure mechanical systems?

1. IT products have a logic behaviour as the most important function. This behaviour is often very complex. Tools to model the behaviour are therefore necessary to assure right quality. Logical behaviour is seldom important in pure mechanical systems.

---

[1] IT = electronics + software

2. Production of IT systems are simpler than production of mechanical systems. Production of software is very cheep and simple. This means that production concerns have very low priority in parts of the system where IT is the dominant design challenge.
3. IT products are changed often, particularly software. They are flexible and easy to modify in comparisation with mechanical systems. This possibility can be a competitive advantage. On the other hand rapid changes can be difficult to handle and design tools are therefore needed.
4. IT make it possible with extended functionality and advanced user interfaces. For example a modern camera has many functions that is impossible without IT.

1.2 What is the difference between mechatronic and pure IT?

1. A mechanical movement exist. Movement is a kind of behaviour, but is different from the logical behaviour that exist in IT systems.
2. Control of movement is an important function of IT systems. This requires an understanding of the nature of the movements, principles for the control, and an understanding of physical relationships.
3. Interface between IT and mechanical systems exists, i.e. sensors and actuators. These interfaces connect different technologies and physical relationships.
4. Production plays an important role in design.
5. Environment plays an important role.

The relation between mechanical systems and IT to realise movement is an important characteristic of mechatronic systems. This relation results in a increased number of design degrees of freedom and more customer needs can be fulfilled. When a company has started to use IT solutions in traditionally mechanical products, the appetite for more functions will quite often increase as a result of the new possibilities in IT.

1.3 Mechatronics as a product development strategy to increase the competitive power

As a result of reduced marked life time and accelerated shorter innovation time for new technology in the products, product development plays an important role in companies. The need for efficiency in this area will increase. Mechatronics is a product oriented discipline. The main goal is to develop competitive products that fulfil the customer´s wishes and demands. A common feature is co-ordination and selection among technologies and disciplines. The

following four factors are important for use of mechatronic as a strategy to increased competitive power:
1. To respond fast to changes that affect the competitive power.
2. To reduce the product development time. Use modularity and flexibility in software.
3. To make use of interplay between technologies to achieve competitive power in products and production. Particularly integrate product aspects such as user-friendliness and increased functionality.
4. To make a plan for new possibilities (marked, production and product technology).

The potential is enormous, integration of electronics, software, and mechanical solutions in products, and systems will be necessary in many new products. A new design methodology is necessary to achieve this.

1.4 The Norwegian Mechatronic Design Methodology program.
The program is focusing on the following five points:
1. Identify and develop design methodology for multi-disciplinary design of mechatronics.
2. Identify and/or develop method, models and tools that can support the designer.
3. Develop a software system based on the methodology and tools described in point 1 and 2.
4. Verify and improve the results by doing projects in co-operation with the industry.
5. To work out teaching material and distribute results.

2 MECHATRONIC DESIGN METHODOLOGY AS A PART OF THE TOTAL PRODUCT DEVELOPMENT PROCESS.

Product development in a company can typically be organised in three levels as (see also fig. 1):
1. **Product planing**
2. **Integrated product development**
3. **Product synthesis**

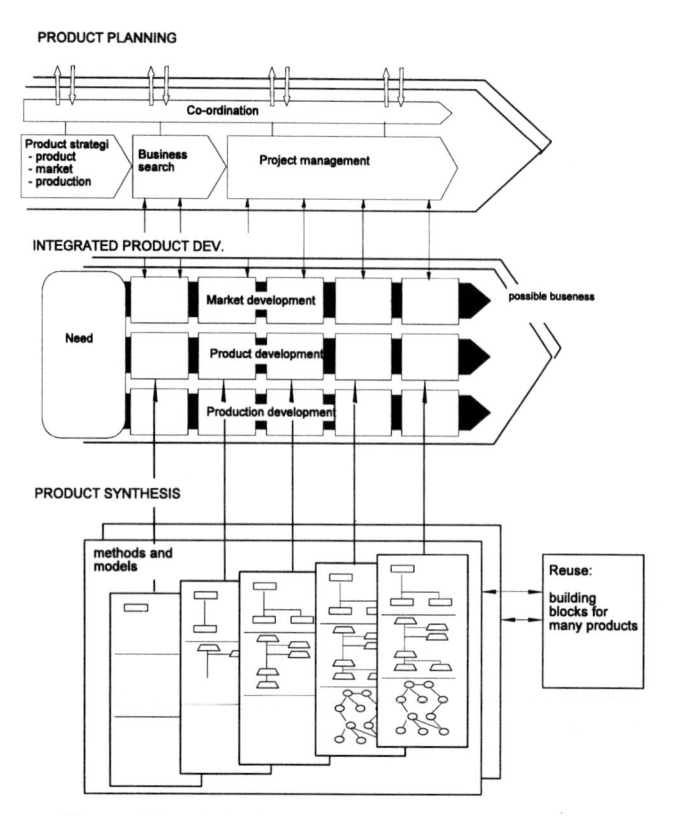

Figure 1 Levels in the product development activity

In this context product planning means initiating, planning and control of a development project. Integrated Product development is the total product development activity, i.e. a simultaneous development of a marked, product and the production. The main goal is to create business through co-operation between these elements.

The product syntheses concerns the technical development of the product, from abstract functions to detailed descriptions of the final product. The mechatronic project is mainly concerned with this level. We want to make a methodology based on structured models that describe different aspects of the product. The models have to be multi-discipline and encourage communications. We have chosen this result oriented approach instead of a more common activity controlled approach. Systematic reuse and how to handle variants are also included.

2.1 Product planning.

To be sure that the development projects are "moving" in the right direction, the product development effort have to be linked to at strategy. The product development strategy concerns all decisions that control the product development activities. The product development strategy has to be linked to the companiy´s overall strategy. This overall strategy can typically include: technology, organisation, business area and so on, see figure 2.

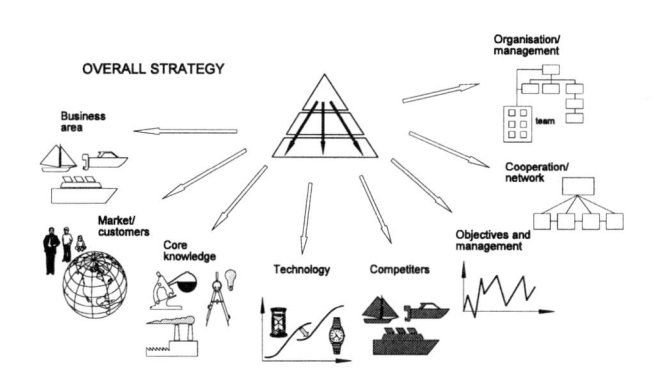

Figure 2 Example of aspects in a companies overall strategy

Based on the overall strategy, the actual situation, and level of ambition, the strategy for the actual development project can be established. This is a complex task and various analyse have to be executed. Important input in this activity is feedback from the marked, production department, and analysis of user properties and competitors. When a business is identified, i.e. a good connections between a marked, a product and production, the design of the product can take place. The following activities are of central importance:

- Start up and follow up of the product development project
- Controll of resources.
- Co-ordination of the actual project
- Controll of results from the product development project.

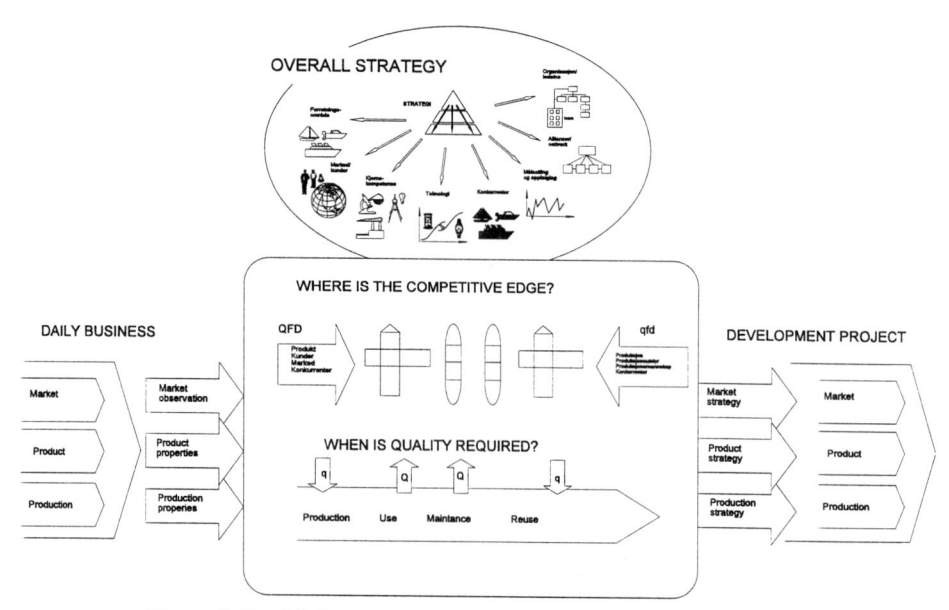

Figure 3 Establishing of product development strategy

2.2 Integrated Product development.

Product development is a creative process involving many disciplines. The three main disciplines are development of marked, product and production. These activities are all necessary for the development of a new product and have to be executed simultaneous in three parallel activities and with business as a common goal, see figure 4. The three main disciplines have their own

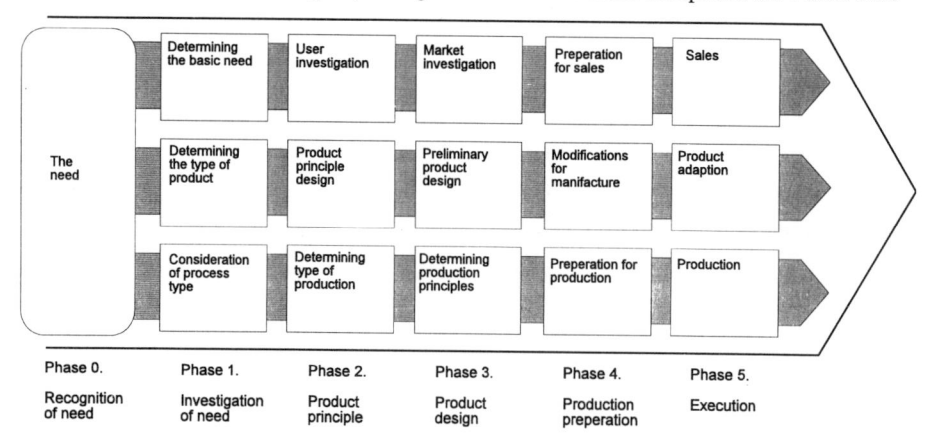

Figure 4 The Integrated Product development model [2]

lives in the company, but have common control and target. As shown in the figure, activities in marked development and production are executed in co-operation with the design of the product.

3 METHODOLOGY FOR MECHATRONICS

The methodology is based on a frame in following levels, see also figure 5.
1. **Methodology**
2. **Common components**
3. **Tools**
4. **Actual product**

Methodology:
The methodology includes a frame of methods, connected models, the relation between these methods, design rules, and strategy for the design work.

We have three types of specifications, a domain specification that relate the product, technology and users, a project specification to control the development project and the use of resources, and a product specification that specifies what type of product we want to develop.

Design models play an important role in design of mechatronic products. In the early phases these models are a way to by information of the final product. Models make it possible for the designer to describe and visualise his thoughts. Multi disciplinary models that can be used and understood of all in the design team are used to reduce the development time and reduce the risk for misunderstandings. Detailed design tasks are based on these interdisciplinary models and are executed by experts.

Common components:
This level gives help for reuse of earlier solutions. It is possible to look for solutions at different levels. The methods in this level make it possible to handle product families, variants and descriptions of the differences that gives these variants.

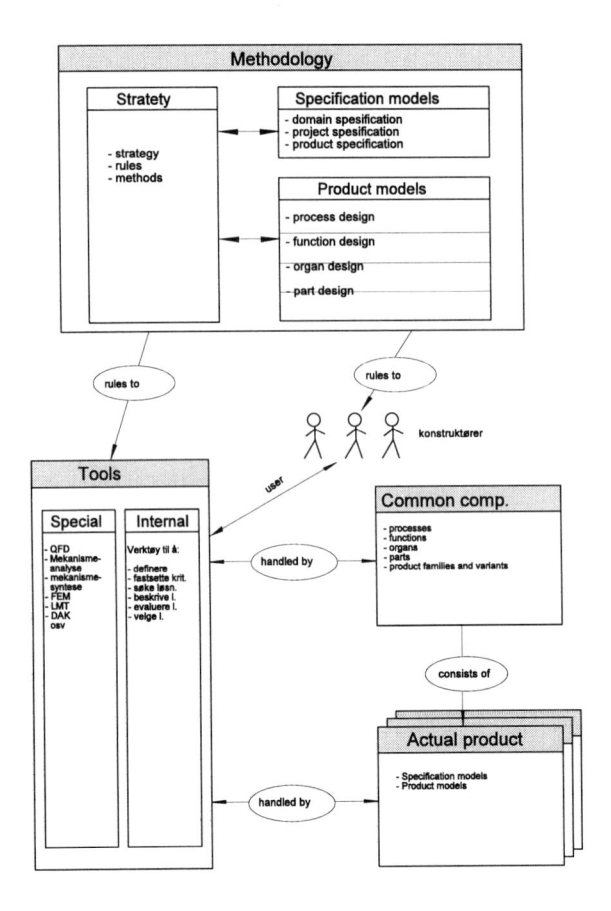

Figure 5 A methodical frame for mechatronic design

Tools:

This level is divided in two types of tools, special and internal tools. Special tools are commercial software tools that are useful for the designer. We will not develop such tools in this project, the important job is to find suitable tools and relate them to the design models. The internal tools are support tools to serve the models and handle communication. All communication between the user and the system are performed through these tools.

Actual product:

The description and models of the designed product are stored her.

4   PRODUCT MODELS.

No single model or representation of a mechatronic system will fully describe all aspects, a "metamodelling" approach is needed [3]. Several models on different levels are needed, for the total project, single design phases, individual design steps and technologies. The challenge is to find suitable models that in total give a complete description of wanted aspects. We want the models in the early design phases to be interdisciplinary and create a basis for further detail design. Two models have vital importance and describe the frame of the methodology approach: the Reference model and the System model.

4.1   Reference model for mechatronic.

The reference model organises models on different levels of abstraction (vertical causality). The reason for using models on different levels of abstraction, is that it is easier to achieve the target of right quality, shorter time to the marked and low cost. At this point we have chosen to analyze following models:

- *The chromosome model* (the theory of domain)[8]
- *The SISU model*, used real time computer systems [7]
- *Common models used for electronic design*

Many similarities exist between these models, but also some important differences. The differences are partly caused by different focus, different design challenges, and partly because of tradition.

Process design:

    The task is to describe working tasks, technology and user. This is done by a structure diagram of transformations (energy, material and information), objects and users. The focus is to analyze which tasks that are performed of the system, by users and environment.

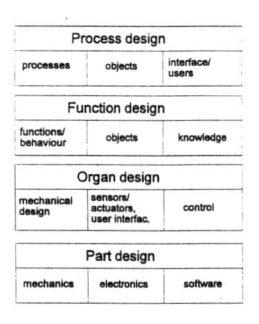

Process design		
processes	objects	interface/ users
Function design		
functions/ behaviour	objects	knowledge
Organ design		
mechanical design	sensors/ actuators, user interfac.	control
Part design		
mechanics	electronics	software

Figure 6 The Reference model

Function design:

    A system of purpose functions needed in the system to create the specified transformations, behaviour, objects, and knowledge

Organic design:
>A structure of organs, each of which realises one or more functions through physical effects. Particular design of mechanical systems, control and interfaces as sensors, actuators, and user interfaces.

Part design:
>- A structure of single components and their properties.

4.2 The system model.

At the first look, mechatronic systems seem to have few structural similarities. Further analysis shows that important similarities exist and a common system model can be established. The system model gives a structure of the problem and identifies important design challenges that exist in common systems. The system model is inspired by The theory of technical systems by Hubka [6], see figure 7. In this context, the system is seen together with the user, other systems, and processes that express the working tasks to be performed. The following models are studied:

- *Theory of technical systems* (Hubka, Eder [6])
- *System model* (Hildre [5])
- *General knowledge about design of mechatronic systems*
- *General knowledge about design computer systems*

The theory of technical systems cover transformation systems where technical systems and humans are working together to do work tasks (the technical process).

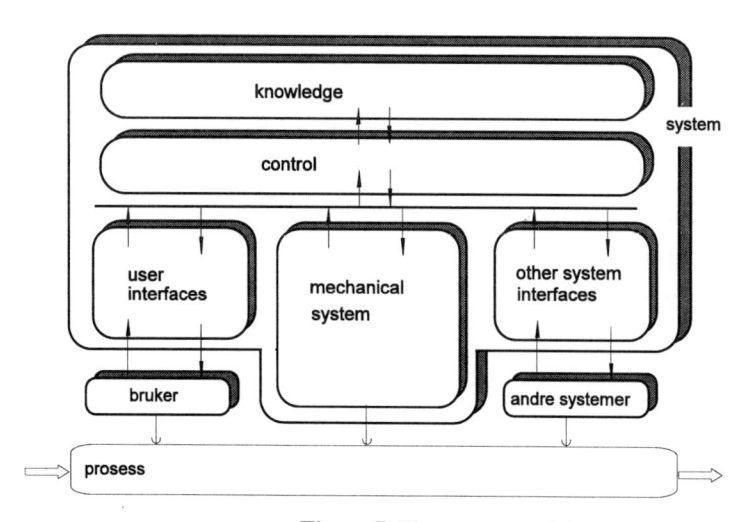

Figure 7 The system model

The technical process is work tasks to be done by one or several technical systems or humans i co-operation to fulfil identified needs. The idea is that analysis of work tasks gives a basic understanding of the need, technology, and distribution of work tasks between human and the technical system. The system can be divided in several part systems. Figure 7 shows a general system model typical for mechatronic systems. Note that the model is independent of technology and is suited to describe the functionality independent of solutions. Figure 8 shows the connection between the system model and the reference model.

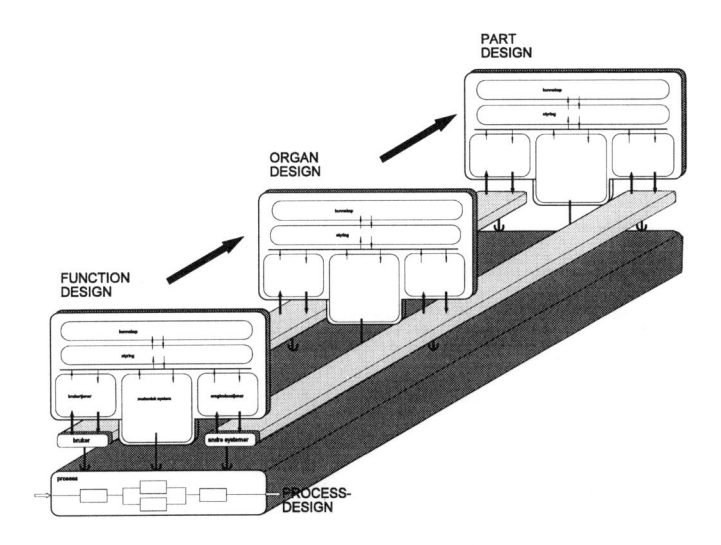

Figure 8 The relation between the Reference model and the System model

4.3 Process design.

Process design is used to describe work tasks executed by systems or humans. The objective is to make a common understanding of the system requirement and borders. The process information is expressed in a form and structure that can be understood of everybody in the project team, the marketing department and customers.

Mechanical part systems that execute work tasks is common in all mechatronic systems. An understanding of the mechanical processes is therefore important. These are:

- *transformations of material in time and space*
- *transformations of energy in time and space*

We have passive and active objects. The active objects execute the transformations on passive objects. For example movements is transportation of material in space. The passive object is the material that is moved and it is moved by active objects.

An information technology system that can:
- *control the mechanical processes*
- *communication between users, the system and other systems*
- *other information services*

The process description is based on a process structure consisting of following elements:
- *active objects described as boxes*
- *energy, material and information flow described as arrows*
- *passive objects (material) at attributes to flow of material*
- *signal and data as attributes to flow of information*
- *type of energy as attribute to objects*
- *knowledge as attribute to objects*
- *relations as symbols between objects*
- *users for active objects*

and process behaviour as:
- *event lists as MSD (massage sequence diagram)*
- *time history diagrams*
- *behaviour of control objects as state diagrams*
- *behaviour of material transformations as text, figures and equations.*

Figure 9 shows an example on process structure diagram.

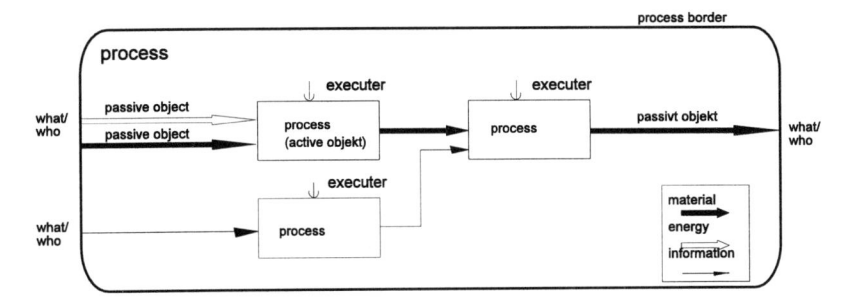

Figure 9 Example on process structure diagram

4.4 Function design

The objective with function design is to give a complete description of functions and the behaviour of the system. The behaviour description require that all relation between a stimuli and response are described clear. To do this it is necessary to split the system into objects with parallel behaviour.

The level of function design is very different for IT design and mechanical design. In design of mechanical systems we usually start working with functions and tries to find principal solutions without any complete description of the behaviour. In design of IT systems a complete description of the controlled behaviour will be made. This discrepant is independent of selection of solutions and technology. The behaviour description make it possible to analyses and simulate the behaviour even before principal solutions is made.

The following descriptions are essential:
- *Functions-structure.* Structure of functions with parallel behaviour and connections between these functions and the surroundings.
- *Conceptual data model.* The knowledge the system needs about processes.
- *Signals through interfaces.*
- *Behaviour for functions.*

In this level a more detailed system model is required, see figure 9.

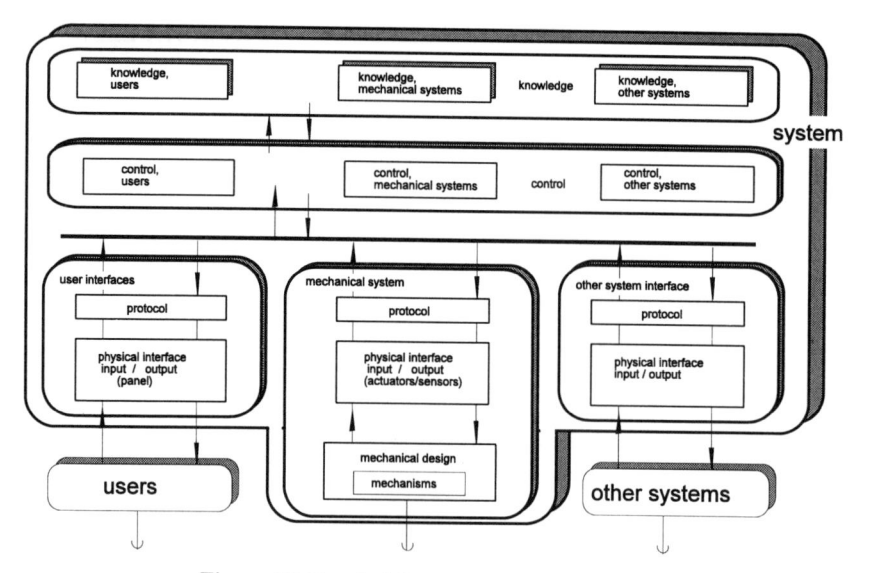

Figure 10 Detailed System model

References

[1] Jacob Buur; A Theoretical Approach to Mechatronic Design, Instituttet for Engineering Design 1990, Technical University of Denmark, 1990

[2] M. Myrup Andreasen, Lars Hein; Integrated Product Development, Universitetsforlaget AS 1986

[3] V. Salminen et al; Metamethodics in the task of definition and conceptual design phases. int, Conf. Of Eng. Design (ICED), Dubrovnic 1990-B

[4] Knut Gulbrandsen, Mekatronikk notat, Specifications, SINTEF 1994- M-20

[5] Hans Petter Hildre, Rolv Bræk, Mekatronikk notat, Mechatronic Methodology, SINTEF 1995, M-16

[6] Vladimir Hubka, W. Ernst Eder, Theory of Technical Systems, Springer Verlag 1984

[7] R. Bræk, Ø. Haugen; Engineering Real Time Systems, an Object-Oriented Methodology using SDL, Prentice Hall, 1993

[8] M. Myrup Andreasen, Design Methodology, Seminar on Design Theory and its Application, The Norwegian Academy of Technical systems, Trondheim May 1991

REASONING AND TRUTH MAINTENANCE OF CAUSAL STRUCTURES IN INTERDISCIPLINARY PRODUCT MODELLING AND SIMULATION

X-T Yan & J E E Sharpe

1 INTRODUCTION

With the increasing availability of advanced technology and the rapidly changing nature of design requirements, modern products are becoming more and more complicated in their functionality. A product's success is heavily dependant upon a high quality design. With such high expectation from the market, the designer faces a great challenge in choosing enabling technologies and producing a product design with highest possible quality and minimum cost in the shortest time to market. There is, therefore, a necessity for powerful tools to assist the designer in addressing such demanding design requirements. Most software tools available today only support a designer in a very limited aspect of the whole design process and it is very difficult if not impossible to use them effectively as a homogenous tool. In addition, the designer is more likely to be hindered by incompatabilities between a set of heterogeneous software.

A complex system needs to be designed using highly integrated and interconnecting technology to fulfil the various requirements, which could be mutually conflicting. That is, in the design of such a system, one often faces the dilemma that one has to satisfy a wide range of design requirements and at the same time allow the system to perform normally in various extreme conditions. The resolution of this is vital to the success of a complex product.

Many researchers have focused their study on the general aspects of supporting a product design, including geometric modelling and functional behaviour modelling. Few systems provide any assistance to the designer regarding a system's dynamic behaviour in a complex situation. The underlying structure of the system behaviour is not as fixed as is assumed by many designers. The causal structure is an important aspect of a total product design and this paper will address the necessity of considering causal structure in a complex system design and demonstrates how Schemebuilder, a knowledge based computer support system for mechatronic product design, supports the causal analysis, modelling and conflict resolution during scheme generation and evaluation, using artificial intelligence techniques.

2 INTERDISCIPLINARY PRODUCT DESIGN

Engineering design is a process of generating a solution or a scheme to a design problem to satisfy a design requirement and its specification. Various processes and activities, such as concept generation, invention, realisation, evaluation, modification, refinement and detail specification and so forth are involved throughout the entire design. A generally well-accepted division of engineering design can be expressed as conceptual design and embodiment design [1,2,3,4,5]. Conceptual design takes the statement of problem expressed in an abstract form and aims at generating broad working solutions to it in the form of schemes, whereas embodiment design concentrates on further specification of conceptual schemes in order to produce, on one hand, a set of detailed specification drawings of components and, on the other hand, to provide feedback to the conceptual stage and verification of the design schemes. Although the design process is broadly divided into these two separate processes, overlapping is inevitable and encouraged in the context of concurrent engineering practice.

Modern electronics and computer technology have helped to remove many of the traditional constraints to design, both in terms of methodology and implementation, and this has given the designer a much greater freedom of choice and decision-making power as to the best solution to any given design problem. This not only provides greater opportunities for designers to produce better designs, but also generates new difficulties as to how the multi-disciplinary approach can be best applied in complex and highly integrated product designs.

Mechatronics has been well established as an interdisciplinary subject and can be considered as the synergistic integration of mechanical engineering, electronics and computer technology. System modelling and simulation technology have traditionally been employed to predict the performance of a design scheme by employing a mathematical method to describe a physical system. These modelling methods were generally developed for a specific discipline and can only cope with systems within that discipline. With the advent of interdisciplinary subjects, such as mechatronics, the lack of flexibility inherent in conventional modelling methods has rendered them useless.

The aim of this research is to assist the designer at both the conceptual and the embodiment design stages in the analysis of the causal structure of a solution scheme and its verification using sophisticated simulation techniques. The method employed is to catalogue commercially available components into a library of generic component class models, the simulation of which is based on bond graph methodology for energetic components and a function block representation for information related components. Components are represented in a knowledge base using class or frame structures and production rules. This paper describes how these techniques have been applied in the system, focussing particularly on the reasoning and truth maintenance of causal structures.

3 COMPONENT MODELLING

3.1 Energetic components

Components in most mechatronic systems can be grouped into two basic classes, namely energetic components and information components. Energetic components are those which are designed to convert energy from one form to another. For instance a d.c. motor converts electrical energy into mechanical rotational energy. Information components are those which utilise negligible energy flows to indicate changes in states of interest. According to the characteristics of the different types of components, appropriate methods of modelling and simulation have been adapted and implemented in Schemebuilder [6]. Bond graph theory

[7,8,9,10] is a unified approach to modelling and simulating energy conservation, transformation and interaction among components in different energy domains. It has been adapted for use within Schemebuilder and deals with the energetic components. A block diagram method has also been employed to model and convert a bond graph model into a dual effort and flow energy co-variable model and this has a number of distinctive advantages [11]. The Yourdon diagram [12,13] has been employed to handle information components and is implemented in Simulink, a block diagram based simulation package developed by Math Works Inc., using C functions.

3.2 Information components

A literature survey has been conducted of the fundamental functions of information related components. Several issues have been identified as being important and requiring special consideration in modelling information components, these are: Number System Selection (binary or decimal), Timing Issue, Order Sequence Versus Parallelism, Address Systems (one, two, three or four address system), Level of Modelling Interest and Operation, e.g. at a digital level, or at machine binary code level for a device, or at routine level, and finally Interaction with the Environment. Based on these considerations some forty fundamental functions in modelling information systems have been identified and these are shown in tables 1.1 to 1.5 in the appendix [14,15]. From these tables an interesting observation can be made; like energetic components, any information system can be decomposed into elementary functions and one can construct more complicated components performing sophisticated functions by using the fundamental functions listed in the tables. Wherever it is possible, an example working principle is also given for each of the functions.

4 CAUSAL STRUCTURE OF A MECHATRONIC PRODUCT

4.1 Analysis of component causality

In an interdisciplinary product, energy is often transferred from one domain to another in order to achieve a function which is otherwise not possible to be realised. The final desired function is normally satisfied by introducing intermediate functions in other domains, which can be realised using a component. This process often results in the selection and assembly of several components in different energy domains as well as in the information domain to meet a design requirement. It is inevitable that these components will interact with each other and the final function requirement is achieved collectively through the aggregation of several intermediate functions embodied by the chosen components. The system as a whole requires co-ordination between components especially among those with logical relationships, communication between the information system components and the energetic components and interaction between energetic components to drive any linked components in order to achieve a needed function at a required magnitude.

In the design of a system with such complexity, it is extremely important and vital to establish and maintain throughout the entire design process a correct causal structure or causality. A causal structure is the relationship of interaction between two elementary components. In the energetic domain, the causality is defined by the direction of the effort co-variable. There exist two types of causalities, namely integral causality and differential causality for energetic components [7].

In the information system domain, it is often true that a component can only perform a designed function in such a way that information itself can only flow in one direction carried by its carrier. Once the information has been transmitted from its sender to its receiver, it remains captured in a storage place for future use or via a buffer for processing. The carrier only carries the information when it is being sent and remains idle or static otherwise. A carrier can carry information in both directions, but the information carried each way may be different. This feature of discontinuity and discreteness (in digital communication) determines that information components only work in their intended flow direction. Due to this feature, the designer very often employs decoupling mechanisms to prevent the information flow in a forbidden direction, and this is especially true in the systems in which the functions are embodied in most cases in an electronic form. This literature study indicates the causality of information systems are relatively simple. The causality assignment of information components at the modelling stage can therefore be simplified by giving the flow direction of information from one component to another in the permitted direction. Information flow in the opposite direction is not allowed in these components.

4.2 Rules of assignment of causal structure

4.2.1 Rules for energetic components. A list of rules for assigning the causality to a component bond graph model are given below:

(1) Assign the required causality to all source elements (SEs determining the effort of a connecting bond, SFs the flow of a connecting bond), and extend the causal implication through the graph as far as possible using the constraint elements, namely common effort junction, common flow junction, gyrator and transformer. This rule is used after any of the following steps of assigning causality;

(2) Assign the preferred (integration) causality to energy storage elements (C determining the effort of energy flow of a connecting bond, I determining the flow of energy flow);

(3) Assign causality to any unassigned R-element in any form;

(4) Assign causality to any remaining unassigned bond according to constraint elements;

(5) Adjust the causality assignment of some elements if there is a contradicting causality assignment.

All these rules are clearly defined in a logical manner and were implemented using production rules and logic combinations written in several attached active value methods in the KEE (Knowledge Engineering Environment) developed by IntelliCorp.

4.2.2 Rules for information components. Most information related components have a fixed direction of information flow, in which the component is designed and expected to perform. In other words, the flow of information is one directional and the components are usually designed to stop the back flow of information as a safety measure to enable them to function correctly. A model of any of these components will have either a fixed one directional input or an output, i.e. the causality of these components are component dependent. The function of these models is simply to allow the passage of signals or information through their linking ports.

4.3 Four quadrant causal structure

The causal structure of an energetic component is determined by the imposition of a flow variable or effort variable on the component, which in turn depends on the way and instance of the component being used. When an energetic component is used in a product, it is bound to be driven by an energy source and at the same time is subject to some type of load. The load can be viewed further as a load flow variable and a load effort variable. Corresponding to the energy from its connected or driving component, the energy variables on the component, from the point of view of the component being driven, can also be described by a driving effort variable and a driving flow variable. The real causal structure of a component is determined by the superposition of these compound energy variables, i.e. it depends on the magnitudes and signs of both the driving and the load energy co-variables. Using the bond graph concepts, the following notation is adopted in the analysis and reasoning of the causal structure of a component.

E_D denotes effort variable of the driving energy source of the component;
F_D denotes flow variable of the driving energy source of the component;
E_L denotes effort variable of the load energy source of the component;
F_L denotes flow variable of the load energy source of the component.

Driving Effort E_D	Load Effort E_L	Driving vs Load Effort	Total Effort $E_T = E_D + E_L$	Causal Structure
$E_D \geq 0$	$E_L \geq 0$	$E_D \geq E_L$	$E_T \geq 0$	⊣ Component ⊢
$E_D \geq 0$	$E_L \geq 0$	$E_D < E_L$	$E_T \geq 0$	⊣ Component ⊢
$E_D \geq 0$	$E_L < 0$	$E_D > E_L$	$E_T \geq 0$	⊣ Component ⊣
$E_D < 0$	$E_L \geq 0$	$E_D < E_L$	$E_T < 0$	⊢ Component ⊢
$E_D < 0$	$E_L < 0$	$|E_D| \geq |E_L|$	$E_T < 0$	⊢ Component ⊢
$E_D < 0$	$E_L < 0$	$|E_D| < |E_L|$	$E_T \geq 0$	⊢ Component ⊣

Table 2 The combination of causal structures between two energetic ports in a component

Each of these energy co-variables is represented by its magnitude as well as its sign. Table 2 illustrates all six possible combinations of the effort energy variables with different signs and magnitude, assuming the effort variable is the independent variable among the two energy co-variables. These six combinations can be further degenerated into four causal structure possibilities, a component, in theory, could experience during its application. This is called a four quadrant causal structure due to the obvious correspondence of the causal relationship effort vs flow in a four quadrant co-ordinate frame illustrated in Figure 1. A similar table can also be constructed for flow variables F_D and F_L when assuming the flow variable is independent in the component energy flow.

Any energetic component having one pair of energetic ports, in theory, possesses the above quadrant causal structure. If it has more pairs of energetic ports, further corresponding four quadrant causal structures exist for it. In practice, some components are designed to drive other components only, not vice versa. This implies that between E_D and E_L, or F_D and F_L an implicit relationship has been assumed and it should not be violated during the use of the component, otherwise physical damage might occur to the component. A mechanical worm gear is a typical example of this type of component, only allowing the flow of energy from its worm to its gear, or the transformation of mechanical rotation energy in one orientation into the same form of energy in another orientation. In the modelling and simulation of this type of component, this feature should be specially addressed to ensure that the correct relationship is specified in its model and maintained during the simulation.

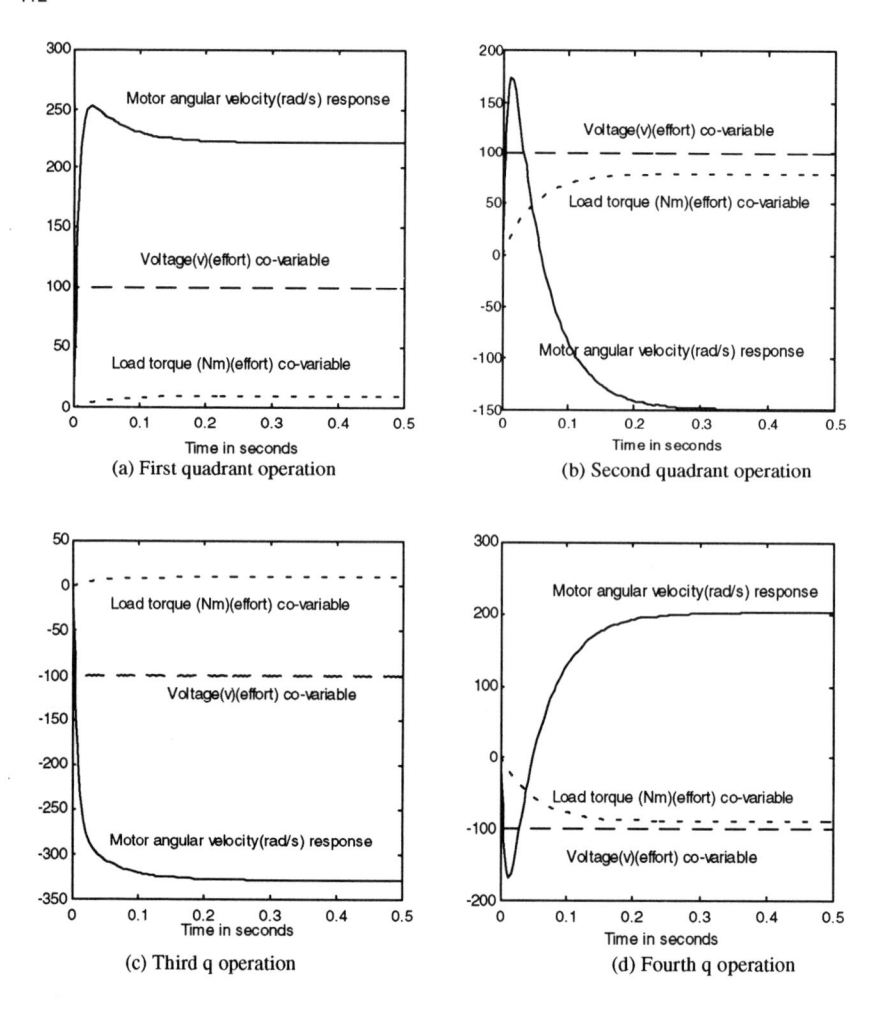

Figure 1 The performance of a d.c. motor in its four quadrant operations

There exists another type of component which is designed to stand all four possible causal structures during its use. A d.c. motor is a typical example of such a component with four quadrant causal structure under a load condition. The four diagrams in Figure 1 show the performance response of a d.c. motor under different load and driving torque combination, (a) showing that driving torque (effort) (from d.c. power conversion) is relatively greater than the load torque and driving the rotor in the intended driving direction (first quadrant causal structure); (b) showing the driving torque (effort) is positive but smaller than the load torque (backward driving) and that compound torque is negative (second quadrant causal structure); (c) showing that the driving torque(effort voltage) is negative with the load torque positive and that compound torque is negative (third quadrant causal structure); and (d) illustrating that the driving torque(effort voltage) is negative with the load torque negative and larger in absolute value than that of driving torque and that compound torque is positive, driving the motor in load driven direction (fourth quadrant causal structure).

The understanding of complicated causal structures is of great importance in the analysis of a component or a system during the course of numerical simulation. This will provide fundamental knowledge for detecting the numerical change from one possible causality to another during simulation. An application of this analysis will be shown in the Fin Actuator system modelling and simulation in section 7.

5 MULTI-LEVEL PRODUCT MODELLING

Modern product development requires a designer to engineer a satisfactory design scheme to the highest possible quality in the shortest time. High quality design needs a product to be looked at from a number of different perspectives to satisfy many constraints including performance, spatial, economic, maintainability and manufacturability requirements as well as environmental considerations. This requires a design support tool be able to provide feedback on any system aspect at any level. This is an extremely important feature for detailed system validation and evaluation, where a more in depth performance evaluation is critical with a multiple states system. Thus a designer needs to be assisted by a design evaluation environment in order to verify a design comprehensively using the required aspect models.

In this research bond graph notion and its model representation have been adopted to unify the analysis of the system dynamic performance, the system spatial and kinematic aspect models, and the system performance aspect models for energetic components. In this way, repetitive description of a component in different aspect models can be avoided. A generalised database with its dataserver has been designed and data for all aspect models can be flexibly stored and accessed. Due to the unified model representation and a single database structure, the database size for these components can be significantly reduced. The data in the generalised database for a component typically consists of energy co-variable effort and flow, (i.e. force and velocity, voltage and current etc.), their integrals (i.e. moment and distance etc.), constants of these variables, spatial dimension specification of a simplified component 3-D solid model, reliability, weight and cost of a component. Some of this data is stored in discrete data form available from the manufacturer's catalogue, some is stored as parametric formula derived from the data analysis of these parameters.

This process of comprehensive evaluation inevitably results in the modification of some design parameters from their initial value specification or even the generation of a new scheme or its variants. This modification and iteration can ultimately generate an optimal design solution to a given problem.

5.1 Component function level modelling

Each component can be represented by a primary or major function and several secondary functions. A primary function is defined as a pure function which the component is designed and intended to perform during its normal use. The primary function can be represented by a high level abstract function which defines the key purpose of the function associated with its principles.

A secondary function is defined as an accompanying function of a component which must exist in order to achieve the designated (primary) function. This is usually the result of constraints imposed upon the component due to the embodiment of primary function. The primary as well as secondary functions are represented in Schemebuilder in bond graph form and are usually described with three levels of complexity, which is thought to be adequate at this stage of the research. Frame structures were used to represent primary and secondary functions.

5.2 Component aspect modelling

The total design of a physical system requires the design engineer to produce a design solution which will not only satisfy the major design specification and requirements, but also optimise the solution to meet the stringent conditions customers might not perceive in their design specifications. A systematic design approach to solving these problems is to provide the designer with support to model the physical system during the entire design process flexibly and comprehensively on a computer. Many researchers and software developers have realised the need to support designers in carrying out some specific or single aspect design by providing software solutions dedicated to these aspects respectively. These tools have proven to be quite successful and powerful in solving problems related to these aspects provided that the designer can follow the procedures the software requires. Each of these tools often acts as an aspect computer modeller to deal with one of many system properties. For example, many solid modelling packages have been produced to facilitate the design of component geometry design and layout design which have been extremely powerful and successful.

From the total design and systematic modelling point of view, it is easy to perceive that one would be much more aided if one could have quick access to as many aspect computer support tools as possible. Thus a designer can use these tools to improve the design efficiency and quality in order to achieve an optimal design. This has been one of Schemebuilder's objectives, which is to provide a designer with an integrated, intelligent environment which support comprehensively the modelling of many aspects of a physical systems. In each of these aspects, the computer acts as an expert advisor and collectively with the designer, forms the knowledge sources for a multi-aspect modelling system.

There are several aspect models which have be identified to address the important aspects in the design of mechatronic systems, including: Component Qualitative Function models, Component Cost and Weight models, Component Spatial models, Component Structural models and Component Energy and Dynamic Behaviour models. Each of these models is an aspect or attribute model of a component class which is defined as an elementary function entity in the design and selection of component for a system. For the entire system, see [6].

In this research, particular attention is given to the application of modelling and simulation technology to mechatronic product design and analysis. A novel system modelling and simulation method was proposed and has been implemented for a generic mechatronic product design based on the underlying function block oriented modelling approach which encompasses bond graph theory, the block diagram method and high level function block representation. This hybrid modelling and simulation method was developed to verify interdisciplinary mechatronic product schemes and has satisfactorily demonstrated that it is feasible to use it as a unified modelling and simulation method for mechatronic systems. A library of component function blocks have consequently been created using the method and collected for mechatronics modelling and simulation. A new product design scheme can be verified by easily aggregating a number of high level function blocks from the library into a simulation model and simulating its performance. The performance of a system can be visualised graphically and design parameters of a scheme can be improved through a comparison between simulation results using different sets of design parameters. The generalised model can be further used to optimise system design parameters and achieve design improvements [11].

6. TRUTH MAINTENANCE OF THE CAUSAL STRUCTURE DURING SCHEME GENERATION AND SIMULATION

A library of bond graph models for energetic components have been created and stored in bond graph library for both the causal structure analysis and the simulation of a component or a system. At the causal structure analysis stage, each of component bond graph models is instantiated from the library with a preferred causal structure. This implies that during the construction of a complete bond graph mode for a system there are possibilities for a designer to build a system bond graph model with inconsistent causality. It is desirable that the software can provide an automatic support to maintain the consistent and logic causality, i.e. truth maintenance of the causality of the system being built.

The acceptable causalities between components are stored in the mechanism of the truth maintenance system and further set of production rules were also implemented. Thus, whenever an illegal causal structure is established by the user, the truth maintenance mechanism will automatically warn the user and an alternative bond graph model of the component will be used instead to maintain a correct causal structure for the designed system.

```
         R1                                    R2
          |                                     |
    ——| 1 ——| 0 ——| TF ——| 0 ——| 1 ——
                |                     |
              Sf=w=0                Sf=v=0
             Structure            Structure
              Support              Support
```

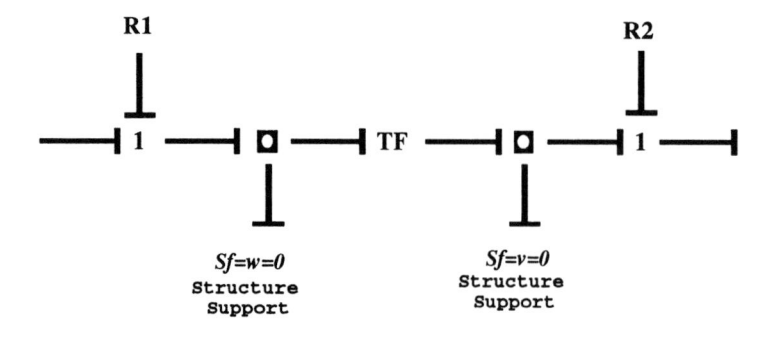

Figure 2 A causal structure of a leadscrew at its normal operation in its bond graph model

6.1 The preferred causality assignment and causal conflict

A set of production rules of the preferred causality for each bond graph elements and information components have been listed in section 4.2. During the modelling of a mechatronic system, bond graph models are employed to model energetic components and corresponding preferred causality hence being assigned to each of its bond graph elements. Figure 2 shows that a preferred causality is assigned to all bond graph elements in a bond graph model for a leadscrew, where the reversed "0" sign indicates the nature of action and reaction effort variable on a structural support. This is the causal structure in its normal operation which shows the leadscrew is driven by a rotational torque (effort) and the load is not bigger than this driving torque.

During some extreme conditions, however, there exists a causal conflict when all the preferred causalities of each bond graph elements are assigned. As alluded in section 4.3, some components are designed to withstand only some of causal structures and if an invalid causal structure is imposed upon them, a causal conflict occurs. Figure 3 shows an example of causal conflict, where a leadscrew is forced by the load for the nut to drive the leadscrew backward. A causal conflict is therefore imposed upon the structure which supports and passes the energy flow. If this conflict is not resolved properly, physical damage might occur to either the thread on the leadscrew or on the nut. This will of course only arise temporally as is the nature of a transient process. It will last until the conflict is resolved by the system or the occurrence of damage. It is important for the designer to take into consideration these causal conflicts and try to avoid them or resolve them in the real-time control of energy flow.

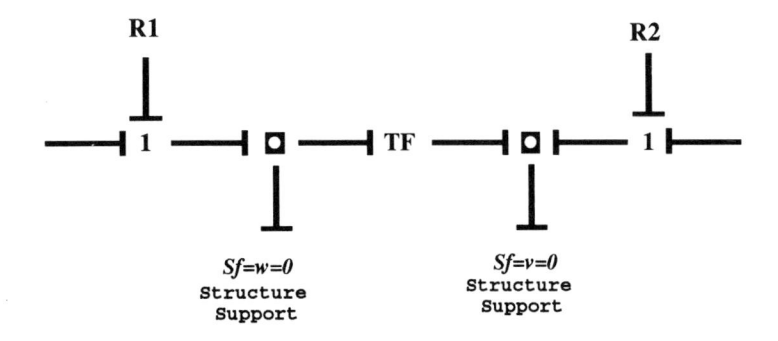

Figure 3 An alternative causal structure representation of a leadscrew at its transient operation in its bond graph model

6.2 Truth maintenance system

Truth maintenance system is a mechanism which establishes a dependent relationship between facts in a knowledge base. All facts about the causal structure of a component are stored in the component bond graph knowledge base, where all possible causal structures of a component at all levels can be retrieved. The truth maintenance mechanism implemented in the system works at two separate levels for both system causal structure analysis and dynamic simulation.

At the analysis level, the user can build a system bond graph model by aggregating the component bond graph models from its library based on the specification of a qualitatively specified system scheme. See [6] for more detail about the qualitative scheme generation. When a component bond graph model is inserted in the system bond graph model, all causal facts associated with the component model are known to the truth maintenance mechanism and a new causality for the linking bond is established which might not be acceptable in the existing bond graph model for the previous component. The truth maintenance mechanism will warn the user and suggest the use of the complimentary bond graph model of the second component.

At the simulation level, A detecting mechanism will check if there is a causal structure change for each of the components in a system. If so, the simulation will be paused and an appropriate component model will be used before resuming the simulation. This will be discussed in the following sections.

7 CAUSAL CHANGE AUTO FLAG IN SIMULATION

Understanding of causal structures is of essential importance in maintaining the energy flow and ensuring the desired direction of such a flow in a system from a component to its adjacent ones. As illustrated in table 2, any two ports of a component can in theory have four possible structures, though some of them are only designed to have two. This implies that there will be at least two possible causal structures for every pair of ports in a component. These are, a causal structure with the effort as the independent input co-variable to the component and the flow variable as the independent output co-variable and one with the flow variable as the independent input co-variable to the component and the effort variable as the independent output co-variable. These two possible structures should be considered at the dynamic modelling stage of any dynamic energetic components and each of them should be modelled separately as they are different and usually opposite.

With this in mind, two separate models for each energetic component at a particular level based on the bond graph modelling techniques have been created and collected in the mechatronic modelling and simulation library and are available for use.

The causal structure of a component which should be used depends on the imposition of the driving effort and the load at an instance of time when the component is used in the system. Some components are designed to withstand only one causality during their normal usage whilst the others may be designed to cope with more than one causal structure. In both cases, a modelling and simulation system should provide support for the designed system to be flexibly modelled and simulated with all their possible causal structures.

7.1 Causal change in the multi-state simulation

The causal structure of any component can change during normal operation and this causes a great difficulty in modelling and especially in simulating any system which contains this type of component. The method used in the modelling and simulation of the Schemebuilder system was generic and applicable to all components. All possible causal structures of a component were created and stored in the library of modelling and simulation. The energy co-variables (effort and flow) of each component energy port are monitored and tested by a causal change detector mechanism, to compare the overall effort and flow value imposed upon the inter-linked pair of ports. Based on the detection of the effort and flow variable, a set of production rules derived from the table 2 are employed to determine which causal structure is imposed on the component at a particular moment and if there is a causal structure change the appropriate component model will be called to meet this change.

7.1.1 An example of causal structure change. The example used in this paper is a fin actuator system of an un-manned target aircraft. A qualitative scheme was generated based on the design requirement and specifications. The fin actuator system is a true mechatronic system and consists of a battery power source, a power amplifier, a d.c. motor, a gearbox, a fin structure, an angular sensor measuring the fin angle, a flight controller and a pre-flight programme(see Figure 4). The scheme was generated on the Schemebuilder's Building Site in component name block form and the causal information of components in their normal in component name block form and the causal information of components in their normal operation (or static causal structure) was also described in the scheme. Based on this information, a simulation model of the fin actuator system was created automatically by analysing the component structure and the static causal structure of each component, instantiating appropriate component models with the corresponding causal structure and linking these objects with appropriate ports and linking direction.

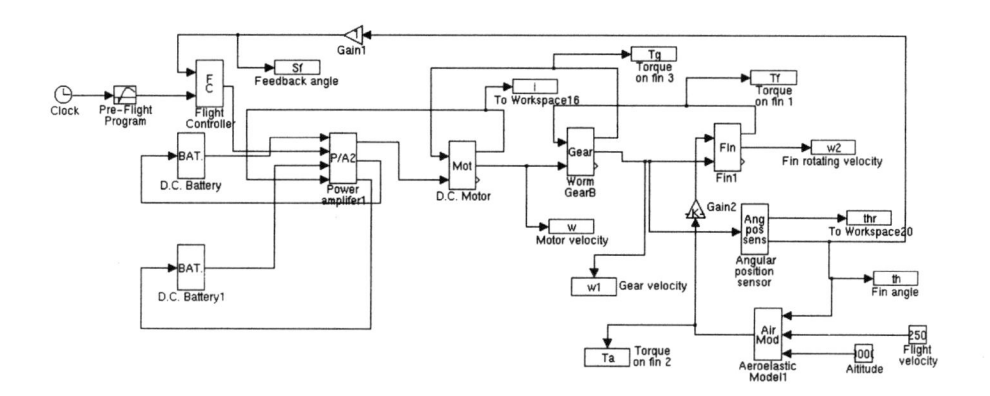

Figure 4 A fin actuator model with its initial causal structure representation

This fin system is subject to the torque load from its surrounding air flow and in normal operation the drive system should enable the fin to rotate in a controlled manner. The overall energy flow direction is therefore from the core of the driving system d.c. motor to the fin structure in normal operation. However, in some extreme conditions, if the torque generated from a gust for instance exceeds the driving torque from the d.c. motor, the energy flows backward temporally and the causal structure is thus changed. Appropriate action needs to be taken to deal with these dynamic causal changes.

7.2 The detection of a causal change

The detection of a causal change is a key aspect in maintaining correct simulation for a complicated system operating in a changing environment. The identification of a causal structure for a component is dealt with by the simulation system in two aspects: static identification and dynamic identification. The initial causal structure of a component is assigned at the scheme generation stage and stored in the KEE knowledge base system. It is easy to retrieve and identify this information just before the creation of a simulation model. This involves the search of assigned causalities of linking bonds or information connections of a component model. Once the correct causal structures have been identified, a simulation model creator mechanism implemented in the KEE can be used to create a simulation model, complying with the requirements of simulation system implemented in Simulink in a text file format together with initial values for each simulation parameter specified in component matching process [6]. This text file is then called in Simulink to initiate the simulation of the design scheme.

The dynamic identification of any causal change is done at system simulation stage and is implemented in the Simulink software. The overall criterion is to observe and monitor the direction of the energy flow between components, which is indicated by the signs of energy co-variables for a component energy port in the simulation system.

7.3 The causal change processing

If there is a change in the flow direction, i.e. the sign of any energy co-variable changes, simulation will be paused to reason if there is a causal change. This will be indicated by the signs of both energy co-variables as illustrated in table 2. If only one of the energy co-variables changes and there is no causal change, the simulation resumes and it carries on until the next possible causal change time instant. If there is a causal change, the mechanism will store all current state variables, energy co-variables and time variables and create a new simulation model of the physical system with the correct new causal structure. This new simulation model can be used to continue the simulation from the same time instant using all state variables, energy co-variables and time variables stored in the previous simulation run. The same set of simulation parameters both for the simulation itself and the physical system are still used in this new causal structure of the system to maintain the consistency of the simulation.

CONCLUSION

Engineering design is a highly competitive intellectual activity and designers face much more pressure and greater challenges in producing rapid as well as high quality design solutions. In this paper, the complicated causal structures of mechatronic systems have been thoroughly analysed both in their static and dynamic operations, and their sophisticated and changing natures have been studied and generalised into four possible causal structures. Due to its generic features, the bond graph method provides a useful tool for carrying out this analysis. A systematic approach to reasoning and dealing with the causal structures has been described. Using a number of AI techniques, the modelling and simulation system automatically supports a designer in determining a sensible causal structure and thereafter maintaining these structures through the truth maintenance system at both the conceptual design and the evaluation stages of mechatronic system design. This enables a designer to concentrate on more intellectual design activities and to make a decision confidently based on the causal analysis and reasoning. A comprehensive study of the information domain was also carried

REFERENCES

[1] French, M.J., "Conceptual design for engineers", Springer-Verlag, second edition, 1985.

[2] Sandor, G. N., "The Seven Stages of Engineering Design", Mechanical Engineering, Vol. 86, No. 4, pp. 21-25, 1964.

[3] Sandgren, E., "A Multi-Objective Design Tree Approach for the Optimisation of Mechanism", Mechanical Machine Theory, Vol. 25, No. 3, pp. 257-272, 1990.

[4] Dixon, J.R., Duffey, M. R., Irani, R., Meunier, K. and Orelup, M., "A Proposed Taxonomy of Mechanical Design Problems", Proceedings of the ASME Computers in Engineering Conference, San Francisco, California, 1988.

[5] Welch, R.V and Dixon, J.R., "Conceptual Design of Mechanical Systems", in Design Theory and Methodology DTM'91 edited by Stauffer, L.A., DE-Vol. 31, pp.61-68, 1991.

[6] Bracewell, R.H., Chaplin, R.V., Langdon, P.M., Li, M., Oh, V.K., Sharpe, J.E.E., and Yan, X.T., "Integrated computer support for interdisciplinary system design", Technical Report EDC 1994/11, Engineering Design Centre, Lancaster University, 1994.

[7] Karnopp, D.C, Margolis, D.L. and Rosenberg, R.C., "System Dynamics, A Unified Approach", Second Edition, John Wiley & Sons, Inc., 1990.

[8] Sharpe, J.E.E and Goodwin, E.M., "The Application of Bond Graphs in Complex Concurrent Multi-disciplinary Engineering Design", International Conference on Bond Graph Modelling, Simulation series Vol. 25, No. 2, pp. 23-28, California, 1993.

[9] Margolis, D., "Bond Graph causality and the derivation of analytical transfer Functions for Low Order Systems", International conference on Bond Graph Modelling, Simulation series Vol. 25, No. 2, pp. 243-248 California, 1993.

[10] Breedveld, P.C., "Multibond Graphs Elements in Physical Systems Theory", J. Frankline Institute, Vol. 319, No. 1/2, pp.51-65, 1985.

[11] Yan, X.T. and Sharpe, J.E.E., "A System Simulation Platform for Mechatronic Product Design", European Simulation Multi-Conference, Barcelona, Spain, pp789-793, June 1994,

[12] Cooling, J.E., "Software Design for Real-time Systems", Chapman and Hall, 1990.

[13] Yourdon, E, "Modern Structured Analysis", Prentice-Hall, Inc., 1989.

[14] Randell, Brian, ed. "The origins of digital computers : selected papers". - 2nd ed. Berlin: Springer-Verlag, 1975.

[15] Norton, H. N., "Handbook of transducers", Prentice-Hall, Inc., 1989.

APPENDIX

These are the results of an initial study with reference to the early development of computer systems and latest sensing technology. It can be seen that it is not too difficult to use the underlying methodology described in the paper to produce similar set of functions and especially embodiments of these using available electronic technology to realise them. However, this remains as future work.

Table 1.1 Fundamental binary functions

Function	Embodiment	Principles
Basic storage	Bi-stable circuits (flip flop), Electronic switch And circuits(gates)	Flip-flop principle
Set, Reset a flip flop	Pulses	
Read a value	Perforated cards, Keyboards A-D converter	
Store a numerical data	Mechanical means, Perforated cards, wheels, brush, Different geometry, Electrical, register, flip flop circuit electronic	
Translation or transmitting signals	Transceiver	Pulse principle
Receiving signals	Receiver	
Record intermediate results	Mechanical means, electronic means	
Record final results, or Print a value	Punchers, printer, display monitor, LCD display window	
Shift a no. to the right	Shifting register (parallel, serial)	Multiplication by 2^{-n}
Shift a no. to the left	Shifting register (parallel, serial)	Multiplication by 2^n
Compare two quantities	coincidence unit in EDSAC	Comparing pulses
Sequence of figures (numbers)	Sequence control tank	
Conditional switch	If (expr.= true) then action1 else action 2	Boolean
Logical control	If (expr. >= 0) then action1 else action 2	Boolean operation + Comparison + Switch

Table 1.2 Fundamental arithmetical operations

Function	Embodiment	Principles
Addition	Three input binary adder; decimal operation, Combination of switching circuits and flip-flops.	Binary operation,
Subtraction	Combination of switching circuits and flip-flops.	Complement operation principle
Multiplication	Combination of switching circuits and flip-flops, Multiplication control unit	Binary operation, Shifting operation,
Division	Combination of switching circuits and flip-flops	Binary operation,
Binary to Decimal Convertion	B-D converter	B-D principle
Decimal to Binary Convertion	D-B converter	D-B principle

Table 1.3 Other fundamental functions

Function	Embodiment	Principles
Warning	Ring bells, textual warning	
Clear a register	Reset	Pulse principle
Initialisation order	Initialisation or reset.	
Supply power	Batteries, a.c. power supplies	
Timing	Program control timer, Secondary timer, tertiary timer, constructed by flip-flops, Timing control tank	

Table 1.4 Fundamental functions of sense

Function	Embodiment	Principles
Thermometers sensors	Thermocouple circuits,	Seeback effect principle, Peltier effect principle and Thomson effect principle.
Force sensors	Capacitive force transducers, Reluctive force transducers, Strain-gauge force transducers, Pizoelectric transducers.	Convertion of a deformation of an elastic element into signal principle, capacitive transduction principle, strain-gauge principle, pizoelectric principle.
Flow rate sensors	Differential-pressure flow sensors, turbine flowmeters, annular-rotor flowmeters, variable-area flowmeters, thermal flowmeters, Ultrasonic flowmeters,	Differential-pressure flow principle, mechanical desplacing principle, heat transfer of mass flow principle, Doppler effect of ultrasonic energy in mass flow.
Altitude sensors	Attitude gyros, gravity-reference attitude transducers, compass-type sensors, Celestial-reference attitude sensors,	Gyroscope principle, gravity-reference sensing principle, magnetic-reference principle, optical-reference principle
Position sensors	Capacitive displacement transducers, inductive displacement transducers, reluctive displacement transducers, potentiometric displacement transducers, strain-gauge sensors, electro-optical displacement sensors, angular and linear encoders (brush-type, optical and magnetic encoders), radar, lidar, sonar.	Contacting sensing principle, capacitive displacement principle, inductive displacement principle, reluctive displacement principle, potentiometric displacement principle, strain-gauge principle, encoder principle, electromagnetic radiation (radar, lidar and sonar) principle.
Velocity sensors	Electromagnetic linear-velocity transducers, Tachometer generators, toothed-rotor electromagnetic tachometers, Electro-optical tachometers.	Electromagnetic flux induction principle, integration of an accelerometer principles, Doppler effect, tachometry principle,
Acceleration sensors	Capacitive accelerometers, Pizoelectric accelerometers, Potentiometric accelerometers, Strain-Gauge accelerometers, Reluctive accelerometers, Servo accelerometers	Capacitive transduction principle, pizoelectric principle, wiper driven potentiometric principle, linear variable differential transformer(LVDT) principle, servo principle.
Pressure sensors	Diaphragms, Capsules, Bellows, Straight tubes, Bourdon tubes.	Diaphragm principles, Bourdon tube principles
Actuation of signals(limbs)	See energetic components.	Working principles for energetic components.

Notes:
1. Each type of sensing functions can be embodied in several ways depending on their working principles.
2. Working principles can be used in more than type of sensing functions as illustrated in the table.

Table 1.5 Device Level Functions, Embodiments & Principles

Function	Embodiment	Principles
Storing a number	Memory organ	
Computing function	Arithmetic organ	Arithmetical + logical principles
Sequence control	Control organ Memory selector + order transmission + number	transfer + arithmetical operations + input control + Output control
Control register (20)	flip-flop register	